Intelligent Reconfigurable Surfaces (IRS)
for Prospective 6G Wireless Networks

IEEE Press
445 Hoes Lane
Piscataway, NJ 08854

Intelligent Reconfigurable Surfaces (IRS) for Prospective 6G Wireless Networks

Edited by

Muhammad Ali Imran
James Watt School of Engineering, University of Glasgow, Glasgow, UK

Lina Mohjazi
James Watt School of Engineering, University of Glasgow, Glasgow, UK

Lina Bariah
Technology Innovation Institute, Abu Dhabi, UAE

Sami Muhaidat
KU Center for Cyber-Physical Systems, Department of Electrical Engineering and Computer Science, Khalifa University, Abu Dhabi, UAE

Tie Jun Cui
State Key Laboratory of Millimeter Waves, Southeast University, Nanjing, China

Qammer H. Abbasi
James Watt School of Engineering, University of Glasgow, Glasgow, UK

The ComSoc Guides to Communications Technologies
Nim K. Cheung, *Series Editor*
Richard Lau, *Associate Series Editor*

Published by John Wiley & Sons, Inc., Hoboken, New Jersey.
Published simultaneously in Canada.

Library of Congress Cataloging-in-Publication Data Applied for:

Hardback ISBN: 9781119875253

Cover Design: Wiley
Cover Image: © natrot/Shutterstock

Set in 9.5/12.5pt STIXTwoText by Straive, Chennai, India

Contents

List of Contributors

Qammer H. Abbasi
James Watt School of Engineering
University of Glasgow
Glasgow UK

Mohammad O. Abualhauja'a
James Watt School of Engineering
University of Glasgow
UK

Hanaa Abumarshoud
James Watt School of Engineering
University of Glasgow
UK

Ian F. Akyildiz
Truva Inc.
Georgia
USA

Shuja Ansari
James Watt School of Engineering
University of Glasgow
UK

Stylianos D. Assimonis
ECIT Institute
Centre for Wireless Innovation
Queen's University Belfast
Northern Ireland
UK

Lina Bariah
Technology Innovation Institute
Masdar City
Abu Dhabi
UAE

Ertugrul Basar
Department of Electrical and
Electronics Engineering
Koç University
Sariyer, Istanbul
Turkey

Qiang Cheng
State Key Laboratory of Millimeter
Waves
Southeast University
Jiangsu Province, Nanjing
China

and

Institute of Electromagnetic Space
Southeast University
Jiangsu Province, Nanjing
China

and

Frontiers Science Center for Mobile Information Communication and Security
Southeast University
Nanjing, China

Michail Christodoulou
Electrical and Computer Engineering Department
Aristotle University
Thessaloniki
Greece

Tie Jun Cui
State Key Laboratory of Millimeter Waves
Southeast University
Jiangsu Province, Nanjing
China

and

Institute of Electromagnetic Space
Southeast University
Jiangsu Province, Nanjing
China

and

Frontiers Science Center for Mobile Information Communication and Security
Southeast University
Nanjing
China

Jun Y. Dai
State Key Laboratory of Millimeter Waves
Southeast University
Jiangsu Province, Nanjing
China

and

Institute of Electromagnetic Space
Southeast University
Jiangsu Province, Nanjing
China

and

Frontiers Science Center for Mobile Information Communication and Security
Southeast University
Nanjing
China

Tri N. Do
The Department of Electrical Engineering
the École de Technologie Supérieure (ÉTS)
Université du Québec
Montréal, QC
Canada

Konstantinos Dovelos
ECIT Institute
Centre for Wireless Innovation
Queen's University Belfast
Northern Ireland
UK

Amal Feriani
Department of Electrical and Computer Engineering
University of Manitoba
Manitoba, Winnipeg
Canada

Harald Haas
LiFi Research and Development Centre (LRDC)
Department of Electronic & Electrical Engineering
The University of Strathclyde
Glasgow
UK

Ekram Hossain
Department of Electrical and Computer Engineering
University of Manitoba
Manitoba, Winnipeg
Canada

Muhammad Ali Imran
James Watt School of Engineering
University of Glasgow
Glasgow
UK

Sotiris Ioannidis
Foundation for Research and Technology – Hellas
Heraklion
Greece

and

School of Electrical and Computer Engineering
Technical University of Chania
Crete
Greece

Muhammad A. Jamshed
James Watt School of Engineering
University of Glasgow
Glasgow
UK

Georges Kaddoum
The Department of Electrical Engineering
the École de Technologie Supérieure (ÉTS)
Université du Québec
Montréal, QC
Canada

Nikolaos Kantartzis
Electrical and Computer Engineering Department
Aristotle University
Thessaloniki
Greece

Jalil R. Kazim
James Watt School of Engineering
University of Glasgow
Glasgow
UK

Christos Liaskos
Computer Science Engineering Department
University of Ioannina
Ioannina
Greece

and

Foundation for Research and Technology – Hellas
Heraklion
Greece

Michail Matthaiou
ECIT Institute
Centre for Wireless Innovation
Queen's University Belfast
Northern Ireland
UK

Lina Mohjazi
James Watt School of Engineering
University of Glasgow
Glasgow
UK

Sami Muhaidat
Department of Electrical Engineering and Computer Science
KU Center for Cyber-Physical Systems
Khalifa University
Abu Dhabi
UAE

and

Department of Systems and Computer Engineering
Carleton University
Ottawa
Canada

Hien Q. Ngo
ECIT Institute
Centre for Wireless Innovation
Queen's University Belfast
Northern Ireland
UK

Thanh L. Nguyen
The Department of Electrical Engineering
the École de Technologie Supérieure (ÉTS)
Université du Québec
Montréal
QC, Canada

Alexandros Pitilakis
Electrical and Computer Engineering Department
Aristotle University
Thessaloniki
Greece

Andreas Pitsillides
Computer Science Department
University of Cyprus
Nicosia
Cyprus

and

Department of Electrical and Electronic Engineering Science
University of Johannesburg (Visiting Professor)
Johannesburg Gauteng
South Africa

Olaoluwa R. Popoola
James Watt School of Engineering
University of Glasgow
Glasgow
UK

Georgios G. Pyrialakos
Electrical and Computer Engineering Department
Aristotle University
Thessaloniki
Greece

James Rains
James Watt School of Engineering
University of Glasgow
Glasgow
UK

Taniya Shafique
Department of Electrical and Computer Engineering
University of Manitoba
Manitoba, Winnipeg
Canada

Hina Tabassum
Department of Electrical Engineering and Computer Science
York University
Ontario, Toronto
Canada

Ageliki Tsioliaridou
Foundation for Research and Technology – Hellas
Heraklion
Greece

Anvar Tukmanov
BT Labs
Adastral Park
Ipswich
UK

Jalil ur Rehman
James Watt School of Engineering
University of Glasgow
Glasgow
UK

Masood Ur-Rehman
James Watt School of Engineering
University of Glasgow
Glasgow
UK

Anil Yesilkaya
LiFi Research and Development Centre (LRDC)
Department of Electronic & Electrical Engineering
The University of Strathclyde
Glasgow
UK

Ibrahim Yildirim
Department of Electrical and Electronics Engineering
Koç University
Sariyer, Istanbul
Turkey

and

Faculty of Electrical and Electronics Engineering
Istanbul Technical University
Sariyer, Istanbul
Turkey

Lei Zhang
James Watt School of Engineering
University of Glasgow
Glasgow
UK

1

Introduction

Muhammad Ali Imran[1], Lina Mohjazi[1], Lina Bariah[3], Sami Muhaidat[2,4], Tei Jun Cui[5], and Qammer H. Abbasi[1]

[1]*James Watt School of Engineering, University of Glasgow, Glasgow, UK*
[2]*Department of Electrical Engineering and Computer Science, KU Center for Cyber-Physical Systems, Khalifa University, Abu Dhabi, UAE*
[3]*Technology Innovation Institute, 9639 Masdar City, Abu Dhabi, UAE*
[4]*Department of Systems and Computer Engineering, Carleton University, Ottawa, Canada*
[5]*State Key Laboratory of Millimeter Waves, Southeast University, Nanjing, China*

The roadmap to beyond the fifth-generation (B5G) wireless networks is envisaged to introduce a new spectrum of fully automated and intelligent data-driven services, such as flying vehicles, haptics, telemedicine, augmented and virtual reality, holographic telepresence, and connected autonomous artificial intelligence (AI) systems [1, 2]. Several unprecedented application environments, including machine-to-people and machine-to-machine communications, are expected to be the driving force of B5G systems. As a result, the number of connected Internet-of-Everything (IoE) devices (e.g. sensors, wearables, implantables, tablets) is anticipated to witness a phenomenal growth in the next few years, reaching up to tens of billions [3]. This poses a fundamental challenge on provisioning a ubiquitous seamless connectivity, while concurrently prolonging the lifetime of a massive number of energy-constrained low-power, low-cost devices.

The unprecedented increase of connected devices resulting from the emergence of IoE has created a major challenge for broadband wireless networks, requiring a paradigm shift towards the development of key enabling technologies for the next generation of wireless networks. Fifth-generation (5G) wireless networks have been identified as the backbone of emerging IoE services and prominently support three use cases: enhanced mobile broadband, ultra-reliable and low-latency communications, and massive machine-type communications. These services are rate- and data-oriented, heterogeneous in nature and defined by a diverse set of key

Intelligent Reconfigurable Surfaces (IRS) for Prospective 6G Wireless Networks, First Edition.
Edited by Muhammad Ali Imran, Lina Mohjazi, Lina Bariah, Sami Muhaidat,
Tie Jun Cui, and Qammer H. Abbasi.

performance indicators. Therefore, enabling them through a single platform while concurrently meeting their stringent requirements in terms of data rate, reliability, and latency is a challenging task [3].

To address these challenges at the physical layer, 5G wireless systems have leveraged the evolution of cutting-edge technologies, including millimetre wave (mmWave) and terahertz (THz) communications. Although mmWave and THz are highly promising in offering unparalleled data rates and significantly reducing the required device size, their present use is limited due to signal degradation at these extremely high-communication frequency bands. Moreover, wireless links suffer from attenuation incurred by high propagation loss, high penetration loss, multipath fading, molecular absorption, and Doppler shift [4]. With the lack of full control over the propagation and scattering of electromagnetic (EM) waves, the wireless environment remains unaware of the time-variant communication, posing fundamental limitations towards building truly pervasive software-defined wireless networks [5].

Motivated by this, reconfigurable meta-surfaces, also known as Intelligent Reconfigurable Surfaces (IRS), have emerged as a low-complexity and energy-efficient solution that aims at turning the wireless environment into a software-defined entity. Reconfigurable meta-surfaces are envisaged to be indispensable in future sixth-generation (6G) wireless systems due to their potential in realizing massive multiple-input multiple-output (MIMO) gains while attaining a notable reduction in energy consumption. The unique design principle of reconfigurable meta-surfaces lies in realizing artificial structures with massive antenna arrays whose interaction with impinging EM waves can be intentionally controlled through connected passive elements, such as phase shifters, in a way that enhances wireless systems' performance in terms of coverage, rate, and so on, giving rise to the concept of smart radio environments (SREs) [6].

IRS have been deemed as a key contributor in putting down the fundamentals of future 6G networks, and therefore, have attracted a considerable attention from the industrial and academic communities. Therefore, this book will equip the reader with the fundamental knowledge of the operational principles of reconfigurable meta-surfaces, resulting in its potential applications in various intelligent, autonomous future wireless communication technologies. The opportunities opened by IRS have spurred, in a short span of time, research in many areas related to wireless communication systems. This includes multi-user resource allocation, beamforming optimization, design of efficient enabling mechanisms, and performance analysis of IRS-assisted wireless networks.

The aim of this book is to offer the readers the opportunity to explore and comprehend the field of IRS from different angles, including the underlying physics, hardware architecture, operating principles, as well as prototype designs. The book will allow the readers to grasp the knowledge of the interplay of IRS and top-notch technologies, accompanied by the evolution of 6G networks, with comprehensively studying the advantages, key principles, challenges, and potential use-cases.

This book is aimed to be a solid foundation for the theoretical investigation and practical implementation of IRS-enabled wireless networks. The book is envisioned to be a concrete reference for students, researchers, university professors, and industrial people working in the field of intelligent surfaces, in which they can exploit it to identify open research problems, and hence steer their research and industrial activities in those directions. With the diverse aspects studied in our book, we look forward to facilitating a smooth comprehension of the preliminary concepts, as well as providing solid answers to more advanced critical concerns raised in the field of IRS.

Chapter 2 discusses the fundamental principles of IRS-aided communications and provides an analysis on the near-field region, wherein the channel modelling and phase shift design problems differ from those in the far-field. Specifically, the chapter highlights the impact of beamfocusing in manipulating the emitted EM waves to achieve desired signal propagation. This chapter also investigates IRS-aided MIMO communications and the relationship between the number of reflecting elements and the achieved energy efficiency gains.

In Chapter 3, the potential of deploying IRSs in merging non-terrestrial networks (NTNs) is explored. This is linked with discussions related to 3GPP standardization guidelines in the context of the various operational aspects, architecture types, and connectivity mechanisms in NTN. Additionally, this chapter highlights how IRSs can be integrated in NTN to enable a typical mobile handset to directly communicate with satellites.

Chapter 4 introduces a new concept called the Internet of MetaMaterial Things (IoMMT), where artificial materials with real-time tunable physical properties can be interconnected to form a network to realize communication through software-controlled EM, acoustic, and mechanical energy waves. After exploring the means for abstracting the complex physics behind these materials, their integration into the IoT world is discussed. The chapter presents two novel software categories for the material things, namely the meta-material Application Programming Interface and Meta-material Middleware, which will be in charge of the application and physical domain.

Chapter 5 overviews the general hardware architecture of IRS that opened a new platform to dynamically manipulate EM waves. This includes describing the design of an IRS structure based on different categorizations. The available IRS modes of operation will be discussed in deployments relevant to wireless communication systems. The chapter also discusses the hardware aspects and features related to the IRS's main operational principles: reconfigurability, interconnection, computing, networking, programmability, and sensing. This chapter reviews the state-of-the-art on advancements in IRS prototype designs for wavefront manipulation and information modulation.

In Chapter 6, the authors discusses practical design considerations for IRS. Specifically, the tunability of the IRS unit-cell elements will be explained. Also, the biasing network of an IRS, which provides a means of control over the individual unit cell reflection characteristics, will be detailed. The chapter will provide a comprehensive treatment on the physical limitations of the IRSs including the trade-off between the bandwidth and phase resolution, the incidence angle response, and the quantization effects.

Chapter 7 explores channel modelling frameworks for facilitating a thorough and accurate evaluation of the system performance of IRS-aided communications operating in the mmWave and sub-6 GHz bands. Specifically, the chapter discusses the channel side limitations of the IRS and sheds light on the important role the channel plays in the IRS implementation. The chapter focuses on discussing small-scale fading and path loss model of IRS-enabled wireless networks, for different scenarios, including far-field and near-field scenarios. Finally, the chapter introduces the open-source, user-friendly, and widely applicable SimRIS Channel Simulator v2.0.

Chapter 8 develops an iterative optimization framework to maximize the data rate of a given user by jointly optimizing the user service mode selection along with phase shifts of the nearest IRS of IRS-assisted users in a large-scale multi-user, multi-base station, multi-IRS network. This chapter presents semi-definite programming (SDP)-based iterative approach for phase optimization, whereas a heuristic approach for mode selection is adopted. In addition, a deep reinforcement learning (DRL) framework is presented with proximal policy optimization (PPO) and double deep policy gradient (DDPG) based solutions to optimize phase shifts.

Chapter 9 investigates the significant role that IRS will play in B5G and 6G wireless networks. Precisely, the chapter discusses the potential of IRS in supporting IRS-assisted multi-user communication, IRS-assisted RF sensing and imaging, IRS-assisted unmanned aerial vehicle (UAV) communication, IRS-assisted wireless power transfer, and IRS-assisted indoor localization. In this context, the authors examine their performance and highlight major performance limiting factors, which open the door for future research directions.

In Chapter 10, the authors study the channel modelling and characterization for multi-IRS-assisted wireless systems. For a distributed multi-IRS (DMI)-assisted system, in which the IRSs have different geometric sizes and are distributively deployed to aid wireless communications, the authors propose a mathematical framework based on the moment-matching method to determine the statistical characterization of the end-to-end (e2e) channel fading of the DMI system. The obtained approximate distributions are employed to derive tight approximate closed-form expressions of the outage probability (OP) and ergodic capacity (EC) of the DMI system.

Chapter 11 presents the fundamental characteristics of the UAVs, major paradigms for integrating the UAVs into the wireless networks, and their possible applications, as well as addressing open problems and challenges in the UAV communications. The chapter sheds light on the possible IRS-assisted UAV systems scenarios. Furthermore, this chapter investigates the performance analysis of the IRS-assisted UAV systems.

Chapter 12 sets the scene for the integration of IRS with optical wireless communication (OWC), as a promising candidate to support blockage-free, and therefore, extended coverage communication. The chapter sheds light on the advantages accomplished by leveraging various RIS functionalities in OWC networks, from a transceiver, as well as propagation environment perspectives. The authors present a case study for an IRS-assisted indoor LiFi system and examine the performance of the considered scenario. The chapter finally highlights some of the challenges and open research directions related to the integration of IRS in OWC.

References

1 Di Renzo, M. Debbah, M. Phan-Huy, DT et al. (2019). Smart radio environments empowered by reconfigurable AI meta-surfaces: an idea whose time has come. *EURASIP J. Wireless Commun. Netw.* 2019 (1): 1–20.

2 Bariah, L. Mohjazi, L. Muhaidat, S. et al. (2020). A prospective look: key enabling technologies, applications and open research topics in 6G networks. *IEEE Access* 8: 174792–174820.

3 Tariq, F. (2020). A speculative study on 6G. *IEEE Wireless Commun.* 27 (4): 118–125.

4 Mohjazi, L., Zoha, A., Bariah, L. et al. (2020). An outlook on the interplay of artificial intelligence and software-defined metasurfaces: an overview of opportunities and limitations. *IEEE Veh. Tech. Mag.* 15 (4): 62–73.

5 Liaskos, C., Nie, S., Tsioliaridou, A. et al. (2018). A new wireless communication paradigm through software-controlled metasurfaces. *IEEE Commun. Mag.* 56 (9): 162–169.

6 Basar, E. Di Renzo, M. De Rosny, J. et al. (2019). Wireless communications through reconfigurable intelligent surfaces. *IEEE Access* 7: 116753–116773.

2

IRS in the Near-Field: From Basic Principles to Optimal Design

Konstantinos Dovelos, Stylianos D. Assimonis, Hien Q. Ngo, and Michail Matthaiou

ECIT Institute, Centre for Wireless Innovation, Queen's University Belfast, Northern Ireland, Belfast, UK

2.1 Introduction

Fifth-generation wireless technologies, such as massive multiple-input multiple-output (MIMO), millimetre wave (mmWave) communication, and ultra-dense networks, are expected to boost data rates into an unprecedented level [1]. Yet, those technologies rely on the postulate that the wireless channel is uncontrollable, and hence its detrimental effect can be mainly compensated by sophisticated transmission and reception schemes. The advent of meta-materials, though, has paved the way for the disruptive paradigm of electromagnetic (EM) meta-surfaces, broadly known as intelligent reconfigurable surfaces (IRSs) [2]. Specifically, an IRS is a planar structure of multiple reconfigurable elements of sub-wavelength size, which can alter the characteristics of an impinging EM wave. Furthermore, IRSs consist mainly of passive components and hence are more power-efficient than MIMO relays [3]. Thanks to those unique wave manipulation capabilities, IRSs are reasonably considered by many experts as the 'next big thing in wireless' [4]. In this chapter, we will first outline the basic principles behind IRSs and then delve into the specifics of near-field communication. Finally, we will conclude the chapter by presenting possible avenues for future research. Table 2.1 summarizes the nomenclature used in this chapter.

Intelligent Reconfigurable Surfaces (IRS) for Prospective 6G Wireless Networks, First Edition.
Edited by Muhammad Ali Imran, Lina Mohjazi, Lina Bariah, Sami Muhaidat,
Tie Jun Cui, and Qammer H. Abbasi.

Table 2.1 Nomenclature adopted in this chapter.

Notation	Description
$\mathbb{E}[\cdot]$	Expectation operator
$\mathcal{CN}(\mu, \sigma)$	Complex Gaussian variable with mean μ and variance σ^2
$\lvert \cdot \rvert$	Absolute value of a complex number
$\arg(\cdot)$	Argument of a complex number

2.2 Basic Principles

2.2.1 IRS Model

In general, an IRS is modelled as a rectangular array of $N = N_1 \times N_2$ reflecting elements of sub-wavelength size, also called unit cells or meta-atoms. Each unit cell is engineered so that it can alter the phase and amplitude of an impinging EM wave. Particularly, the unit cell usually comprises of a metal patch on the top of a grounded dielectric substrate [5, 6]. Furthermore, a semi-conductor device[1] is embedded into the metal patch to dynamically control the overall element response through a biasing voltage. Owing to the sub-wavelength size of unit cells, each one can be represented by an equivalent lumped RLC circuit. To this end, the impedance of the (n, m)th IRS cell is given by [7]

$$Z(C_{n,m}) = \frac{j\omega L_1 \left(j\omega L_2 + \frac{1}{j\omega C_{n,m}} + R \right)}{j\omega(L_1 + L_2) + \frac{1}{j\omega C_{n,m}} + R} \tag{2.1}$$

In (2.1), L_1, L_2, $C_{n,m}$, R, and ω are the bottom-layer inductance, top-layer inductance, effective capacitance, effective resistance, and angular frequency of the incident wave, respectively. Note that the semi-conductor device varies the effective capacitance $C_{n,m}$, thereby reflecting the incident wave due to the impedance discontinuity between the free-space impedance Z_0 and element impedance $Z(C_{n,m})$. The *reflection coefficient* of the (n, m)th unit cell is therefore defined as follows:

$$\Gamma_{n,m} \triangleq \frac{Z(C_{n,m}) - Z_0}{Z(C_{n,m}) + Z_0} \tag{2.2}$$

1 In practice, a positive-intrinsic-negative diode or a varactor diode can be used as the semiconductor device [6].

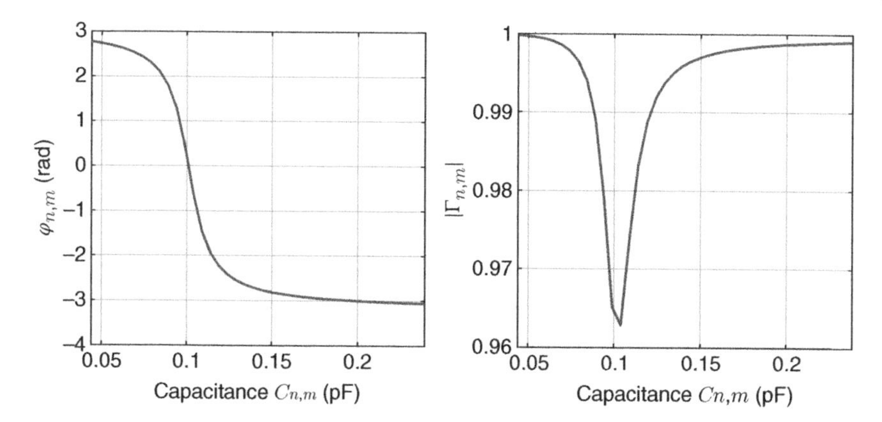

Figure 2.1 Phase and amplitude of the reflection coefficient versus $C_{n,m}$ for $R = 0.1\ \Omega$, $L_1 = 0.25$ nH, and $L_2 = 0.07$ nH

As seen from Figure 2.1, the absorption losses are negligible. For this reason, it is assumed that $|\Gamma_{n,m}| \approx 1$. Then, the reflection coefficient of each unit cell is expressed as $\Gamma_{n,m} = e^{j\varphi_{n,m}}$, where $\varphi_{n,m} \triangleq \arg\left(\frac{Z(C_{n,m})-Z_0}{Z(C_{n,m})+Z_0}\right) \in [-\pi, \pi]$.

2.2.2 Signal Model of IRS-Aided System

The IRS is deployed between the transmitter (Tx) and receiver (Rx) to assist communication. Consider that the Tx and Rx are equipped with a single antenna each. The received baseband signal can then be written as follows:

$$y = \left(\sum_{n=0}^{N_1-1}\sum_{m=0}^{N_2-1} h^t_{n,m} h^r_{n,m} e^{j\varphi_{n,m}} + h_d\right) x + \tilde{n} \tag{2.3}$$

where $h^t_{n,m} \in \mathbb{C}$ is the channel from the Tx to the (n,m)th cell, $h^r_{n,m} \in \mathbb{C}$ is the channel from the (n,m)th cell to the Rx, $h_d \in \mathbb{C}$ is the direct channel between the Tx and Rx, x denotes the transmitted data symbol with average power $\mathbb{E}[|x|^2] = P_t$, and $\tilde{n} \sim \mathcal{CN}(0, \sigma^2)$ is the additive noise with power spectral density σ^2. Under perfect channel state information at the Rx, the received signal-to-noise ratio (SNR) is expressed as follows:

$$\text{SNR} = \frac{P_t}{\sigma^2}\left|\sum_{n=0}^{N_1-1}\sum_{m=0}^{N_2-1} h^t_{n,m} h^r_{n,m} e^{j\varphi_{n,m}} + h_d\right|^2 \tag{2.4}$$

The SNR is maximized when the signals propagating through the IRS and the direct link are coherently combined at the Rx. To achieve this, the phase shift induced by the (n,m)th element must be

$$\varphi_{n,m} = -\left(\arg(h^t_{n,m}) + \arg(h^r_{n,m})\right) + \arg(h_d) \tag{2.5}$$

which gives

$$\text{SNR} = \frac{P_t}{\sigma^2}\left|\sum_{n=0}^{N_1-1}\sum_{m=0}^{N_2-1}|h^t_{n,m}||h^r_{n,m}|e^{j\arg(h_d)} + |h_d|e^{j\arg(h_d)}\right|^2$$

$$= \frac{P_t}{\sigma^2}\left(\sum_{n=0}^{N_1-1}\sum_{m=0}^{N_2-1}|h^t_{n,m}||h^r_{n,m}| + |h_d|\right)^2 \geq \frac{P_t}{\sigma^2}|h_d|^2 \tag{2.6}$$

From (2.6), it is evident that the SNR is larger than that obtained without the IRS. In short, the IRS can boost the signal power at the Rx, which is vital when the direct channel is weak or blocked.

2.3 Near-Field Channel Model

The channel coefficients depend on the operating frequency and deployment scenario. For example it is customary to assume that $h^t_{n,m}$ and $h^r_{n,m}$ are $\mathcal{CN}(0,1)$, i.e. Rayleigh fading, to represent rich scattering in the sub-6 GHz band [8, 9]. In this chapter, we consider mmWave IRS-aided channels whose multi-path scattering is limited [10]. Note that communication in the mmWave band (30–300 GHz) is very attractive for 5G-and-beyond networks thanks to the abundant spectrum available at extremely high frequencies [11]. In addition to sparse scattering, an IRS operating at high frequencies is expected to be near the base station so that propagation losses are minimal. Consequently, near-field effects are of particular relevance. To study those effects, we adopt a *geometric* channel model that captures the key features of line-of-sight (LoS) propagation in the vicinity of the IRS. In particular, we focus on the radiating near-field of the IRS called Fresnel zone. Recall that the Fresnel zone includes all distances r from the IRS satisfying [12]

$$0.62\sqrt{\frac{L^3_{\max}}{\lambda}} < r \leq \frac{2L^2_{\max}}{\lambda} \tag{2.7}$$

where λ and $L_{\max}$ denote the wavelength and maximum IRS dimension, respectively.

From Table 2.2, we evince that a mmWave IRS can have a Fresnel region of tens of metres. Thus, the far-field assumption of plane waves breaks down as the IRS size grows.

2.3.1 Spherical Wavefront

Both the Tx-IRS and IRS-Rx links are LoS. The IRS geometry is depicted in Figure 2.2, where $L_x \times L_y$ is the area of each unit cell and $\mathbf{p}_{n,m} = (nL_x, mL_y, 0)$ is the

Table 2.2 Fresnel region versus IRS size at $f = 60$ GHz.

IRS size	Physical IRS size (m^2)	Fresnel region (m)
$50\lambda \times 50\lambda$	0.25×0.25	$(1.096, 25]$
$100\lambda \times 100\lambda$	0.5×0.5	$(3.1, 100]$
$120\lambda \times 120\lambda$	0.6×0.6	$(4.075, 144]$

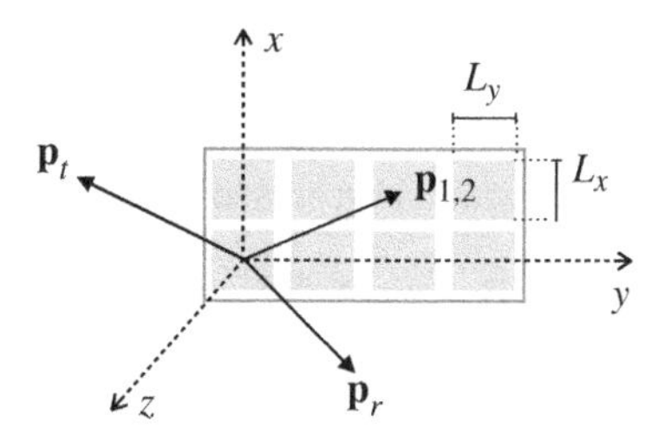

Figure 2.2 Illustration of the IRS geometry used in the channel model.

position vector of the (n, m)th cell. Furthermore, the Tx and Rx position vectors, $\mathbf{p}_t$ and $\mathbf{p}_r$, are described in Cartesian coordinates as follows:

$$\mathbf{p}_t = (r_t \cos\phi_t \sin\theta_t, r_t \sin\phi_t \sin\theta_t, r_t \cos\theta_t) \tag{2.8}$$

$$\mathbf{p}_r = (r_r \cos\phi_r \sin\theta_r, r_r \sin\phi_r \sin\theta_r, r_r \cos\theta_r) \tag{2.9}$$

where r, ϕ, and θ denote the corresponding radial distances, azimuth angles, and polar angles, respectively. The cascaded channel $h_{n,m} \triangleq h^t_{n,m} h^r_{n,m}$ can now be decomposed as [13]

$$h_{n,m} = \sqrt{\mathrm{PL}_{n,m}} e^{-jk(r_t(n,m)+r_r(n,m))} \tag{2.10}$$

where $\mathrm{PL}_{n,m}$ is the path loss through the (n, m)th IRS cell, $k = 2\pi/\lambda$ is the wavenumber, whilst

$$\begin{aligned} r_t(n,m) &\triangleq \|\mathbf{p}_t - \mathbf{p}_{n,m}\| \\ &= r_t\sqrt{1 + \frac{(nL_x)^2}{r_t^2} - \frac{2\cos\phi_t \sin\theta_t nL_x}{r_t} + \frac{(mL_y)^2}{r_t^2} - \frac{2\sin\phi_t \sin\theta_t mL_y}{r_t}} \end{aligned} \tag{2.11}$$

and

$$\begin{aligned} r_r(n,m) &\triangleq \|\mathbf{p}_r - \mathbf{p}_{n,m}\| \\ &= r_r\sqrt{1 + \frac{(nL_x)^2}{r_r^2} - \frac{2\cos\phi_r \sin\theta_r nL_x}{r_r} + \frac{(mL_y)^2}{r_r^2} - \frac{2\sin\phi_r \sin\theta_r mL_y}{r_r}} \end{aligned} \tag{2.12}$$

are the distances from the Tx and Rx to the (n, m)th cell, respectively. Note that the phase variations in (2.10) depend on $r_t(n, m)$ and $r_r(n, m)$, which represent the spherical wavefront of the incident and reflected waves, respectively.

2.3.2 Path Loss

To derive the path loss of the IRS-aided link, we focus on an arbitrary IRS element and omit the subscript 'n, m'. The Tx and Rx are in the far-field of the *individual element*, which implies that $r_t, r_r > 2L_{\max}^2/\lambda$. Thus, a plane wavefront is assumed across the IRS element. For simplicity, we consider a transverse electric incident wave which is linearly polarized along the x-axis. The electric field of the incident plane wave is expressed as follows:

$$\mathbf{E}_i = E_i e^{-jk(y \sin\theta_t - z\cos\theta_t)} \mathbf{e}_x \tag{2.13}$$

where $\mathbf{e}_x \triangleq (1,0,0)^T$ denotes the unit vector along the x-axis. Next, the scattered field $\mathbf{E}_s$ at the Rx location $\mathbf{p}_r$ is determined by leveraging *physical optics* [14]. Assuming that the IRS is a perfect electric conductor, the squared magnitude of the scattered electric field is given by [14, Ch. 13]

$$\|\mathbf{E}_s\|^2 = \left(\frac{L_x L_y}{\lambda}\right)^2 \frac{|E_i|^2}{r_r^2} F(\theta_t, \phi_r, \theta_r) \mathrm{sinc}^2(X) \mathrm{sinc}^2(Y) \tag{2.14}$$

$$\approx \left(\frac{L_x L_y}{\lambda}\right)^2 \frac{|E_i|^2}{r_r^2} F(\theta_t, \phi_r, \theta_r) \tag{2.15}$$

where $F(\theta_t, \phi_r, \theta_r) \triangleq \cos^2\theta_t(\cos^2\theta_r \cos^2\phi_r + \sin^2\phi_r)$, while $X \triangleq \frac{\pi L_x}{\lambda} \sin\theta_r \cos\phi_r$ and $Y \triangleq \frac{\pi L_y}{\lambda}(\sin\theta_r \sin\phi_r - \sin\theta_t)$. The approximation in (2.15) follows from $\mathrm{sinc}(X) \approx 1$ and $\mathrm{sinc}(Y) \approx 1$ for $X \approx 0$ and $Y \approx 0$, which holds for $L_x \leq \lambda$ and $L_y \leq \lambda$. This is also verified in Figure 2.3. Note that each IRS element has sub-wavelength size in order to act as an isotropic scatterer. The power density of the scattered field at the Rx location is determined as follows:

$$S \triangleq \frac{\|\mathbf{E}_s\|^2}{2Z_0} = \left(\frac{L_x L_y}{\lambda}\right)^2 \frac{P_t G_t}{4\pi r_t^2 r_r^2} F(\theta_t, \phi_r, \theta_r) \quad (\mathrm{W/m}) \tag{2.16}$$

where the relationship $\frac{|E_i|^2}{2Z_0} = \frac{P_t G_t}{4\pi r_t^2}$ was used, with G_t denoting the antenna gain of the Tx. Considering the Rx aperture $A_r = G_r \frac{\lambda^2}{4\pi}$ gives the received power

$$P_r = S_s A_r = P_t \frac{G_t G_r}{(4\pi r_t r_r)^2} (L_x L_y)^2 F(\theta_t, \phi_r, \theta_r) \tag{2.17}$$

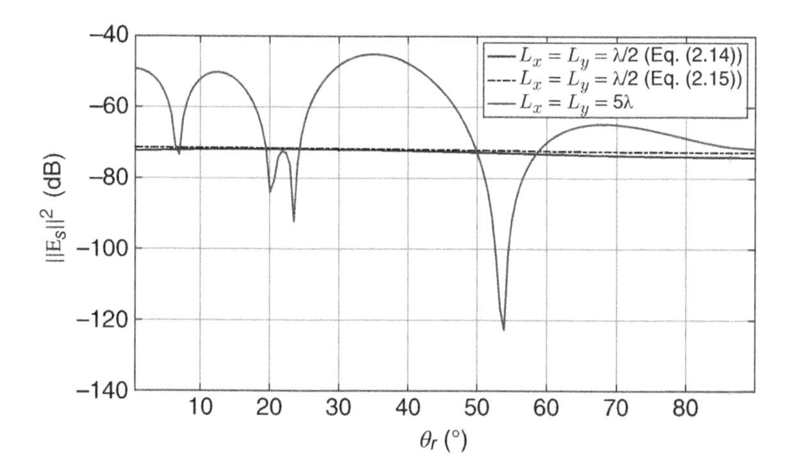

Figure 2.3 Squared magnitude of the scattered field versus observation angle θ_r for incident angle $\theta_t = 30°$ and scattering plane $\phi_r = 60°$; $|E_i|^2 = 1$, carrier frequency $f = 60$ GHz, and $r_r = 4$ m.

Therefore, the path loss of the Tx-IRS-Rx link through the (n, m)th element is given by

$$\text{PL}_{n,m} = \frac{G_t G_r (L_x L_y)^2}{(4\pi r_t(n,m) r_r(n,m))^2} F(\theta_t, \phi_r, \theta_r) \tag{2.18}$$

According to (2.18), the path loss of the IRS-aided link hinges on the reciprocal of the product $(r_t(n,m)r_r(n,m))^2$ rather than of the sum $(r_t(n,m) + r_r(n,m))^2$, as in the case of specular reflection [15]. As a result, the path attenuation of each cascaded channel $h_{n,m}$ is expected to be very high in general.

2.4 Phase Shift Design

In this section, we discuss the phase shift design that maximizes the SNR, and hence the achievable rate, in the near-field. Specifically, we show that conventional far-field beamforming can become highly suboptimal when the Tx operates close to the IRS. Instead, the optimal IRS configuration makes use of the spherical wavefront to focus the incident EM wave towards a specific point of space, a capability that is not possible with beamforming.

2.4.1 Beamfocusing

Based on the spherical wave model, the phase profile that maximizes the SNR is given by

$$\varphi_{n,m} = k(r_t(n,m) + r_r(n,m)) \tag{2.19}$$

which is referred to as *beamfocusing*. This is because the IRS acts like a lens focusing the incident EM wave towards a specific point of space rather than towards a direction (θ, ϕ) [16]. Under beamfocusing, the SNR becomes

$$\mathrm{SNR} = \frac{P_t}{\sigma^2}\left|\sum_{n=0}^{N_1-1}\sum_{m=0}^{N_2-1}\sqrt{\mathrm{PL}_{n,m}}\right|^2 \geq \frac{P_t}{\sigma^2}\min\{\mathrm{PL}_{n,m}\}N^2 \tag{2.20}$$

and, hence, it increases quadratically with the number N of IRS elements.

2.4.2 Conventional Beamforming

Typical IRS processing [17] relies on the far-field assumption, whereby the spherical wavefront degenerates into a plane wavefront. This enables the use of the parallel-ray approximations (i.e. Figure 2.4)

$$r_t(n,m) \approx r_t - nL_x\cos\phi_t\sin\theta_t - mL_y\sin\phi_t\sin\theta_t \tag{2.21}$$

$$r_r(n,m) \approx r_r - nL_x\cos\phi_r\sin\theta_r - mL_y\sin\phi_r\sin\theta_r \tag{2.22}$$

Mathematically speaking, (2.21) and (2.22) follow from the first-order Taylor expansion $(1+x)^a \approx 1+ax$ of (2.11) and (2.12), respectively. In the so-called beamforming strategy', the phase shifts are designed as follows:

$$\begin{aligned}\varphi_{n,m} = -k(&nL_x\cos\phi_t\sin\theta_t + mL_y\sin\phi_t\sin\theta_t \\ &+ nL_x\cos\phi_r\sin\theta_r + mL_y\sin\phi_r\sin\theta_r)\end{aligned} \tag{2.23}$$

which depend solely on the angular information (θ_t, ϕ_t) and (θ_r, ϕ_r). As a result, beamforming can be highly suboptimal in the Fresnel zone. To analytically characterize the reduction in the SNR, we first consider that the IRS is deployed close

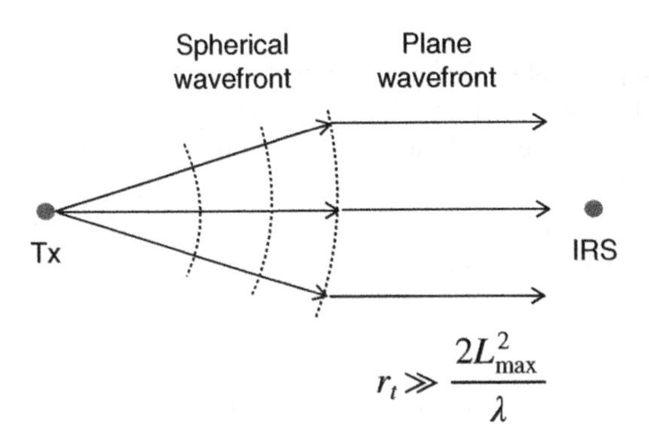

Figure 2.4 Illustration of the parallel-ray approximation in the far-field region.

to the Tx, whilst the Rx is in the far-field of the IRS.[2] Akin to (2.20), the SNR is lower bounded as follows:

$$\mathrm{SNR} \geq \frac{P_t}{\sigma^2} \min\{\mathrm{PL}_{n,m}\} N^2 G \tag{2.24}$$

where $G \in [0,1]$ is the normalized power gain defined as follows:

$$G \triangleq \frac{\left|\sum_{n=0}^{N_1-1} \sum_{m=0}^{N_2-1} e^{-jk(r_t(n,m)+r_r(n,m))} e^{j\varphi_{n,m}}\right|^2}{N_1^2 N_2^2} \tag{2.25}$$

We next exploit the Fresnel approximation of the Tx distance

$$\begin{aligned} r_t(n,m) \approx r_t &+ \frac{(nL_x)^2(1-\cos^2\phi_t \sin^2\theta_t)}{2r_t} - nL_x \cos\phi_t \sin\theta_t \\ &+ \frac{(mL_y)^2(1-\sin^2\phi_t \sin^2\theta_t)}{2r_t} - mL_y \sin\phi_t \sin\theta_t \end{aligned} \tag{2.26}$$

to recast G under the phase shift design (2.23) as follows:

$$G \approx \frac{\left|\sum_{n=0}^{N_1-1} e^{-jk\frac{(nL_x)^2(1-\cos^2\phi_t \sin^2\theta_t)}{2r_t}}\right|^2}{N_1^2} \frac{\left|\sum_{m=0}^{N_2-1} e^{-jk\frac{(mL_y)^2(1-\sin^2\phi_t \sin^2\theta_t)}{2r_t}}\right|^2}{N_2^2} \tag{2.27}$$

which admits the approximation:

$$\begin{aligned} G &\approx \frac{\left|\sum_{n=0}^{N_1^2-1} e^{-jk\frac{nL_x^2(1-\cos^2\phi_t \sin^2\theta_t)}{2r_t}}\right|^2}{N_1^4} \frac{\left|\sum_{m=0}^{N_2^2-1} e^{-jk\frac{mL_y^2(1-\sin^2\phi_t \sin^2\theta_t)}{2r_t}}\right|^2}{N_2^4} \\ &= \left|D_{N_1^2}\left(\frac{2\pi}{\lambda}\frac{L_x^2(1-\cos^2\phi_t \sin^2\theta_t)}{2r_t}\right)\right|^2 \left|D_{N_2^2}\left(\frac{2\pi}{\lambda}\frac{L_y^2(1-\sin^2\phi_t \sin^2\theta_t)}{2r_t}\right)\right|^2 \end{aligned} \tag{2.28}$$

where $D_N(x) = \frac{\sin(Nx/2)}{N\sin(x/2)}$ denotes the Dirichlet sinc function. The validity of the Fresnel approximation (2.26) is shown in Figure 2.5.

In addition, the accuracy of the approximate closed-form expression (2.28) is evaluated in Figure 2.6, which exhibits a very good match with the exact formula. Most importantly, we observe that beamforming substantially decreases the power

2 In fact, deploying the IRS close to one of the link ends yields the maximum SNR, compared to placing it somewhere in between [18].

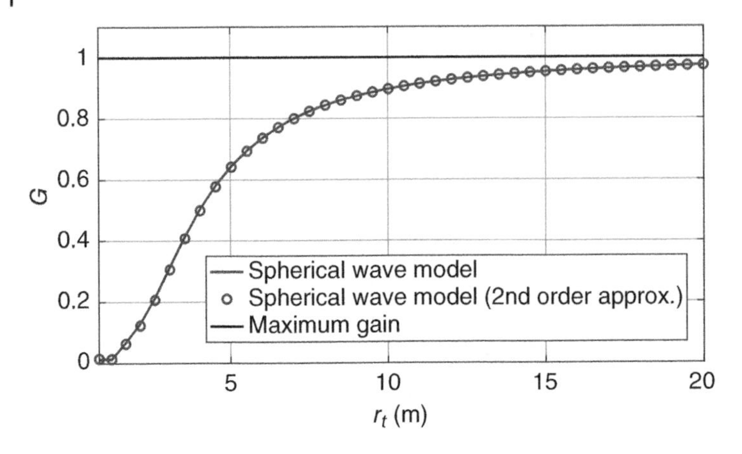

Figure 2.5 Normalized power gain versus distance r_t for an 100×100-element IRS with $L_x = L_y = \lambda/2$, $\mathbf{p}_t = (0.4, 0.4, z)$, $0.5 \leq z \leq 10$, and $0.755 \leq r_t \leq 10.016$ m.

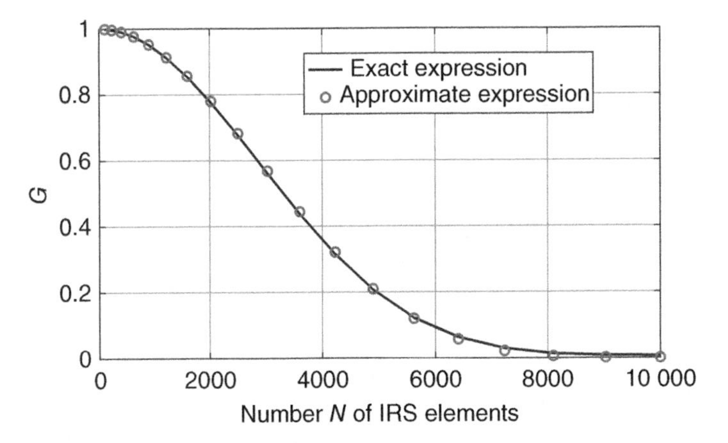

Figure 2.6 Normalized power gain versus number N of IRS elements for $L_x = L_y = \lambda/2$, $\mathbf{p}_t = (0.4, 0.4, 1)$, and $r_t = 1.15$ m.

gain when the Tx operates in the near-field of an electrically large IRS. Lastly, capitalizing on (2.28), we have the asymptotic result $G \to 0$ as $N \to \infty$. This implies that for a finite yet large number N of IRS elements, the total power gain N^2G tends to zero as N grows, which is demonstrated in Figure 2.7. In conclusion, the decrease in the power gain cannot be compensated by increasing the number of IRS elements, and hence, beamfocusing is the optimal mode of operation in the Fresnel zone.

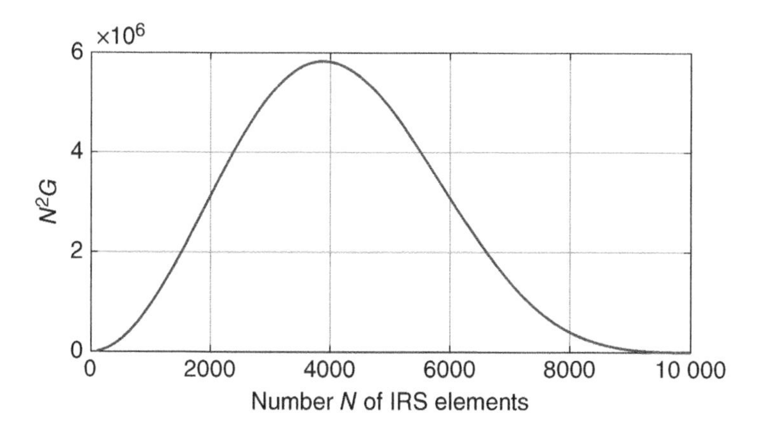

Figure 2.7 Power gain versus number N of IRS elements for $\mathbf{p}_t = (0.4, 0.4, 1)$, $r_t = 1.15$ m, and $L_x = L_y = \lambda/2$.

2.5 Energy Efficiency

So far we have restricted our discussion to a single-antenna Tx and Rx. We now extend our analysis to the MIMO case, where both the Tx and Rx are equipped with multiple antennas. Specifically, we investigate under which conditions IRS-aided MIMO can outperform MIMO in terms of energy efficiency (EE).

2.5.1 MIMO System

Consider a MIMO system where the Tx and Rx are equipped with N_t and N_r antennas, respectively. The Tx seeks to communicate a single data stream to the Rx through a LoS channel, whose path loss is calculated using the Friss transmission formula:

$$\mathrm{PL}_{\mathrm{MIMO}} = G_t G_r \left(\frac{\lambda}{4\pi r_d} \right)^2 \tag{2.29}$$

where $r_d = \|\mathbf{p}_r - \mathbf{p}_t\|$. For an adequately small N_r and N_t, far-field propagation can be assumed. In this case, the LoS channel is rank-one [19, Ch. 7]. Hence, analogue beamforming and combining yield the received SNR:

$$\mathrm{SNR}_{\mathrm{MIMO}} = \frac{N_r N_t P_t \mathrm{PL}_{\mathrm{MIMO}}}{\sigma^2} \tag{2.30}$$

For a hybrid array architecture with a single radio-frequency chain, the power consumption of the MIMO system is[3]

$$P_{\mathrm{MIMO}} = P_t + N_r(P_{\mathrm{PS}} + P_{\mathrm{PA}}) + N_t(P_{\mathrm{PS}} + P_{\mathrm{PA}}) \tag{2.31}$$

3 The power consumption of signal processing is neglected.

where P_{PS} and P_{PA} denote the power consumption values for a phase shifter and a power amplifier, respectively.

2.5.2 IRS-aided MIMO System

The Tx and Rx perform beamforming and combining to communicate a single stream through the N-element IRS. Due to the directional transmissions, the Tx–Rx link is very weak, and hence is ignored. The received SNR of the IRS-aided MIMO system is simply given by

$$\mathrm{SNR}_{\mathrm{IRS}} = \frac{N_t N_r N^2 P_t \mathrm{PL}_{\mathrm{IRS}}}{\sigma^2} \tag{2.32}$$

where $\mathrm{PL}_{\mathrm{IRS}}$ denotes the path loss of the IRS-aided link in (2.18) calculated for an arbitrary IRS element. The phase of each IRS element is controlled by a varactor diode, which consumes a negligible power compared to a typical phase shifter. Thus, the power consumption of each reflecting element is set to $P_{\mathrm{e}} = 0$ [20], and the total power expenditure of the IRS-assisted MIMO system is $P_{\mathrm{IRS}} \approx P_{\mathrm{MIMO}}$.

Proposition 2.1 *The IRS-aided MIMO system with N_t/α and N_r/α antennas, where α is a positive integer, attains a higher SNR than MIMO with N_t and N_r antennas for*

$$N^{\star} \geq \alpha \frac{\lambda}{L_x L_y} \frac{r_t r_r}{\sqrt{F(\theta_t, \phi_r, \theta_r)} r_d} \tag{2.33}$$

Proof: Using (2.30) and (2.32), the IRS-aided system achieves a higher SNR for $N^{\star} \geq \sqrt{\alpha^2 \mathrm{PL}_{\mathrm{MIMO}}/\mathrm{PL}_{\mathrm{IRS}}}$, which gives the desired result after basic algebra. ∎

According to Proposition 2.1, we can decrease the number of Tx and Rx antennas by a factor α to reduce the power consumption as follows:

$$P_{\mathrm{IRS}}(N_t/\alpha, N_r/\alpha) = P_t + \frac{N_r}{\alpha}(P_{\mathrm{PS}} + P_{\mathrm{PA}}) + \frac{N_t}{\alpha}(P_{\mathrm{PS}} + P_{\mathrm{PA}}) \approx P_{\mathrm{MIMO}}/\alpha \tag{2.34}$$

while keeping the achievable rate fixed. Consequently, the EE gain with respect to MIMO is approximately equal to α. Let B denote the transmit bandwidth. The achievable rate is finally calculated as follows:

$$R = B \log_2(1 + \mathrm{SNR}_{\mathrm{IRS}}) \quad \text{(bit/s)} \tag{2.35}$$

whilst the EE is given by $\mathrm{EE} \triangleq R/P_{\mathrm{IRS}}$. In this numerical experimental, we consider $N_t = N_r = 100$ antennas for the MIMO system without an IRS. From Figure 2.8, we verify that the IRS-assisted MIMO system, with $N_t = N_r = 50$ antennas, offers a twofold EE gain. Specifically, the IRS provides an alternative communication

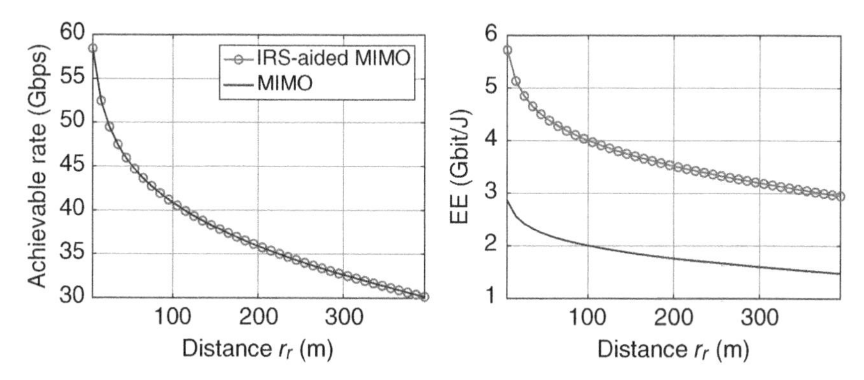

Figure 2.8 Achievable rate and EE versus distance r_r for $\alpha = 2$ and a fixed IRS location at $(0,0,0)$. In the MIMO system, $N_t = 100$ and $N_r = 100$ antennas. The other parameters are as follows: $G_t = G_r = 20$ dBi, $P_t = 10$ dBm, $P_{\mathrm{PS}} = 42$ mW and $P_{\mathrm{PA}} = 60$ mW, $\sigma^2 = -174$ dBm/Hz, $B = 2$ GHz, $f = 60$ GHz, $L_x = L_y = \lambda/2$, $\mathbf{p}_t = (0, -0.6, 1)$ with $r_t = 1.16$ m, and $\mathbf{p}_r = (0, r_r, 1)$.

link in addition to LoS, where the Tx and Rx employ a smaller number of antennas to communicate with each other, hence reducing the power consumption of the system. However, note that the suggested benefits hold when: (i) the power expenditure of IRS elements is negligible compared to that of conventional phase shifters; and (ii) reflection losses are small [21].

Remark 2.1 In our EE analysis, we assumed $P_e \approx 0$, which implies that the IRS can have an arbitrarily large number of IRS elements without increasing the power consumption of the system. Since this assumption may not be realistic, we derive the necessary condition for having an EE gain when $P_e > 0$. Particularly, the power consumption of each IRS element must satisfy

$$P_{\mathrm{MIMO}} > P_{\mathrm{IRS}} \Rightarrow P_{\mathrm{MIMO}} > N^{\star} P_e + P_{\mathrm{MIMO}}/\alpha \Rightarrow P_e < \frac{P_{\mathrm{MIMO}}(1 - 1/\alpha)}{N^{\star}} \tag{2.36}$$

2.6 Optimal IRS Placement

We now study the impact of the IRS position on the number $N^{\star}$ of reflecting elements. For the deployment considered in Figure 2.8, r_t is small, and hence $r_r^2 \approx r_t^2 + r_d^2$. Further, $\phi_r = \pi/2$ which gives $F(\theta_t, \phi_r, \theta_r) = \cos^2\theta_t$. Then, (2.33) reduces to

$$N^{\star} = \alpha \frac{\lambda}{L_x L_y} \frac{r_t r_r}{\cos\theta_t \sqrt{r_r^2 - r_t^2}} \tag{2.37}$$

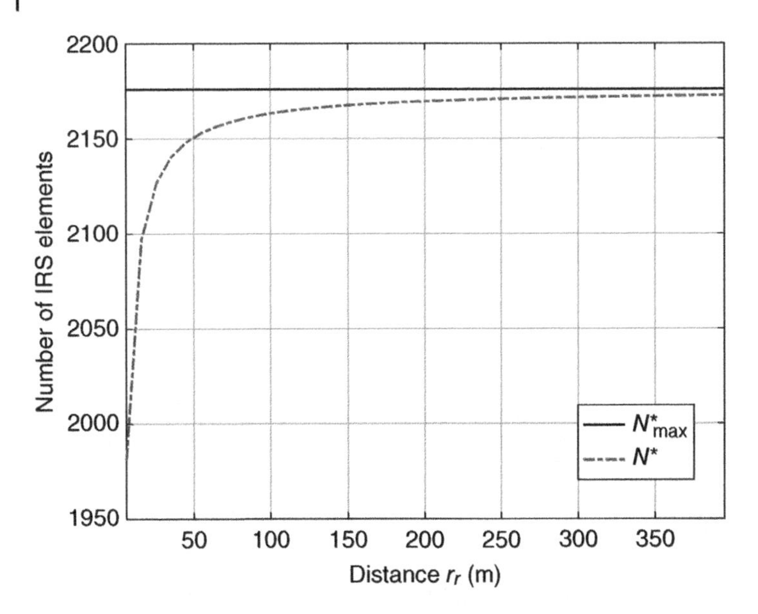

Figure 2.9 Number of IRS elements versus distance r_r for $\alpha = 2$ and a fixed IRS location at (0,0,0). The other parameters are as follows: $\mathbf{p}_t = (0, -0.6, 1)$, $\mathbf{p}_r = (0, r_r, 1)$, $f = 60$ GHz, and $L_x = L_y = \lambda/2$.

which takes the asymptotic value:

$$N^{\star}_{\max} = \alpha \frac{\lambda}{L_x L_y} \frac{r_t}{\cos \theta_t} \tag{2.38}$$

as $r_r \to \infty$; this follows from $\sqrt{r_r^2 - r_t^2} \approx r_r$ for $r_r \gg r_t$. As evinced, $N^{\star}$ is bounded for a fixed IRS position near the Tx. We stress that due to symmetry, the same result holds when the IRS is near the Rx. For instance, $N^{\star}_{\max} = 2176$ elements in Figure 2.9. In contrast, when the IRS is deployed always in the middle of the link ends, $N^{\star}$ increases as $O(r_t r_r)$. This scaling law is depicted in Figure 2.10. In light of this finding, we conclude that the IRS should be deployed close to one of the link ends to effectively mitigate the path loss with a reasonable number of reflecting elements.

2.7 Open Future Research Directions

The area of IRSs has been extensively explored over that last years. Nevertheless, there is still a plethora of research problems that are worth investigation. In this section, we will delineate those problems along with some representative work from the related literature.

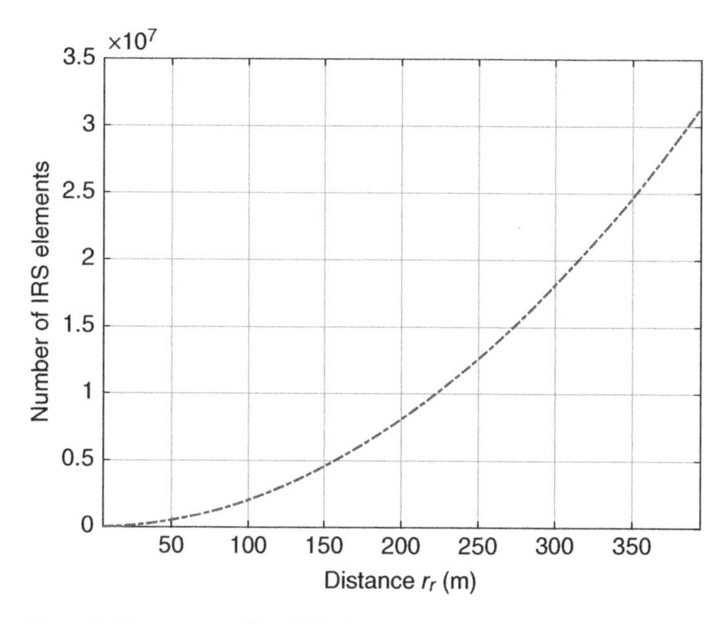

Figure 2.10 Number $N^{\star}$ of IRS elements versus distance r_r for $\alpha = 2$ and a varying IRS position. The other parameters are as follows: $\mathbf{p}_t = (0, -0.6, 1)$, $\mathbf{p}_r = (0, r_r, 1)$, IRS at $(0, (r_r - y_t)/2, 1)$ with $y_t = -0.6, f = 60$ GHz, and $L_x = L_y = \lambda/2$.

1. **Channel estimation**: One critical question is how the IRS acquires timely channel state information used in passive beamforming without possessing sensing units. A common approach is to have a channel estimation stage preceding data transmission, wherein the receiver sends pilots through the IRS. However, this entails overheads that can become overwhelmingly high even for a moderate number of IRS elements, and especially for fast fading channels [22].
2. **Path loss modelling**: Scattering from an IRS can be described using surface equivalence principles, such as physical optics [23]. In this context, the IRS is treated as a thin sheet of infinite extent whose surface currents determine the scattered field. However, an IRS is composed of closely spaced discrete elements. Thus, mutual coupling effects kick in, which can affect the overall surface current distribution and phase shift design [24]. The impact of mutual coupling on surface currents is yet not clear.
3. **Wideband reflection**: As the IRS size and transmission bandwidths grow, the propagation delay over the IRS becomes larger and can exceed the symbol period. In this case, we have a spatially wideband channel that requires frequency-dependent beamforming even in LoS conditions [25]. Existing solutions rely on optimizing the reflection coefficients of narrowband unit cells

over a wide range of frequencies [26]. Thus, how to design an IRS architecture similar to true-time-delay of phased arrays is a very interesting direction for future work.

2.8 Conclusions

In this chapter, we first provided a general description of IRS-aided communication, which is deemed a very promising candidate for 6G. We then focused our analysis on the near-field region, wherein the channel modelling and phase shift design problems differ from those in the far-field. Specifically, the spherical wavefront of the emitted EM waves plays a key role, rendering beamfocusing the optimal model of operation. In this context, we showed that conventional beamforming can significantly decrease the SNR, and hence should be avoided. We finally investigated IRS-aided MIMO, where large EE gains can be attained in the near-field with a reasonable number of reflecting elements.

References

1 Boccardi, F., Heath, R.W., Lozano, A. et al. (2014). Five disruptive technology directions for 5G. *IEEE Commun. Mag.*, 52 (2): 74–80.

2 Di Renzo, M. Zappone A., Debbah M., et al. (2020). Smart radio environments empowered by reconfigurable intelligent surfaces: how it works, state of research, and the road ahead. *IEEE J. Sel. Areas Commun.*, 38 (11): 2450–2525.

3 Zhang, J. Björnson, E. Matthaiou, M. et al. (2020). Prospective multiple antenna technologies for beyond 5G. *IEEE J. Sel. Areas Commun.*, 38 (8): 1637–1660.

4 Matthaiou, M. Yurduseven, O. Quoc Ngo, H. et al. (2021). The road to 6G: ten physical layer challenges for communications engineers. *IEEE Commun. Mag.*, 59 (1): 64–69.

5 Wu, Q. and Zhang, R. (2020). Towards smart and reconfigurable environment: intelligent reflecting surface aided wireless network. *IEEE Commun. Mag.*, 58 (1): 106–112.

6 Gros, J.-B., Popov, V., Odit, M.A. et al. (2021). A reconfigurable intelligent surface at mmWave based on a binary phase tunable metasurface. *IEEE Open J. Commun. Soc.*, 2: 1055–1064.

7 Zhu, B.O., Zhao, J., and Feng, Y. (2013). Active impedance metasurface with full 360 reflection phase tuning. *Sci. Rep.* 3: 3059–3064.

8 Wu, Q. and Zhang, R. (2019). Intelligent reflecting surface enhanced wireless network via joint active and passive beamforming. *IEEE Trans. Wireless Commun.*, 18 (11): 5394–5409.

9 Huang, C., Zappone, A., Alexandropoulos, G.C. et al. (2019). Reconfigurable intelligent surfaces for energy efficiency in wireless communication. *IEEE Trans. Wireless Commun.*, 18 (8): 4157–4170.

10 Wang, P., Fang, J., Yuan, X. et al. (2020). Intelligent reflecting surface-assisted millimeter wave communications: joint active and passive precoding design. *IEEE Trans. Veh. Technol.*, 69 (12): 14960–14973.

11 Rappaport, T.S. Xing, Y. Kanhere, O. et al. (2019). Wireless communications and applications above 100 GHz: opportunities and challenges for 6G and beyond. *IEEE Access*, 7: 78729–78757.

12 Balanis, C.A. (1996). Antenna cheory. In: *Analysis and Design*. Wiley. 960.

13 Dovelos, K., Assimonis, S.D., Ngo, H.Q. et al. (2021). Intelligent reflecting surfaces at terahertz bands: channel modeling and analysis. *Proceedings of IEEE ICC*, June 2021.

14 Balanis, C.A. (2012). *Advanced Engineering Electromagnetics*, 2e. Wiley.

15 Han, C., Bicen, A.O., and Akyildiz, I.F. (2015). Multi-ray channel modeling and wideband characterization for wireless communications in the terahertz band. *IEEE Trans. Wireless Commun.*, 14 (5): 2402–2412.

16 Yurduseven, O., Assimonis, S.D., and Matthaiou, M. (2020). Intelligent reflecting surfaces with spatial modulation: an electromagnetic perspective. *IEEE Open J. Commun. Soc.*, 1: 1256–1266.

17 Abeywickrama, S., Zhang, R., Wu, Q., and Yuen, C. (2020). Intelligent reflecting surface: practical phase shift model and beamforming optimization. *IEEE Trans. Commun.*, 68 (9): 5849–5863.

18 Wu, Q. Zhang, S. Zhang, B. et al. (2021). Intelligent reflecting surface aided wireless communications: a tutorial. *IEEE Trans. Commun.* 69 (5): 3313–3351.

19 Tse, D. and Viswanath, P. (2005). *Fundamentals of Wireless Communication*. New York: Cambridge University Press.

20 Tang, W. Chen, M.Z. Chen, X. et al. (2021). Wireless communications with reconfigurable intelligent surface: path loss modeling and experimental measurement. *IEEE Trans. Wireless Commun.*, 20 (1): 421–439.

21 Chou, S.-K., Yurduseven, O., Ngo, H.Q., and Matthaiou, M. (2021). On the aperture efficiency of intelligent reflecting surfaces. *IEEE Wireless Commun. Lett.*, 10 (3): 599–603.

22 Zappone, A., Di Renzo, M., Xi, X., and Debbah, M. (2021). On the optimal number of reflecting elements for reconfigurable intelligent surfaces. *IEEE Wireless Commun. Lett.*, 10 (3): 464–468.

23 Dovelos, K., Assimonis, S.D., Ngo, H.Q. et al. (2022). Electromagnetic modeling of holographic intelligent reflecting surfaces at the terahertz band. *Proceedings of ASILOMAR*.

24 Gradoni, G. and Di Renzo, M. (2021). End-to-end mutual coupling aware communication model for reconfigurable intelligent surfaces: an electromagnetic-compliant approach based on mutual impedances. *IEEE Wireless Commun. Lett.*, 10 (5): 938–942.

25 Dovelos, K., Matthaiou, M., Ngo, H.Q., and Bellalta, B. (2021). Channel estimation and hybrid combining for wideband terahertz massive MIMO systems. *IEEE J. Sel. Areas Commun.*, 39 (6): 1604–1620.

26 Cai, W., Li, H., Li, M., and Liu, Q. (2020). Practical modeling and beamforming for intelligent reflecting surface aided wideband systems. *IEEE Commun. Lett.*, 24 (7): 1568–1571.

3

Feasibility of Intelligent Reflecting Surfaces to Combine Terrestrial and Non-terrestrial Networks

Muhammad A. Jamshed, Qammer H. Abbasi, and Masood Ur-Rehman

James Watt School of Engineering, University of Glasgow, Glasgow, UK

3.1 Introduction

Rapid evolution has been observed in the field of mobile communication, where every decade represents a new generation of wireless communication [1]. Although a significant improvement has been achieved in terms of high data rates, lower latency, and ultra-reliability, the telecommunication industry only focuses on providing high-end wireless communication services to densely populated areas or metropolitan cities [2, 3]. Consequently, all the research efforts of industry and academia are mainly focused on developing methods that are more effective in enhancing the capacity of metropolitan cities-based cellular infrastructure [4]. This focus has resulted in the development of networks that are not suitable to fulfil the business needs of the telecommunication industry in providing communication facilities to inhabitants of very less-populated areas [5]. It is noted that the business part of any project must be considered before adopting a technology.

Every telecommunication operator offers homogeneous services, i.e. the cost of mobile data rate is fixed across the whole country. These plans are designed based on the clientâs needs to capture more users and develop a profitable business [6]. However, the telecommunication vendors are facing pressure to reduce these prices further due to a rapid change in market dynamics. This price reduction will significantly reduce the telecommunication vendor's interest in deploying telecommunication infrastructure in rural areas as the profit margins are null or negative [7]. Nowadays, a typical telecommunication vendor can get a profit by deploying a traditional terrestrial network in locations with over 10k subscribers [8]. Although to cover areas with 1k subscribers, the telecommunication operators are relying on technologies that are designed and optimized for

Intelligent Reconfigurable Surfaces (IRS) for Prospective 6G Wireless Networks, First Edition.
Edited by Muhammad Ali Imran, Lina Mohjazi, Lina Bariah, Sami Muhaidat,
Tie Jun Cui, and Qammer H. Abbasi.

low-density locations, yet the main challenges arise to cover the location with less than 1k subscribers or in a region with scattered houses. To overcome this issue, new ways and reinvention of technology are required.

In order to address these issues, the 3rd generation partnership project (3GPP) in Release 17, is working on the standardization of non-terrestrial networks (NTN) to enable 5G-based terrestrial networks to support NTN [9, 10]. Based on an initial version of Release 17, the NTN will consist of low earth orbit (LEO) satellites constellation, whereas the unmanned aerial vehicles (UAVs) and high-altitude platforms (HAPs) are considered as a special use case of NTN networks [9]. Moreover, in 3GPP Release 17, the new standards are introduced to support satellites. This support can play a major role in reducing the total telecommunication infrastructure cost. Moreover, it can also allow satellite-based narrow-band communication from a typical mobile handset (enabled to support non-terrestrial connectivity solutions (NTCS)) and can improve network interoperability. Based on TELECOM INFRA PROJECT (TIP) [8], a typical architecture of NTCS supporting 5G-based terrestrial networks is shown in Figure 3.1.

The signals from these satellites are weakened and dispersed by the indoor environment. As a result, the NTN are unable to provide an effective indoor coverage [11]. Moreover, the power and delay are the critical factors that restrict a mobile handset to obtain reliability, seamless coverage, and low latency by using NTN [12]. Based on the recent theoretical and experimental analysis, the intelligent reflecting surfaces (IRSs) have proven to be much effective in tunning wireless signals to provide an additional channel gain between each transmitter and

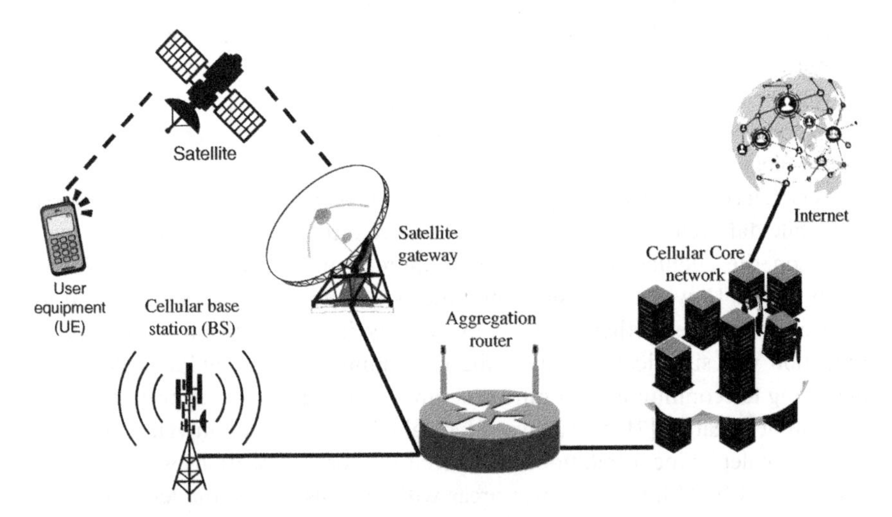

Figure 3.1 Potential architecture of enabling a typical mobile handset to directly communicate with NTN.

receiver pair [13, 14]. In this chapter, we have explored the possibilities of using IRSs in merging the NTN in terrestrial networks. The remainder of the chapter is organized as follows: in Section II, the background of IRSs has been discussed. Section II also discusses the opportunities of using IRS in wireless networks. Based on 3GPP guidelines, Section III discusses different aspects of NTN, various architectures of NTN, and connectivity solution proposed by 3GPP. Section IV discusses how IRSs can be integrated with NTN to enable a typical mobile handset to directly communicate with satellites. Finally, Section IV concludes the chapter.

3.2 Intelligent Reflecting Surfaces

With the commercialization of the 5G wireless communication systems, nowadays, researchers are exploring various means to further improve cellular coverage, achieving reliable, and faster data rates [15]. In recent literature, the IRSs have shown some significant results in achieving such improvements [16]. In this section, we have briefly discussed the background, architecture, and opportunities of using IRSs in a wireless network.

3.2.1 Background and Architecture

The IRS is a flat surface composed of a large number of passive reflecting components, each of which may independently produce a controlled amplitude and/or phase shift on the input signal [14, 17]. The NTT DoCoMo and MetaWave demonstrated the first use of IRS in reflecting the wireless data, while operating at 28 GHz [18]. IRSs are capable of controlling the propagation of electromagnetic waves by intelligently varying the magnetic and electric field of each IRS unit [19]. An array of IRS units can incur some independent changes, i.e. phase, amplitude, frequency, polarization, etc., to engineer the propagation of wireless signals [14, 17]. Mostly, the research outputs, consider the phase shift to smartly tune the signal reflection, to aid the transmissions between the sender and the receiver [20].

IRS is not only theoretically appealing but it also has several advantages. Unlike standard active antenna arrays, the IRSs passively reflect the impinging signals, mitigating the need of radio frequency (RF) transmission chains [21]. As a result, it can be operated with lower hardware and energy cost [21]. IRS operates in full-duplex mode, with no antenna noise amplification or self-interference, hence, providing competitive advantages over traditional active relays [14]. IRS may also be readily placed on or removed from environment objects for deployment or replacement because of its low profile and conformal shape. A typical

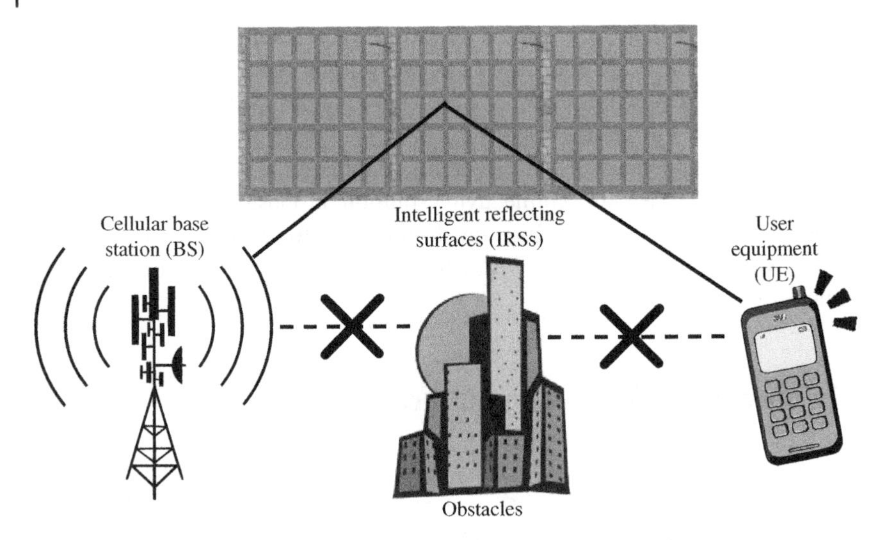

Figure 3.2 Illustration of a typical IRS to aid the transmissions between the sender and the receiver.

configuration of IRS in enhancing the coverage of cellular network is shown in Figure 3.2 [22]. Based on research carried out so far, the IRSs have proven to be much more effective to achieve seamless coverage, ultra-reliable, and faster data rates.

3.2.2 Intelligent Reflecting Surfaces in Wireless Networks

IRSs functions as an auxiliary device in wireless networks and can be seamlessly incorporated into them, giving it a lot of flexibility and compatibility. IRS is well suited to be widely applied in wireless networks to considerably improve spectral and energy efficiency while being cost-effective. Therefore, IRS is expected to result in a fundamental paradigm shift in wireless system designs, such as moving from the existing multiple-input-multiple-output (MIMO) networks without IRS to the new IRS-aided small and moderate MIMO systems [23]. IRS-aided MIMO system uses the large aperture to create fine-grained reflected beams via smart passive reflections, allowing the base station (BS) to be equipped with significantly fewer antennas without compromising the users' quality-of-service (QoS) [24]. As a result, system hardware costs and energy usage may be drastically lowered, particularly for wireless systems that will migrate to higher frequency bands in the future.

The signal propagation between transmitters and receivers can be reconfigured by deploying IRSs to achieve desired realizations, hence providing a new paradigm to effectively combat the issues of interference and channel fading [17]. This can also enable the wireless network in achieving an improvement in capacity and reliability. Integrating IRSs into a wireless network will transform an existing heterogeneous network with solely active components into a new hybrid design with both active and passive components cooperating intelligently. Because IRSs are far less expensive than their active counterparts, they may be deployed more densely in wireless networks at even lower costs, all while avoiding the need for complex interference control between IRSs owing to their passive reflection and resulting local coverage. It is foreseen that the new IRS-aided wireless network will be promising to achieve a cost effective and sustainable capacity growth. Despite the benefits of IRS, it also faces some challenges, such as channel estimation, reflection optimization, and deployment from communication design perspectives [25].

3.3 Non-terrestrial Networks

Recently, the satellite industry has gained popularity and shown some promising technological advancements to be integrated with terrestrial-based 5G networks. One of the main reasons for this addition by 3GPP is a planned initiative by OneWeb to launch 648 LEO satellite constellation by the end of 2022. Moreover, any non-terrestrial flying objects, such as HAPs, UAVs also falls under the umbrella of NTN, as shown in Figure 3.3 [26]. The major focus of the telecommunication industry is to provide high-end communications services to densely

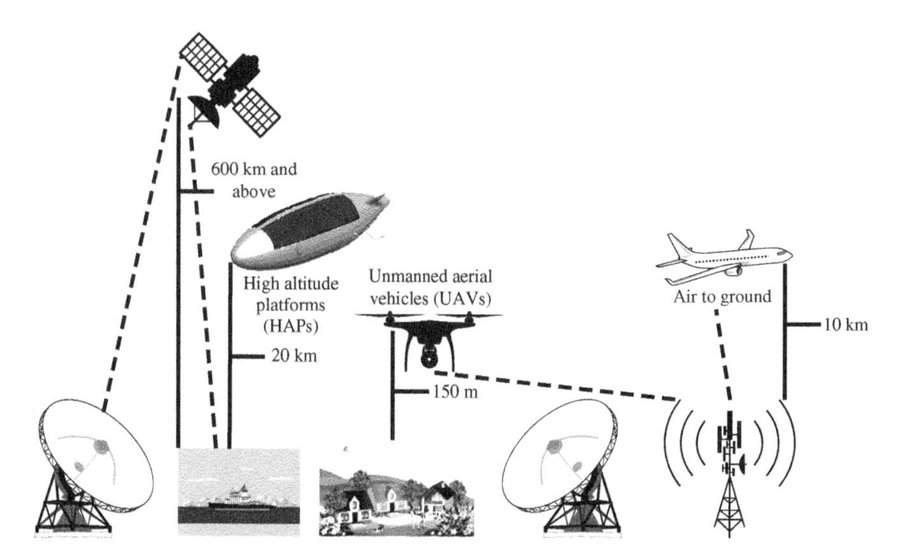

Figure 3.3 Illustration of different NTN platforms. Source: Dzianis Rakhuba/Adobe Stock; sannare/Adobe Stock.

populated areas [2, 3] to capture more users and develop a profitable business. However, the market dynamics are pushing telecommunication vendors to reduce the cost of data rates. This reduction in data rate prices will make the telecommunication vendors completely lose their interest in providing cellular communication services in rural areas. In order to address these issues, the 3GPP in Release 17, is working on the standardization of NTN to enable 5G-based terrestrial networks to support NTN. The NTN is always a part of 3GPPs vision, but the commercial reasons have always delayed the implementation and trials. Over the past few years, tremendous growth in wireless communication as well aerospace industry (including the hype of UAVs, HAPs, and LEO satellite constellation), have made 3GPP to finally work towards the standardization of NTN in 5G-based wireless networks. Figure 3.4 illustrates an initial vision of TIP, to integrate the NTN in 5G networks [8]. In the following, we have briefly discussed different aspects of NTN, various architectures of NTN, and connectivity solutions proposed by 3GPP.

3.3.1 Non-terrestrial Networks: 3GPP Vision

In Release 15, 3GPP started to integrate new radio (NR) in NTN, with the main focus on channel models and deployment scenarios. The main objectives in Release 15 were to define some reference deployment scenarios of NTN and some parameters, i.e. frequency bands, orbital altitude, etc. In Release 16, 3GPP set objective to identify necessary enabling features to support NR in NTN, especially looking into satellites. Finally, in Release 17, the 3GPP has set objectives to

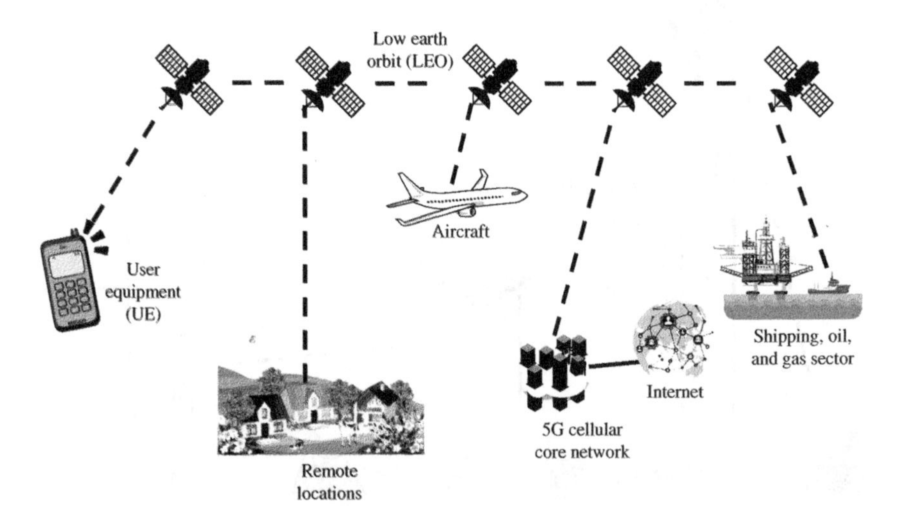

Figure 3.4 Use cases of NTN, using LEO satellite constellations. Source: sannare/Adobe Stock.

start a work item on NTN in NR. The requirements involve defining necessary enhancements of protocols, architecture, RF upgradation, physical layer aspects, etc. The NTN defines a set of architecture options in accordance with TR 38.821 [27] and 38.811 [28]. Two categories have been identified, in the non-terrestrial segment:

1. Satellite-based platforms.
2. UAVs and HAPs.

Table 3.1 summarizes the six macro-scenarios identified in the 3GPP framework [29]. In these scenarios, the user service link can be either in S-band or in Ka-band. The LEO constellations reference scenario can be differentiated based on the type of on-ground coverage provided, i.e. steerable, or fixed beams. The regenerative payload scenarios D1 and D2 are used for the inter-satellite link, whereas transparent payload scenarios A, C2, and D2 have been given higher priority. Recently, the possibility of implementing steerable beams has received increased attention. Finally, for the time being, scenario B has been de-emphasized. The various architecture options can be broadly classified based on the type of payload the satellite is equipped with, i.e. regenerative or transparent, in which the satellite contains a full or a partial Next generation node B (gNB). Based on 3GPP Release 17, Figure 3.5, summarizes the NTCS and is divided into two parts:

1. Direct user equipment (UE)/mobile handset access.
2. Indirect UE access, using satellite as a relay.

Figures 3.6 and 3.7, show 3GPP based NTCS reference architecture based on direct UE access. It is noted that Figure 3.6 refers to transparent satellite-based NTN architecture, whereas Figure 3.7 refers to regenerative-based satellite NTN architecture with a distributed unit on satellite. The gNB is conceptually situated at the gateway (GW) in the transparent scenario, while it is implemented on-board with payloads in the regenerative case. The standard NR-Uu air interface is used in both scenarios to implement the user access connection between the satellite and the on-ground UE; using this air interface, it is vital to appropriately examine the impact of satellite channel impairments on the PHY and MAC operations.

Table 3.1 NTN reference scenarios.

System	Regenerative	Transparent
Fixed antenna beams LEO satellites	D2	C2
Steerable antenna beams LEO satellites	D1	C1
Geosynchronous equatorial orbit (GEO)	B	A

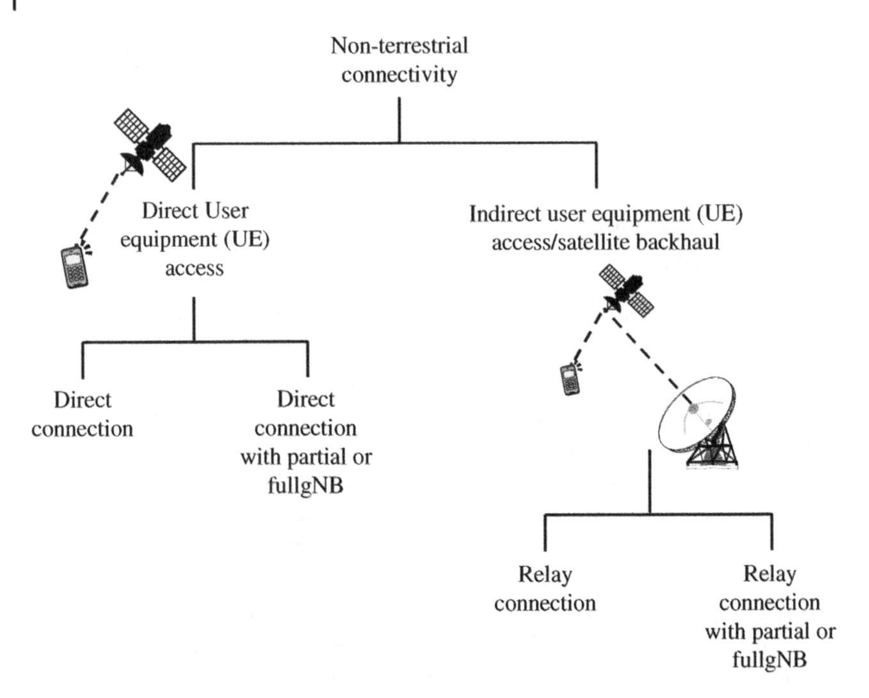

Figure 3.5 Hierarchical representation of NTCS.

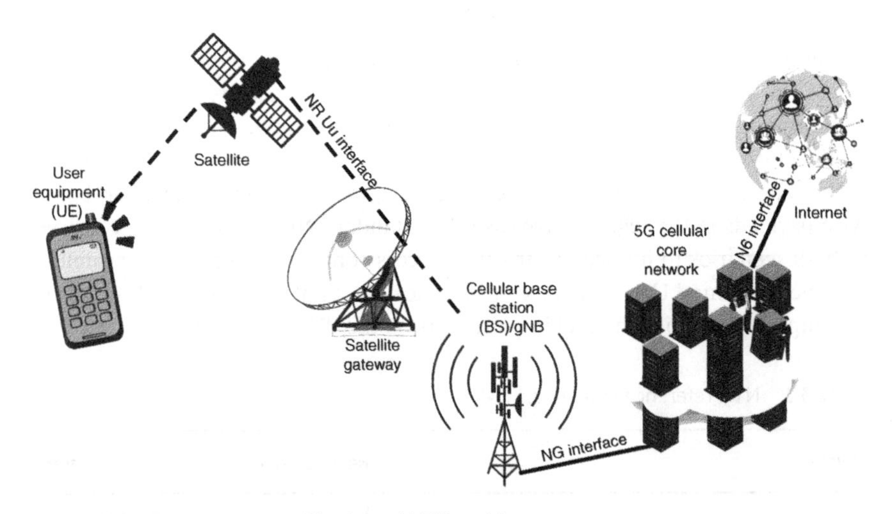

Figure 3.6 Transparent satellite-based NTN architecture.

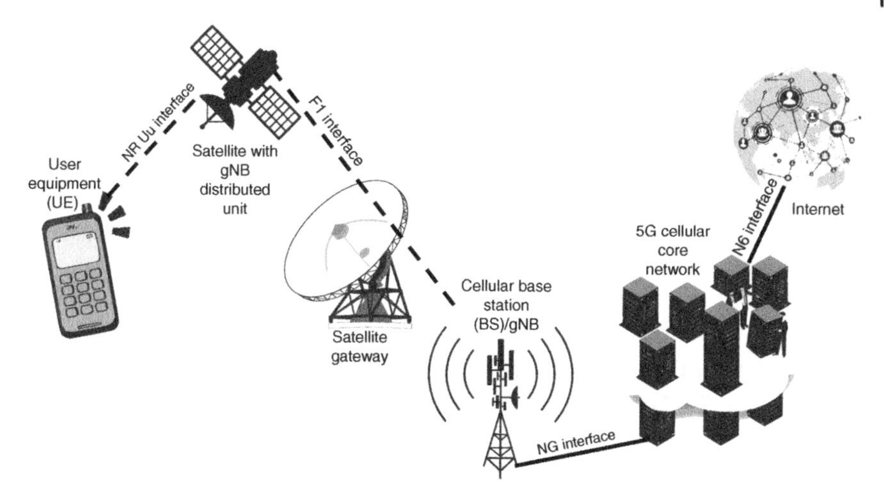

Figure 3.7 Regenerative-based satellite NTN architecture with distributed unit on satellite.

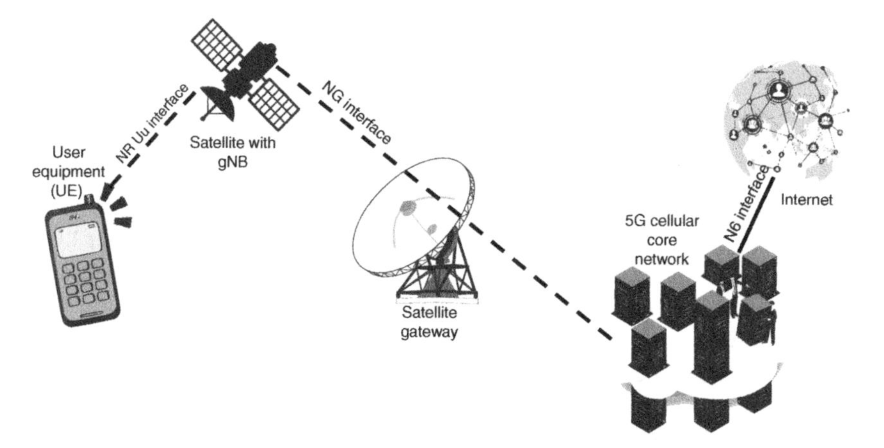

Figure 3.8 Regenerative-based satellite NTN architecture with gNB on satellite.

Figure 3.8 shows the solutions based on the relay network, where the relay network is linked to a donor next generation node B (DgNB), which is the entity that provides a connection to the next-generation core network (NGCN). Based on operational factors, the relay node appears to users as a typical gNB while being viewed as a UE by the DgNB, which justifies the deployment of the NR-Uu interface on both links and is reflected in the relay node attach method. Because of the introduction of a potentially large number of on-ground relays acting as gNBs, the architecture with relay nodes and transparent payloads is more complex than the direct access scenarios; these entities must be managed by several DgNBs. Thus, the overall system cost may increase. However, because the relay node may terminate protocols up to layer 3 (5G NR layer 3 [9]), no changes to the user service connection are required. In this situation, the impact of normal satellite channel impairments must consequently be evaluated exclusively on the backhaul link. The payload cost increases when an on-board relay node is installed, but the benefit of terminating the protocols up to layer 3 on-board the satellite is still maintained [29].

3.4 Revamping Non-terrestrial Networks Using Intelligent Reflecting Surfaces

3.4.1 Satellites for Communication: Background

The satellites phones can also be referred to as wireless handsets, but instead of relying on cellular BSs, they use satellite signals to communicate. The satellite phones are usually suited for a remote location, where the signals of the cellular tower are unavailable. As a result, they are the communication devices of choice in regions where cell service is limited or non-existent. Terminals (an alternate name for satellite-based cell phones), like mobile phones, have all of the standard phone functionality. Although they are larger and heavier than mobile phones, they do not have the practically limitless possibilities of a smartphone. Satellite phone systems function in a variety of ways, depending on the technology used by each organization. Some firms prefer GEO satellites, while others prefer LEO satellites. Each arrangement, or constellation, has its own set of advantages and disadvantages. GEO satellites circle the Earth as it rotates, allowing them to stay in a relatively constant position in the sky. They circle the Earth at a high height of roughly 22 000 miles (35 000 kilometres), constantly centred over the equator [30]. These are massive, powerful satellites, and each one can cover a significant part of the Earth's surface. A business may cover the majority of the globe with a constellation of only three or four satellites. As a result, these satellites are built to manage massive amounts of data, so they can handle video streaming, file sharing, messaging, television, and much more. GEO configurations are used by Inmarsat and Thuraya, two well-known firms [31, 32].

GEO satellites have a disadvantage in that their high orbits, result in large transmission delays of roughly 250 ms one way, or a quarter-second round-trip. As a result, we may have to wait a few seconds for someone to respond to our inquiries when we are conversing with them. We can also hear a jarring echo, which can be aggravating. In addition, because of their modest size, these networks are more susceptible to disruptions. When a satellite needs maintenance (or fails), an entire region of the planet may lose service until the problem is rectified. GEO satellites do not give much coverage for the poles because they orbit largely above the equator. One of the most significant disadvantages of GEO systems is their size. To communicate with these satellites, a device, the size of a laptop computer, is needed with a directional antenna taking up most of that space. To get the optimum reception, a calibration of the antenna is required, while aiming towards the satellite.

Companies such as Globalstar, OneWeb, and Iridium operate LEO satellites, which have far lower orbits of up to 930 miles (1500 km). In comparison to GEO satellites, the LEO satellites are much smaller in size and lighter in weight. Because LEO satellites travel at such low altitudes, a network may require as many as 60 satellites to cover the whole Earth. We might be within range of two or more of LEO satellites at any given moment, since they speed around the world at roughly 17 000 miles (27 359 km) per hour, completing an orbit in about two hours [33]. LEO satellites are noted for their outstanding call quality, short call latency (about 50 ms one way), and durability. They also require less battery power, consuming less energy than a GEO-capable satellite phone. However, the data transmission speeds of LEO are much slower than GEO (about 9600 bits per second). Moreover, they have the advantage of not requiring a huge antenna on a terminal, hence, reducing the size of a satellite phone.

3.4.2 Indoor Connectivity Using Intelligent Reflecting Surfaces

Recently, the satellite industry has gained popularity and shown some promising technological advancements to be integrated with terrestrial-based 5G networks. The NTN signals are unable to provide indoor coverage, as the signals from these satellites are attenuated and scattered. A mobile phone needs a line-of-sight (LoS) view of the satellite, whether using a GEO or LEO system. LoS can be a problem with a GEO constellation, especially if in a dense forest or hilly terrain. With a LEO constellation, though, the phone has several chances to connect as multiple satellites pass overhead, albeit the window of opportunity may be restricted to a few minutes. Although the LEO satellite constellation and well-developed research in UAVs and HAPs, are paving way for mobile operators to rely on the already available infrastructure and provide 5G-based cellular services to rural areas, yet the problem of indoor coverage requires some innovate approach to enable a typical low-cost smartphone to communicate with NTN. For the scope of this chapter, we have focused on the direct UE connectivity aspect of NTCS.

Based on the recent theoretical and experimental analysis, the IRSs have proven to be much effective in tunning wireless signals to improve the spectral and energy efficiencies. In [34], the author showed that an IRS-aided wireless

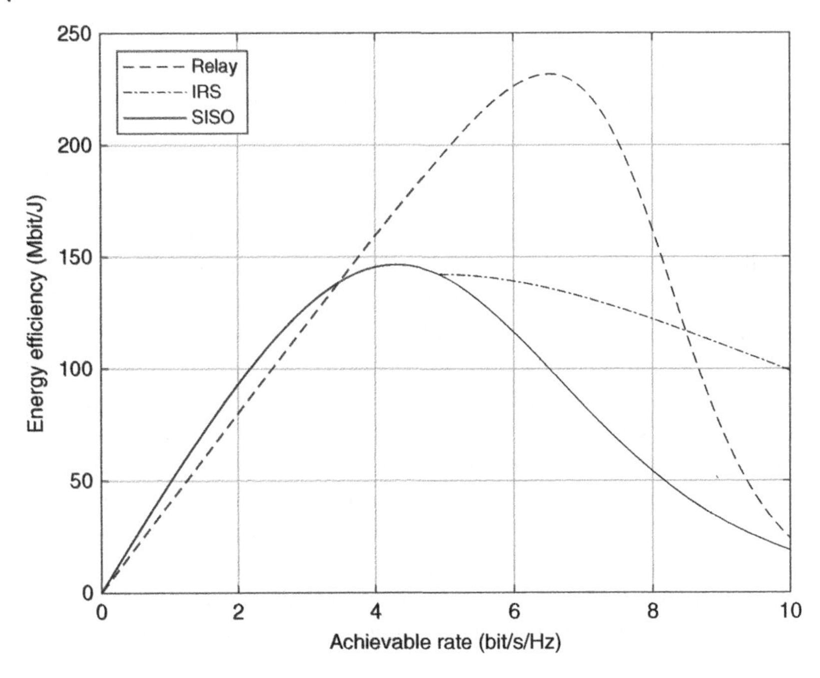

Figure 3.9 Comparison of energy efficiency versus data rate for IRS and decode-and-forward relaying.

network can improve the energy efficiency aspect of a wireless network and can provide higher data rates in comparison to a relay network (see Figure 3.9). IRS may be thought of as a supplementary technique for overcoming cost, security, power, and energy efficiency challenges by allowing more control over a single propagation via software-controllable reflections and obviating the requirement for signal decoding [35]. In [36], the authors demonstrate that, even by employing a 1-bit resolution of IRS, the energy efficiency of a wireless network can significantly be improved. Similarly, in [37], the authors can demonstrate a 300% increase in the energy efficiency of a wireless network equipped with IRS, in comparison to traditional relay networks. The passive nature of IRS provides strong physical-layer security for a wireless network while comparing it with traditional relay networks [38]. In multiple antenna-enabled wireless networks, the IRS can provide a much higher secrecy rate [39]. In NTN, the IRS have shown some impressive results to enable the co-existence of terrestrial and NTNs [40]. In [41], it has been shown that an IRS place near a mobile handset user, having a low-channel gain, can effectively maximize the sum rate of the network. In [42], the authors have shown that an IRS placed near the satellite user can enable a secure cooperative transmission between terrestrial and NTNs.

3.5 Conclusion

In this chapter, we have studied the feasibility of using IRS in merging the NTN in terrestrial networks. We have specifically focused on the direct UE connectivity aspect of NTCS. Based on an initial version of 3GPP Release 17, the NTN will consist of LEO satellites constellation, whereas the UAVs and HAPs are considered as a special use case of NTN networks. We have also explained a brief background of different NTN use case and connectivity aspects. Similar to global positioning system (GPS), the NTN-based signals are unable to provide indoor coverage, as the signals from these satellites are attenuated. Based on the recent theoretical and experimental analysis, the IRSs have proven to be much effective in tunning wireless signals to improve the spectral and energy efficiencies. These useful properties of IRS in improving channel parameters will make IRS a prominent contender in enabling the low-power UEs to directly communicate with NTN.

References

1 Kano, S. (2000). Technical innovations, standardization and regional comparisona case study in mobile communications. *Telecommun. Policy* 24 (4): 305–321.

2 Shapiro, P. (1976). Telecommunications and industrial development. *IEEE Trans. Commun.* 24 (3): 305–311.

3 Grubesic, T.H. and Murray, A.T. (2004). Waiting for broadband: local competition and the spatial distribution of advanced telecommunication services in the united states. *Growth Change* 35 (2): 139–165.

4 Lee, J., Tejedor, E., Ranta-aho, K. et al. (2018). Spectrum for 5G: global status, challenges, and enabling technologies. *IEEE Commun. Mag.* 56 (3): 12–18.

5 Yaacoub, E. and Alouini, M.-S. (2020). A key 6G challenge and opportunity-connecting the base of the pyramid: a survey on rural connectivity. *Proc. IEEE* 108 (4): 533–582.

6 Sacco, F.M. (2020). The evolution of the telecom infrastructure business. In: *Disruption in the Infrastructure Sector*, 87–148, Springer. Gatti, S.; Chiarella, C.

7 Arisar, M.M., Lianju, N., Zhongyuan, S., and Jokhio, I.A. (2019). A comprehensive investigation of telecom business models and strategies. *2019 Cybersecurity and Cyberforensics Conference (CCC)*, pp. 129–135, IEEE.

8 TELECOM INFRA PROJECT. https://telecominfraproject.com/ntcs/ (accessed 17 February 2022).

9 3GPP Release 17. https://www.3gpp.org/release-17 (accessed 17 February 2022).

10 Au, E. (2020). A short update on 3GPP release 16 and release 17 [standards]. *IEEE Veh. Technol. Mag.* 15 (2): 160.

11 Hameed, A. and Ahmed, H.A. (2018). Survey on indoor positioning applications based on different technologies. *2018 12th International Conference on Mathematics, Actuarial Science, Computer Science and Statistics (MACS)*, pp. 1–5, IEEE.

12 Ruan, Y., Jiang, L., Li, Y., and Zhang, R. (2020). Energy-efficient power control for cognitive satellite-terrestrial networks with outdated CSI. *IEEE Syst. J.* 15 (1): 1329–1332.

13 Özdogan, O., Björnson, E., and Larsson, E.G. (2019). Intelligent reflecting surfaces: physics, propagation, and pathloss modeling. *IEEE Wireless Commun. Lett.* 9 (5): 581–585.

14 Gong, S., Lu, X., Hoang, D.T. et al. (2020). Toward smart wireless communications via intelligent reflecting surfaces: a contemporary survey. *IEEE Commun. Surv. Tutorials* 22 (4): 2283–2314.

15 Khan, M.F., Bhatti, F.A., Habib, A. et al. (2017). Analysis of macro user offloading to femto cells for 5G cellular networks. *2017 International Symposium on Wireless Systems and Networks (ISWSN)*, pp. 1–6, IEEE.

16 Yu, X., Xu, D., Sun, Y. et al. (2020). Robust and secure wireless communications via intelligent reflecting surfaces. *IEEE J. Sel. Areas Commun.* 38 (11): 2637–2652.

17 Wu, Q., Zhang, S., Zheng, B. et al. (2021). Intelligent reflecting surface-aided wireless communications: a tutorial. *IEEE Trans. Commun.* 69 (5): 3313–3351.

18 NTT DOCOMO and Metawave Announce Successful Demonstration of 28GHz-Band 5G Using World's First Meta-Structure Technology. https://www.businesswire.com (accessed 17 February 2022).

19 Zou, Y., Gong, S., Xu, J. et al. (2020). Wireless powered intelligent reflecting surfaces for enhancing wireless communications. *IEEE Trans. Veh. Technol.* 69 (10): 12369–12373.

20 Nemati, M., Park, J., and Choi, J. (2020). RIS-assisted coverage enhancement in millimeter-wave cellular networks. *IEEE Access* 8: 188171–188185.

21 Hu, S., Rusek, F., and Edfors, O. (2018). Beyond massive MIMO: the potential of data transmission with large intelligent surfaces. *IEEE Trans. Signal Process.* 66 (10): 2746–2758.

22 Zhao, J. (2019). A survey of intelligent reflecting surfaces (IRSS): towards 6G wireless communication networks. *arXiv preprint arXiv:1907.04789*.

23 Gong, S., Xing, C., Zhao, X. et al. (2021). Unified IRS-aided MIMO transceiver designs via majorization theory. *IEEE Trans. Signal Process.* 69: 3016–3032.

24 Wu, Q. and Zhang, R. (2019). Intelligent reflecting surface enhanced wireless network via joint active and passive beamforming. *IEEE Trans. Wireless Commun.* 18 (11): 5394–5409.

25 Shafi, M., Molisch, A.F., Smith, P.J. et al. (2017). 5G: A tutorial overview of standards, trials, challenges, deployment, and practice. *IEEE J. Sel. Areas Commun.* 35 (6): 1201–1221.

26 Lin, X., Rommer, S., Euler, S. et al. (2021). 5G from space: an overview of 3GPP non-terrestrial networks. *IEEE Commun. Stand. Mag.* 5 (4): 147–153.

27 T.3.3GPP. https://portal.3gpp.org (accessed 17 February 2022).

28 3.T.38.811. https://portal.3gpp.org (accessed 17 February 2022).

29 Vanelli-Coralli, A., Guidotti, A., Foggi, T. et al. (2020). 5G and beyond 5G non-terrestrial networks: trends and research challenges. *2020 IEEE 3rd 5G World Forum (5GWF)*, pp. 163–169, IEEE.

30 Hofmann, C.A. and Knopp, A. (2019). Ultranarrowband waveform for IoT direct random multiple access to geo satellites. *IEEE Internet Things J.* 6 (6): 10134–10149.

31 Inmarsat. https://www.inmarsat.com/en/index.html (accessed 17 February 2022).

32 Thuraya. https://www.thuraya.com/en/about-us (accessed 17 February 2022).

33 Ali, I., Al-Dhahir, N., and Hershey, J.E. (1999). Predicting the visibility of LEO satellites. *IEEE Trans. Aerosp. Electron. Syst.* 35 (4): 1183–1190.

34 Björnson, E., Özdogan, O., and Larsson, E.G. (2019). Intelligent reflecting surface versus decode-and-forward: how large surfaces are needed to beat relaying? *IEEE Wireless Commun. Lett.* 9 (2): 244–248.

35 Tan, X., Sun, Z., Koutsonikolas, D., and Jornet, J.M. (2018). Enabling indoor mobile millimeter-wave networks based on smart reflect-arrays. *IEEE INFOCOM 2018-IEEE Conference on Computer Communications*, pp. 270–278, IEEE.

36 Huang, C., Alexandropoulos, G.C., Zappone, A. et al. (2018). Energy efficient multi-user MISO communication using low resolution large intelligent surfaces. *2018 IEEE Globecom Workshops (GC Wkshps)*, pp. 1–6, IEEE.

37 Huang, C., Zappone, A., Alexandropoulos, G.C. et al. (2019). Reconfigurable intelligent surfaces for energy efficiency in wireless communication. *IEEE Trans. Wireless Commun.* 18 (8): 4157–4170.

38 Chen, J., Liang, Y.-C., Pei, Y., and Guo, H. (2019). Intelligent reflecting surface: a programmable wireless environment for physical layer security. *IEEE Access* 7: 82599–82612.

39 Dong, L. and Wang, H.-M. (2020). Secure MIMO transmission via intelligent reflecting surface. *IEEE Wireless Commun. Lett.* 9 (6): 787–790.

40 Dong, H., Hua, C., Liu, L., and Xu, W. (2021). Towards integrated terrestrial-satellite network via intelligent reflecting surface. *ICC 2021-IEEE International Conference on Communications*, pp. 1–6, IEEE.

41 Dong, H., Hua, C., Liu, L. et al. (2021). Weighted sum-rate maximization for multi-IRS aided integrated terrestrial-satellite networks. *2021 IEEE Global Communications Conference (GLOBECOM)*, pp. 1–6, IEEE.

42 Xu, S., Liu, J., Cao, Y. et al. (2021). Intelligent reflecting surface enabled secure cooperative transmission for satellite-terrestrial integrated networks. *IEEE Trans. Veh. Technol.* 70 (2): 2007–2011.

4

Towards the Internet of MetaMaterial Things: Software Enablers for User-Customizable Electromagnetic Wave Propagation

Christos Liaskos[1,2], Georgios G. Pyrialakos[3], Alexandros Pitilakis[3], Ageliki Tsioliaridou[2], Michail Christodoulou[3], Nikolaos Kantartzis[3], Sotiris Ioannidis[4,2], Andreas Pitsillides[5,6], and Ian F. Akyildiz[7]

[1]*Computer Science Engineering Department, University of Ioannina, Ioannina, Greece*
[2]*Foundation for Research and Technology – Hellas, Heraklion, Greece*
[3]*Electrical and Computer Engineering Department, Aristotle University, Thessaloniki, Greece*
[4]*School of Electrical and Computer Engineering, Technical University of Chania, Crete, Greece*
[5]*Computer Science Department, University of Cyprus, Nicosia, Cyprus*
[6]*Department of Electrical and Electronic Engineering Science, University of Johannesburg (Visiting Professor), Johannesburg Gauteng, South Africa*
[7]*Truva Inc., Georgia 30022, USA*

4.1 Introduction

Electromagnetic (EM) waves are everywhere around us, carrying signals that span the spectrum from radiowaves to beyond the visible. In the past 30 years, wireless mobile telecommunications have been forming information channels only in small slices of the spectrum, where EM waves interact 'conveniently' with the natural materials and the man-made environment. But what if we had *meta*materials [1], i.e. composite materials with engineered sub-wavelength features, that macroscopically exhibit new tailored properties that enhance a wireless environment's performance and extend potential applications therein? And, what if these meta-materials could moreover adaptively reconfigure themselves [2, 3], to improve a specific channel in real-time, and/or sense the ambient environment [4]?

Hypersurfaces (HSFs) or Intelligent Reconfigurable Surfaces (IRSs) are the embodiment of these principles, currently deploying applications in 'classical' EM bands (e.g. microwave) and showcasing applications in previously inconvenient ones (e.g. terahertz). Naturally, they constitute one of the key enables of prospective next-generation mobile networks (6G), and all the systems and man-made sectors that stand to benefit from superfast, energy efficient,

Intelligent Reconfigurable Surfaces (IRS) for Prospective 6G Wireless Networks, First Edition.
Edited by Muhammad Ali Imran, Lina Mohjazi, Lina Bariah, Sami Muhaidat,
Tie Jun Cui, and Qammer H. Abbasi.

and versatile wireless communications. One of these sectors is the emerging Internet-of-Things (IoT), denoting the interconnection of every electronic device and the smart, orchestrated automation it entails [5].

Vehicles, smart phones, sensors, home, and industrial appliances of any kind expose a functionality interface expressed in software, allowing for developers to create end-to-end workflows. As an upshot, smart buildings and even smart cities that automatically adapt, e.g. power generation, traffic, and heat management to the needs of residents, have been devised in recent years. This current IoT potential stems from exposing and controlling a high-level functionality of an electronic device, such as turning on/off lights and air-conditioning units based on the time of day and temperature.

This chapter proposes the expansion of the IoT to the level of physical material properties, such as electrical and thermal conductivity, mechanical elasticity, and acoustic absorption. This novel direction is denoted as the *Internet of MetaMaterial Things (IoMMT)* and can have ground-breaking potential across many industrial sectors, as outlined in this chapter. There are two key enablers for the proposed IoMMT:

4.1.1 Key Enabler 1

The first key enabler of the proposed IoMMT are the meta-materials, the outcome of recent research in physics that has enabled the creation of artificial materials with real-time tunable physical properties [1, 6]. Meta-materials are based on the fundamental idea stating that the physical properties of matter stem from its atomic structure. Therefore, one can create artificially structured materials (comprising sufficiently small elementary 'units' of composition and geometry) to yield any required energy manipulating behaviour, including types not found in natural materials. Meta-materials manipulating EM energy were the first kind of meta-materials to be studied in depth, mainly due to the relative ease of manufacturing as low-complexity electronic boards [6–15].

Going beyond EM waves, the collectively termed elastodynamic meta-materials can manipulate acoustic, mechanical, and structural waves, whereas thermodynamic and quantum-mechanic meta-materials have also been postulated [16]. Elastodynamic meta-materials, empowered by recent advances in nano- and micro-fabrication (e.g. additive manufacturing/3D printing), can exhibit effective/macroscopic non-physical properties such as tunable stiffness and absorption/reflection, extreme mass-volume ratios, negative sonic refraction, etc. [17]. Their cell-size spans several length scales, depending on the application: acoustic cloaking/anisotropy/isolation, ultra-lightweight and resilient materials, devices for medical/surgical applications and food/drug administration, micro-electromechanical switches (MEMS), anti-seismic structures, etc. Tunability of

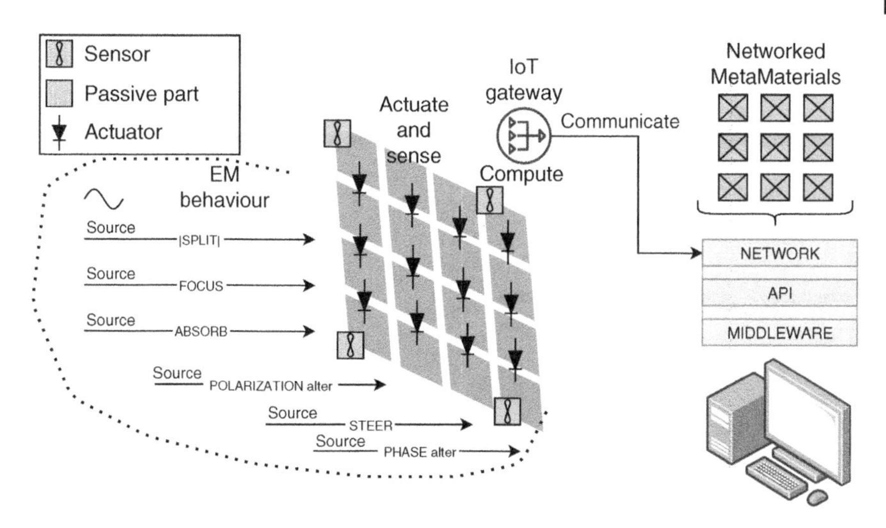

Figure 4.1 Networked meta-material structure and possible energy wave interactions. Source: Adapted from Liaskos et al. [18].

elastodynamic meta-materials can be achieved with electric, magnetic, optical, thermal, or chemical stimuli.

In a nutshell, their operation is as follows: impinging EM waves create inductive currents over the material, which can be modified by tuning the actuator elements within it (e.g. simple switches) accordingly. The Huygens principle states that any EM wavefront departing from a surface can be traced back to an equivalent current distribution over a surface [6]. Thus, in principle, meta-materials can produce any custom departing EM wave as a response to any impinging wave, just by tuning the state of embedded switches/actuators. Such EM interactions are shown in Figure 4.1 (on the right side). The same principle of operation applies to mechanical, acoustic, and thermal meta-materials [19].

4.1.2 Key Enabler 2

The second key enabler of the IoMMT is the concept of networked meta-materials. These will come with an application programming interface (API), an accompanying software middleware and a network integration architecture that enable the hosting of any kind of energy manipulation over a meta-material in real time (e.g. steering, absorbing, splitting of EM, mechanical, thermal, or acoustic waves), via simple software callbacks executed from a standard PC (desktop or laptop), while abstracting the underlying physics. The goal is to constitute the IoMMT directly accessible to the IoT and software development industries, without caring for the intrinsic and potentially complicated physical principles. Regarding

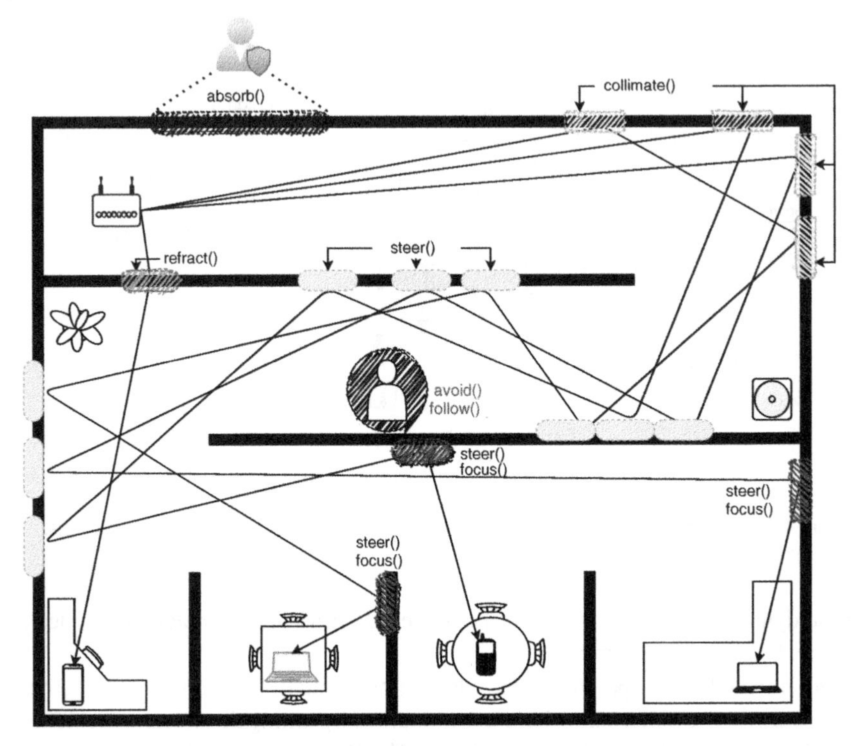

Figure 4.2 The programmable wireless environment introduced in [20] is created by coating walls with networked meta-materials. This allows for customized wireless propagation-as-an-app per communicating device pair, introducing novel potential in data rates, communication quality, security, and wireless power transfer. Source: Adapted from Liaskos et al. [20].

the IoMMT potential, large-scale deployments of EM meta-materials in indoor setups have introduced the ground-breaking concept of programmable/intelligent wireless environment (Figure 4.2) [20]. By coating all major surfaces in a space (e.g. indoors) with EM meta-materials, the wireless propagation can be controlled and customized via software. As detailed in [20], this can enable the mitigation of path loss, fading, and Doppler phenomena, while also allowing waves to follow improbable air-routes to avoid eavesdroppers (a type of physical-layer security). In cases where the device beamforming and the EM meta-materials in the space are orchestrated together, intelligent wireless environments can attain previously unattainable communication quality and wireless power transfer [20]. Extending the EM case, we envision the generalized IoMMT deployed as structural parts of products, as shown in Figure 4.3:

- EM interference and unwanted emissions can be harvested by IoMM-coated walls and be transformed back to usable EM or mechanical energy.

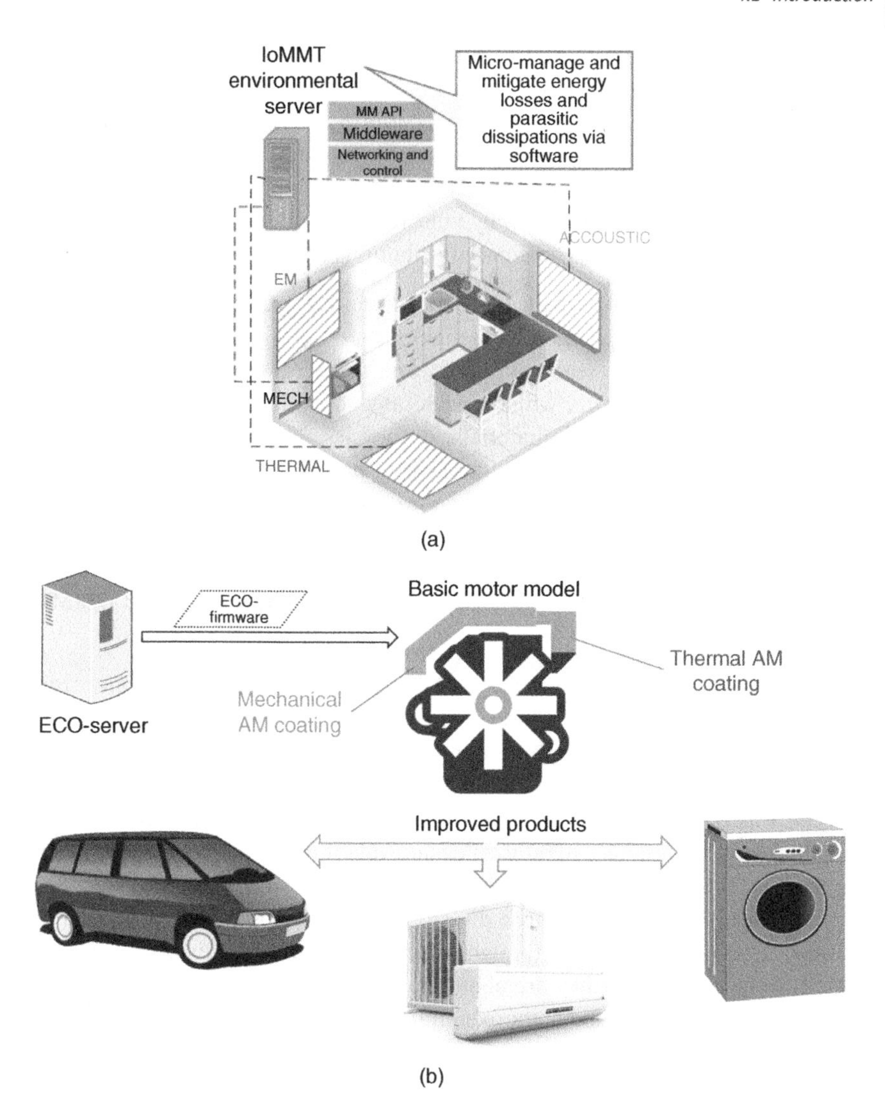

Figure 4.3 Envisioned applications of the IoMMT in smart houses and products. (a) Conceptual IoMMT deployment within a smart. (b) Conceptual IoMMT deployment within single products. Source: Akyildiz and Jornet [21].

- Thermoelectric and mechanical meta-materials can micro-manage emanated heat and vibrations from devices, such as any kind of motor, to recycle it as energy while effectively cooling it. The same principle can be applied to a smart household or a noisy factory.
- The acoustic meta-materials can surround noisy devices or be applied on windows to provide a more silent environment, but to also harvest energy which can be added to a system such as a smart-household.

Assuming a central controller to optimize a given IoMMT deployment allows for further potential. For instance, one can allow for quickly 'patching' of overlooked physical aspects (e.g. poor ecological performance) of IoMM-enabled products during operation, without overburdening the product design phase with such concerns. The 'patching' may also be deferred in the form of 'eco-firmware', distributed via the Internet to ecologically tune a single product or horizontal sets of products. In this context, the principal contributions of the chapter are as follows:

- We propose the concept of the IoMMT and discuss its architecture and interoperability with existing network infrastructures.
- We define two novel categories of software: the *Meta-material API* and the *Meta-material Middleware*, which enable any software developer to interact with a set of networked meta-materials, in a physics-agnostic manner. We establish the data models, workflows, and testbed processes required for profiling and, subsequently, componentizing meta-materials.
- We present an implemented and experimentally verified version of the meta-material API and the Meta-material Middleware for the EM case.
- We highlight promising, new applications empowered by the featured IoMMT concept.

In this aspect, the potential of our IoMMT paradigm is the first to offer true control over the energy propagation within a space, in every physical domain, i.e. for any physical material property and corresponding information-carrying wave. For instance, control over the equivalent RLC parameters of an electric load controls the power that can be delivered to it by an EM wave. Moreover, the presented software is a mature prototype platform for the development of IoMMT applications. This constitutes a major leap towards a new research direction. On the other hand, other research directions have proposed and explored the Internet of NanoThings [21]. Although similarly named, these directions are not related to the IoMMT, as they are about embedding nano-sized computers into materials in order to augment the penetration level of applications (e.g. sense structural, temperature, humidity changes within a material, rather than just over it), and not to control the energy propagation within them. The remainder of this chapter is organized as follows: devoting a section to each of the principal contributions of our work. In Section 4.2 we provide the related work overview and the necessary prerequisite knowledge for networked meta-materials. In Section 4.3, we present the architecture for integrating the IoMMT in existing Software-Defined Networks (SDNs) and systems. In Section 4.4, we present the novel meta-material API, and Section 4.5 follows with the description of the Meta-material Middleware and its assorted workflows. In Section 4.6, we present the implemented version of the software for the EM meta-material case, along with a description of the employed evaluation test bed. Finally, novel realistic applications enabled with our new paradigm are discussed in Section 4.7, and we conclude the chapter in Section 4.8.

4.2 Pre-requisites and Related Work

Meta-materials are simple structures that are created by periodically repeating a basic structure, called a *cell* or a *meta-atom* [6]. Some examples across physical domains are shown in Figure 4.4. The planar (2D) assemblies of meta-atoms, known as *meta-surfaces*, are of particular interest currently [26, 27]. For instance, EM are currently heavily investigated by the electromagnetic/high-frequency community, for novel communications, sensing, and energy applications [28–30].

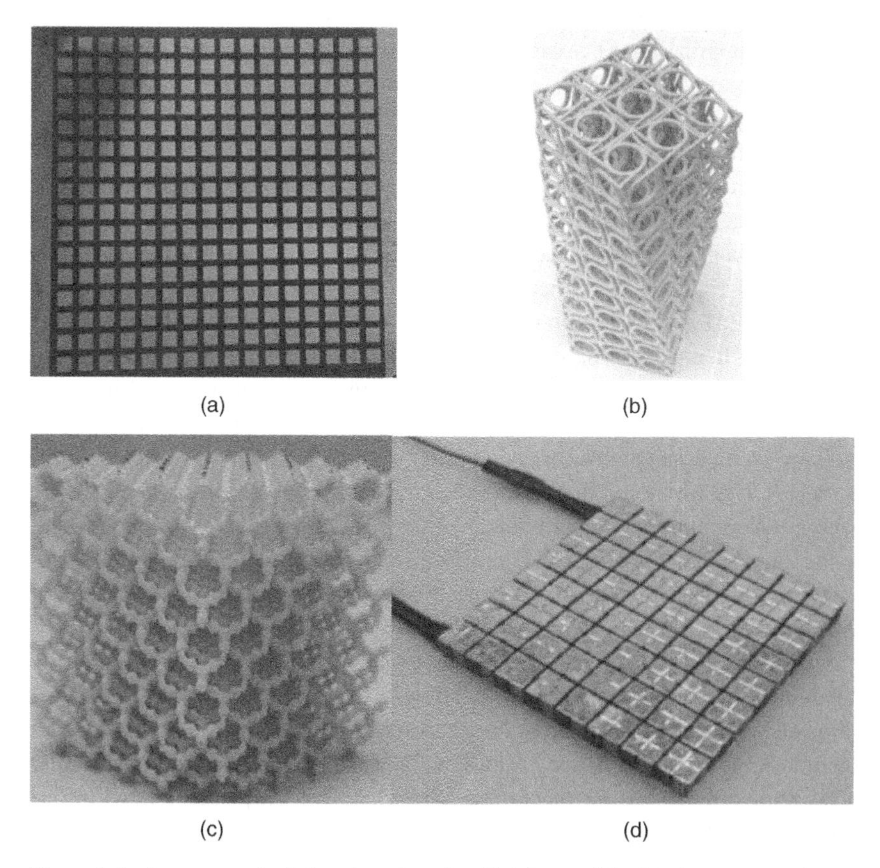

Figure 4.4 Energy manipulation domains of artificial materials: (a) electromagnetic. Source: Ref. [22], (b) mechanical [23]. Source: Photo: T. Frenzel/KIT, (c) acoustic. Source: Haberman and Guild [24], and (d) thermoelectric. Source: Dresselhaus et al. [25].

A notable trait of meta-materials is that they are simple structures and, therefore, there exists a variety of techniques for generally low-cost and scalable production [6]. The techniques such as printed circuit broads, flexible materials such as Kapton, 3D printing, Large Area Electronics, bio-skins, and microfluidics have been successfully employed for manufacturing [6].

In each physical domain, a properly configured meta-material has the capacity to steer and focus an incoming energy wave towards an arbitrary direction or even completely absorb the impinging power. In the EM case, this capability can be exploited for advanced wireless communications [2, 20, 31–34], offering substantially increased bandwidth and security between two communicating parties.

The potential stemming from interconnected meta-materials has begun to be studied only recently [18]. The perspective networking architecture and protocols [18, 20], meta-material control latency models [35], and smart environment orchestration issues have been recently studied for the EM case [36, 37].

Notably, a similarly named concept, i.e. the Internet of NanoThings [21], was recently proposed to refer to materials with embedded, nano-sized computing and communicating elements. In general, these materials are derived from miniaturizing electronic elements and placing them over or embedding them into fabrics and gadgets, to increase their application-layer capabilities. For instance, this could make a glass window become a giant, self-powered touchpad for another IoT device. Originally, the concept of software-defined meta-surfaces was based on the nano-IoT as the actuation/control enabler [2]. Nano-devices can indeed act as the controllers governing the state of the active cells, offering manufacturing versatility and extreme energy efficiency. Nonetheless, until nano-IoT becomes a mainstream technology, other approaches can be adopted for manufacturing software-defined meta-surfaces, as reported in the related physics-oriented literature [19]. It is also noted that nano-IoT as a general concept is about embedding nano-sized computers into materials in order to augment the penetration level of applications (e.g. sense structural, temperature, humidity changes within a material, rather than just over it), and not specifically to control the energy propagation within them.

In contrast, our work refers specifically to the case of meta-materials and the capabilities they offer for the manipulation of energy across physical domains. Moreover, our chapter introduces the software enablers for this direction, which has not been proposed before. Additionally, our chapter focuses more on the networking approaches for meta-materials, which has only been treated in our previous work [18], and only for the EM case. Finally, the work of Chen et al. [38] also advocates for the use of meta-material in any physical domain for distributed energy harvesting, e.g. in a smart house or a city. However, software enablers and networking considerations are not discussed or solved in [38]. Moreover, the energy manipulation type is restricted to harvesting which can be viewed as a subset of our proposed IoMMT potential.

4.2.1 Meta-materials: Principles of Operation, Classification, and Supported Functionalities

A conceptual metamaterial is illustrated in Figure 4.5 [6, 39–42]. Basically, a meta-material consists of periodically repeated meta-atoms arranged in a 3D grid layout, with the meta-surfaces being a sub-case. In particular, unit cells comprise passive and tunable parts, required in reconfigurable meta-materials as well as optional integrated sensory circuits, which can extract information of the incident energy wave. Furthermore, tunable parts are crucial for meta-materials, as they enable re-configurability and switching between different functions. For illustration, in EM meta-materials at microwave frequencies, the tunable parts embedded inside the unit cells can be voltage-controlled resistors (varistors) and/or capacitors (varactors), micro-electromechanical switches (MEMS), to name a few [6, 34].

On the other hand, in mechanical and acoustic meta-materials, the tunable parts can be micro-springs with a tunable elasticity rate [43, 44]. The meta-atoms

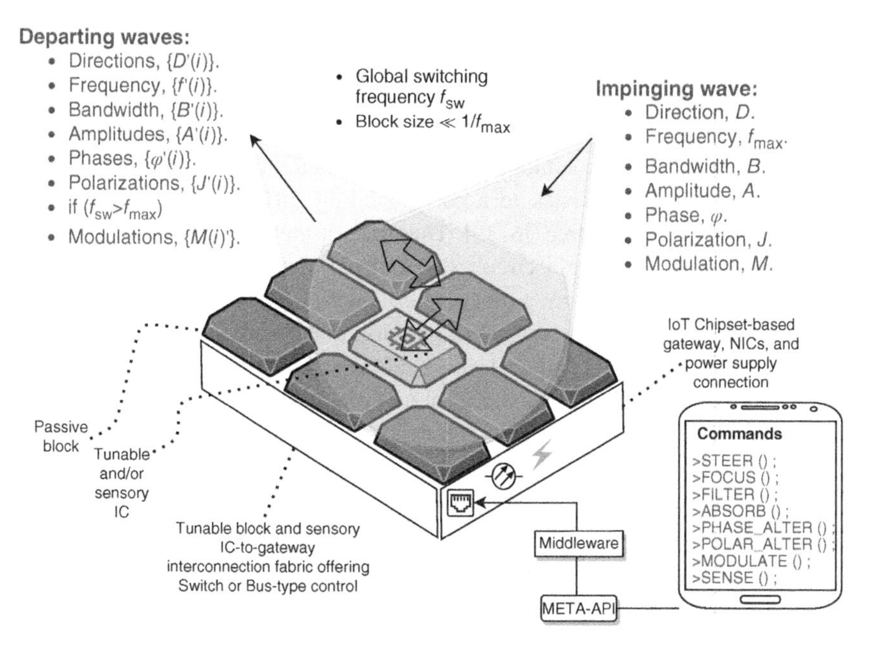

Figure 4.5 Overview of the metasurface/metamaterial structure and operating principles. Source: Adapted from Refs. [6, 34].

may also form larger groups, called *super-atoms* or *super-cells*, repeated in specific patterns that can serve more complex functionalities, as discussed later in this chapter. Lastly, the software-defined meta-materials include a *gateway* [20], i.e. an on-board computer, whose main tasks are to (i) power the whole device and (ii) control (get/set) the state of the embedded tunable elements, (iii) interoperate with the embedded sensors, and (iv) interconnect with the outside world, using well-known legacy networks and protocol stacks (e.g. Ethernet).

The relative size of a meta-atom compared to the wavelength of the excitation (impinging wave) defines the energy manipulation precision and efficiency of a meta-material. For example, EM meta-surfaces share many common attributes with classic antenna-arrays and reflect-arrays. Antenna arrays can be viewed as independently operating antennas, being very effective for coarse beam steering as a whole. Reflect-arrays typically consist of smaller elements (still sub-wavelength), permitting more fine-grained beam steering and a very coarse polarization control. Meta-materials comprise orders of magnitude smaller meta-atoms and may also include tunable elements and sensors. Their meta-atoms are generally considered tiny with regard to the exciting wavelength, hence, allowing full control over the form of the departing energy wave.

Regardless of their geometry and composition, the operating principle of meta-materials remains the same. As depicted in Figure 4.5, an impinging wave of any physical nature (e.g. EM, mechanical, acoustic, thermal) excites the surface elements of a meta-material, initiating a spatial distribution of energy over and within it. We will call this distribution 'exciting-source'. On the other hand, well-known and cross-domain principles state that any energy wavefront, which we demand to be emitted by the meta-material as a response to the excitation, can be traced back to a corresponding surface energy distribution denoted as 'producing-source' [6, 19]. Therefore, a meta-material configures its tunable elements to create a circuit that morphs the exciting-source into the producing-source. In this way, a meta-material with high meta-atom density can perform any kind of energy wave manipulation that respects the energy preservation principle. Arguably, the electromagnetism constitutes a very complex energy type to describe and, as a consequence, manipulate in this manner, as it is described by two dependent vectors (electric and magnetic field) as well as their relative orientation in space, i.e. polarization (mechanical, acoustic, and thermal waves can be described by a single scalar field in space). As such, incoming EM waves can be treated in more ways than other energy types. The common types of EM wave manipulation via meta-materials, reported in the literature [6], can designate a set of high-level functionality types as follows:

- **Amplitude**: Filtering (band-stop, band-pass), absorption.
- **Polarization**: Waveplates (polarization conversion, modulation).

- **Wavefront**: Steering (reflecting or refracting), splitting, focusing, collimating, beamforming, scattering.
- **Bandwidth**: Filtering.
- **Modulation**: Requires embedded actuators that can switch states fast enough to yield the targeted modulation type [45].
- **Frequency**: Filtering, channel conversion.
- Doppler effect mitigation and non-linear effects [18].

Additionally, sensing impinging waves may be considered one of the above functionalities and, as an outcome, the embedded sensors can extract information of any of the above parameters related to the incident wave.

In this aspect, the role of the contributed meta-material API is to model these manipulation types into a library of software callbacks with appropriate parameters. Then, for each callback and assorted parameters, the Meta-material Middleware produces the corresponding states of the embedded tunable elements that indeed yield the required energy manipulation type. In other words, a meta-material coupled with an API and a Meta-material Middleware *can be viewed as a hypervisor that can host meta-material functionalities* upon user request [18].

In the following, we focus on EM meta-materials which, as described, yield the richest API and most complex Meta-material Middleware. The expansion to other energy domains is discussed via derivation in Section 4.7.

4.3 Networked meta-materials and SDN workflows

Many meta-materials deployed within an environment can be networked through their gateways [46–49]. This means that they may become centrally monitored and configured via a server/access point in order to serve a particular end objective [4, 50–53].

An example is given in Figure 4.6, where a set of meta-materials is designed with the proper commands for energy wave steering and focusing, in order to route the energy waves exchanged between two wireless users, thus avoiding obstacles or eavesdroppers. Other applications include wireless power transfer and wireless channel customization for an advanced quality of service (QoS) [20, 29]. Such a space, where energy propagation becomes software defined via meta-materials is called a programmable wireless environment (PWE) [18].

As shown in [20], the PWE architecture is based on the SDN principles. The PWE server is implemented within an SDN controller [54]; the south-bound interface abstracts the meta-material hardware, treating meta-material devices as networking equipment that can route energy waves (e.g. similar to a router, albeit

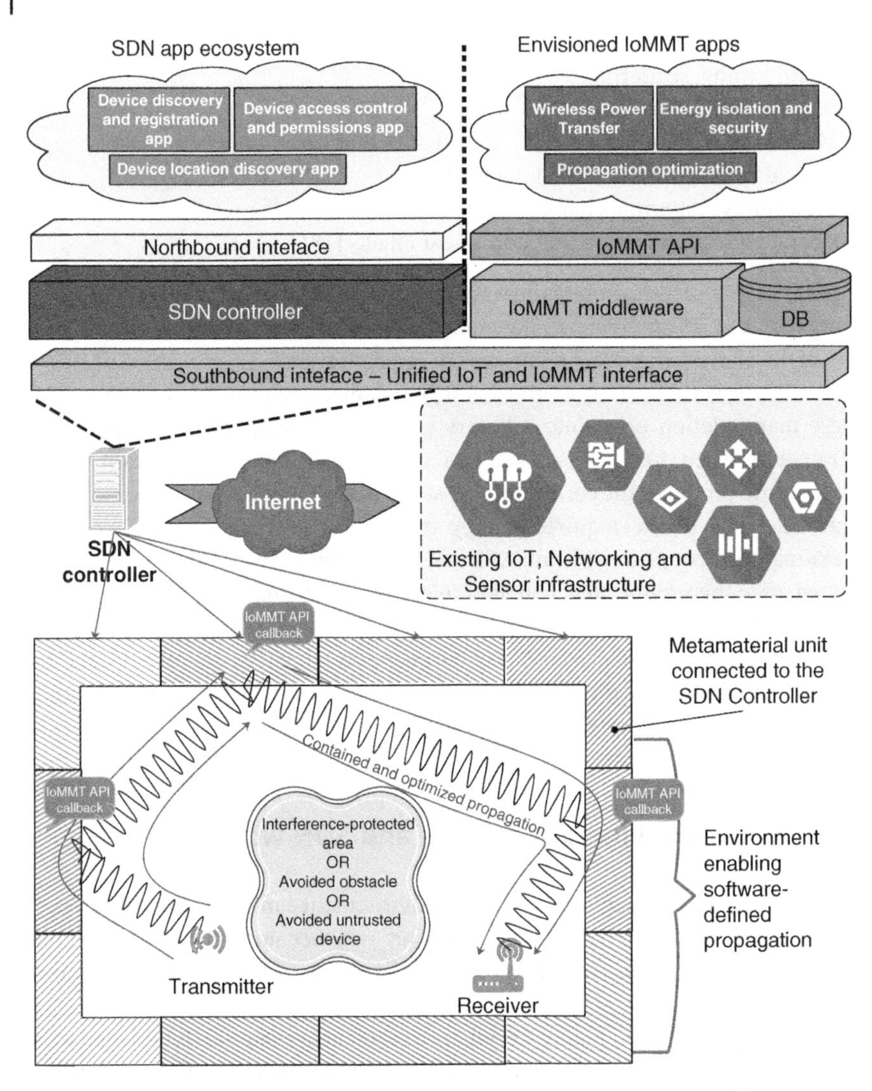

Figure 4.6 SDN schematic display of the system model and the entire workflow abstraction.

with a more extended, and unique parameterization). Thus, the meta-material API constitutes a part of the north-bound SDN interface, atop of which the security, QoS, and power transfer concepts can be implemented as SDN controller applications. On the other hand, the Meta-material Middleware is part of the SDN middleware, translating metamaterial API callbacks into meta-material hardware directives.

A notable trait of the Meta-material Middleware is that it is divided into *two parts*, in terms of system deployment [55]:

1. The meta-material manufacturing stage component, a complex, offline process requiring special meta-material measurement and evaluation setups (discussed in Section 4.5).
2. The meta-material operation stage component, which operates in real time based on a codebook. This codebook is a database populated once by the manufacturing stage component and contains a comprehensive set of configurations for all meta-material API callbacks, supported by a given meta-material.

The operation stage component simply retrieves configurations from the codebook and optionally combines them as needed, using an interleaving process described in Section 4.5.

Notably, other studies propose the use of online machine learning as a one-shot process, which can be more practical when response time is not a major concern [56]. However, in this work, we propose the afore-mentioned separation in deployment, to ensure the fastest operation possible overall, thus covering even the most- demanding cases.

It is noted that SDN is not a choice due to restrictions, but rather a choice due to compatibility. In the software-defined meta-surfaces presented in this chapter, a key point is the abstraction of physics via an API that allows networking logic to be reused in PWEs, without requiring a deep understanding of physics. SDN has (among other things) already introduced this separation of control logic from the underlying hardware and its administrative peculiarities. Therefore, we propose an integration of PWE within SDN to better convey the logical alignment of the two concepts.

4.4 Application Programming Interface for Meta-materials

In the following, we consider a meta-material in the form of a rectangular *tile*. The term 'tile' is used to refer to a practical meta-material product unit, which can be used to cover large objects such as walls and ceilings in a floorplan.

A software process can be initiated for any meta-material tile supporting a unique, one-to-one correspondence between its available switch element configurations and a large number of meta-material functionalities [3, 57, 58]. The meta-material tiles in this work incorporate tunable switch elements, which dictate the response of each individual cell, locally. In this way, providing an arrangement of all the tile cells allows the tuning of the 'concerted' meta-material response of the entire tile.

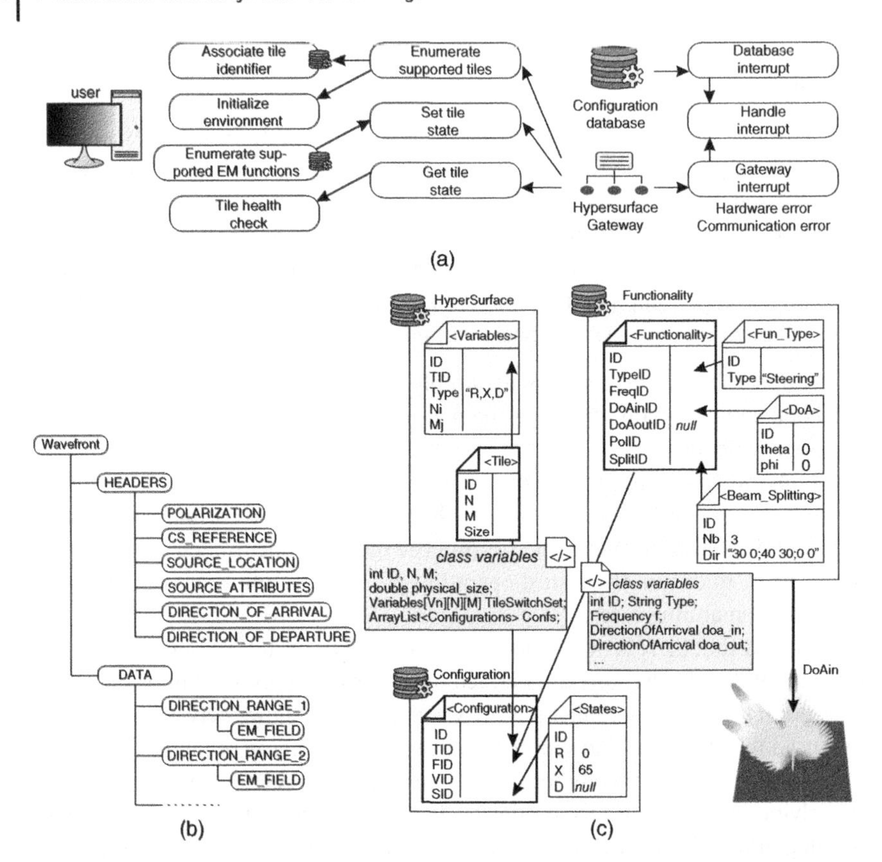

Figure 4.7 (a) Case diagram of the main functions supported by the three basic entities. Tasks highlighted with the database icon indicate that a set of data are to be retrieved from the Configuration Database. (b) Wavefront description in data object format. (c) A simplified overview of the structure of the Configuration Database. The <Tile> table hosts all information regarding a tile's physical implementation. The <Functionality> table combines a set of meta-material parameters to define new functionalities. Both tables are combined in the <Configuration> table with a set of entries from the <States> table to compose a new configuration that supports the functionality FID on tile TID (VID refers to an entry in the table <Variables>).

In this section, we present the API that grants access to the tile's meta-material applications by defining an abstract representation of the meta-material, its switch element configurations, and their respective functionalities. Specifically, the API resides between a user, operating a common PC (desktop, smartphone, etc.) and a tile gateway, linked to the network of switch element controllers. The case diagram of the proposed concept, presented in Figure 4.7a, involves the following main entities:

- A Configuration Database which stores all information regarding the tiles, the switch element configurations, and their corresponding functionalities.

- A User which initiates all API callbacks though either the source code or button click events in the Graphical User Interface. (GUI).
- The HyperSurface Gateway which represents the electronic controller of the hardware [59–61].
- An Interrupt handling service which acts as a persistent daemon, receiving and dispatching commands to the Hypersurface Gateway.

4.4.1 Data Structures of the Meta-material API

In the configuration Database (DB), each tile is associated with an element array S that represents all possible arrangements of switch element states on the meta-material under study. Each switch element is represented by either a discrete or a continuous variable, creating a mathematical space of $V_1 \times N \times M + \cdots + V_n \times N \times M$ dimensions, where V_i is the number of elements of the same type (e.g. capacitors) and N, M the number of unit cells towards the two perpendicular directions. Furthermore, every object in this space corresponds to a different state of S and therefore a different configuration. As an example, a tile with two controllable resistive and one diode elements per unit cell is parameterized by a $2 \times N \times M + 1 \times N \times M$ array, where the first and second sets span a continuous $[R_{\min}, \ldots, R_{\max}]$ and a discrete [0,1] range, respectively. In this case, 0 and 1 correspond to the OFF and ON states of the diode. This representation will then acquire the following form:

$$F \longleftarrow [(d_1, d_2, i_1)_{n=1}, \ldots, (d_1, d_2, i_1)_{n=N\times M}] \tag{4.1}$$

where d_1, d_2 are double-type variables, i_1 is an integer variable, and F is the appointed functionality. The primitive data types of all variables should be selected so as to minimize the total parameter space of combined states without any loss of relevant information. This lays a better optimized communication and computational burden to both the API and the Compiler, especially during the compilation process where a sizable amount of mathematical computations is required. Accordingly, all functionalities are, also, associated with their own representation and classified pertinent to their own type and defining parameters. For instance, a complex beam-splitting and polarization control operation is parameterized by a discrete variable corresponding to the number of outgoing beams, their directivity amplitudes, and an appropriate number of (θ, φ) pairs, indicating the steering angles. The most complex functionality can be generally described by a custom- scattering pattern and represented herein by a collection of variables that indicate the reflected power towards all directions within the tile's viewing area.

It is, also, worth noting that the data objects being passed as arguments in the callbacks are primarily descriptions of wavefronts. A simplified data structure is illustrated in Figure 4.7b. Hence, a wavefront is described by a type (string identifier), such as 'Planar', 'Elliptic', 'Gaussian', 'Custom'. For each type, a series of headers define the location and attributes of the creating source (for impinging wavefronts only) as well as the coordinate system origin with respect to which all distances are measured. Moreover, the direction of arrival and departure are arrays that can be used to define multiple impinging or departing wavefronts at the same time. Notably, the information within the headers may be sufficient to produce any value of the wavefront via simply analytical means. In such cases, the data part can be left empty. In custom wavefronts, the data are populated accordingly. A mechanism for defining periodicity is supplied via the notion of ranges (i.e. coordinate ranges where the energy field is approximately equal), to potentially limit the size of the overall data object.

The parameters that represent the functionalities and configurations of a tile constitute the set of variables that are exposed to the programmer through the meta-material API. They are organized in a unified manner within the Database, as shown in Figure 4.7c which provides an illustration of the unique association between all primary tables. Particularly, the <Tile> table stores all information of a tile's hardware implementation, such as the number of variables per unit cell and the type of switch elements. The <Functionality> table stores the representation scheme described in Section 4.2.1 for all available meta-material functionalities. Each parameter associated with a functionality is organized in a separate table, including a table that stores an identification variable representing the type of functionality. This table ID enumerates all possible operations supported by the tile, including full power absorption, wavefront manipulation (steering, splitting, etc.), and wavefront sensing. Finally, the <Configuration> table combines, in an exclusive manner, both primary tables (<Tile> and <Functionality>) to link each stored functionality with a specific set of switch element states, acquired from the pool of available entries in the secondary table <States>.

4.4.2 API Callbacks and Event Handling

Using the Database as a reference point, the API is responsible for interpreting a configuration array to the proper set of hardware commands, when a suitable callback is executed. In general, an API callback can refer to a number of common requests such as

- Detect the number and type of accessible tiles in the environment.
- Get the current state of all switch elements or set them to a specific configuration.
- Check the health status and handle interrupts from the tiles or the Database.

Prior to any other callback, the API follows an initiation process, while the software detects all presently active and connected (discoverable) tiles by broadcasting a corresponding network message. The tiles report their location and a unique identifier, e.g. a fixed value, that associates all tiles with the same hardware specifications. The API validates the support of the active tiles by checking if the identifier exists in the tile list present in the database. It, then, retrieves the switch element arrays that correspond to these tiles and remains idle until a new 'get' or 'set' request is received for a currently active or new configuration, respectively.

This means that the API is now open to receive new functionality requests from a user, physically operating the software, or generate its own requests by reacting to unexpected changes in the environment of devices linked to the MS network. When a new functionality request is received, the API retrieves one of the available configurations from the Database and translates it to a proper set of element states on an active meta-material tile. The corresponding API callback process is illustrated in Figure 4.8. In particular, the Caller (user) executes a meta-material Function Deployment request, which, in turn, invokes the Configuration Resolver, identifying a tile that supports the requested functionality. Next, the resolver queries the Database and returns a configuration that matches the intended meta-material application, looking for a proper entry in the <Configuration> table. The API creates a string command, using the tile identifier and a hardware representation of the element state variables, which is, then, conveyed to the tile Gateway using the corresponding protocol. This notifies the intra-tile control network to assign the switch element states to their suitable values. Finally, feedback from a successful or failed configuration setup is received from the tile, notifying the user [62]. The state of the newly set configuration is evaluated through either the identification of failed or unresponsive switches, or by activating the sensing app, in controlled conditions, as a self-diagnosing tool for the tile.

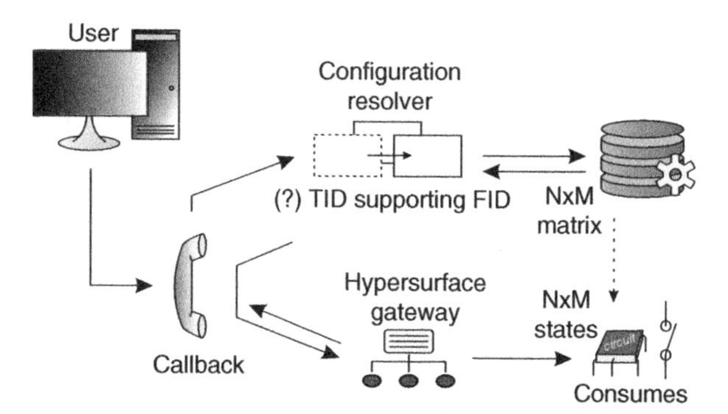

Figure 4.8 A metamaterial Function Deployment request initiates an API callback on the tile (TID). The Configuration Resolver seeks an appropriate configuration that supports the selected functionality (FID). The matrix of states is conveyed to the HyperSurface Gateway, where it is translated to a set of corresponding hardware states.

A more advanced API callback may involve the assignment of a secondary or supplementary functionality, on top of an already existing operation. For instance, the Meta-material Middleware may receive independent requests from different users to steer the wavefront of several point power sources (i.e. the users' cell-phones) towards the direction of a single nearby network hotspot. This can be handled by the API in many ways. In the case where several tiles are present in the environment, each tile can be repurposed to host a separate functionality, distributing all users to their own active tiles. When this is not feasible, the API can divide a single tile into separate areas and associate the respective element switches to different configurations (the division is usually performed in equal-sized rectangular patterns, but interlacing can, also, be used). Lastly, two functionalities can be combined into a single one, when a corresponding physical interpretation exists. For example, two separate steering operations, from the same source, may be combined into a single dual-splitting operation, expressed by a single unified pattern on the meta-material.

In other cases, the configuration resolver may need to combine several functionalities to produce a special new application. This occurs when a functionality is parameterized by a continuous variable (e.g. a steering angle), while the Meta-material Middleware can evaluate and store only a finite number of entries in the database. In such a scenario, the resolver seeks the two closest matching entries through an appropriate minimization function (e.g. for a beam-steering operation, the minimum distance between the requested and currently stored steering angles), whereas performing an interpolation of the switch state values.

To ensure a seamless operation, the API is reinforced with a set of specialized algorithms to handle unexpected failures in the hardware or communication network. So a tile must include all necessary identification capabilities (e.g. the ability to identify power loss in its switch elements), and be able to notify the Meta-material Middleware in case of failure. The API is, then, responsible for handling these errors by ensuring no loss of the current functionality [55]. For illustration, if a group of switch elements is stuck to an unresponsive state, the Meta-material Middleware may instantly seek the closest matching functionality with a fixed configuration for the faulty elements. In more severe cases of demanding human intervention, like an unresponsive network (after a certain timeout), the API is also responsible for informing an available user.

4.5 The Meta-material Middleware

For the meta-material to be reconfigured between different functionalities, a physical mechanism for locally tuning each unit cell response must be infused [18].

In the context of the present work, we assume that the response of the unit cells is controlled by the variable *impedance loads* connected to the front-side metallization layer of the meta-material, where structures such as the resonant patch pair resides [6]. The loads are complex valued variables, comprising resistors and capacitors or inductors. The value of the ith load, $Z_i = R_i + jX_i$, comprises two parameters: its resistance ($R_i > 0$) and reactance ($X_i = -(\omega C_i)^{-1}$ or $X_i = +\omega L_i$), for capacitive and inductive loads, respectively. The loads are, thus, electromagnetically connected to the surface impedance of the 'unloaded' unit cell and by tuning their values we can regulate the unit cell response, e.g. the amplitude and phase of its reflection coefficient. The latter is naturally a function of frequency and incoming ray direction and polarization. When the meta-material unit cells are properly 'orchestrated' by means of tuning the attached (R_i, X_i) loads, the desired functionality (global response) of the meta-material is attained.

In the most rigorous approach, the meta-material response can be computed by full-wave simulations, which implement Maxwell's laws, given the geometry and meta-material properties of the structure as well as a complex vector excitation, i.e. the impinging wave polarization and wavefront shape (phase and amplitude profile). The full-wave simulation captures the entire physical problem and, hence, does not require a meta-material-level abstraction for the structure. Frequency-domain solvers, which assume linear media and harmonic excitation (i.e. the same frequency component in both the excitation and the response), are the prime candidates for full-wave simulation. They typically discretize the structure's volumes or surfaces at a minimum of $\lambda/10$ resolution, formulate the problem with an appropriate method (e.g. the finite-element or the boundary-element method) and, then, numerically solve a large sparse- or full-array system to compute the response, in our case, the scattered field. Conversely, time-domain solvers assume a pulsed excitation, covering a pre-defined spectral bandwidth and iteratively propagate it across the structure, solving Maxwell's equations to compute its response; they, typically, require a dense discretization of the structure, e.g. a minimum of $\lambda/20$ resolution. From this process, it becomes evident that meta-material with a wide aperture, i.e. spanning over several wavelengths along the maximum dimension, require high-computational resources in the full-wave regime, scaling linearly, when parametric simulations need to be performed to optimize the structure and/or the response. For instance, broadband simulation for the response of a unit cell of volume $(\frac{\lambda}{5})^3$ on a contemporary desktop computer could take several minutes, especially if the cell includes fine sub-wavelength features. The memory and CPU resources scale-up linearly for meta-surfaces comprising hundreds of thousands of unit cells. Moreover, full-wave simulations do not explicitly unveil the underlying principles that govern the meta-material functionality.

4.5.1 Functionality Optimization Workflow: Meta-material Modelling and State Calibration

In this section, we establish the optimization workflow of Figure 4.9 that drives the calibration process of the meta-material via an appropriate approximation model. Here calibration denotes the matching of actual active element states (e.g. the states of a tunable varactor) to the corresponding model parameter

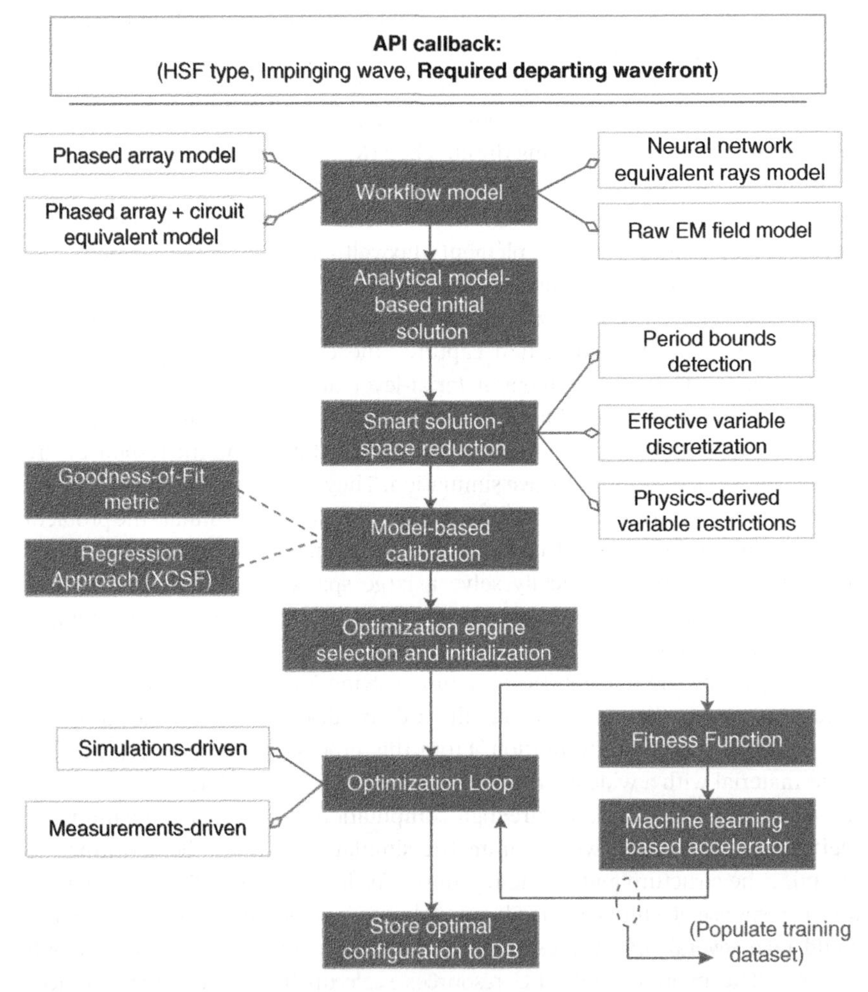

Figure 4.9 The Meta-material Middleware functionality optimization workflow. The workflow seeks to match an analytical meta-material model and its parameters to a specific parameterized API callback. A selected analytical model is first calibrated. Then, an iterative process (simulation or measurement-based) optimizes the input parameters of the model that best yield the API callback.

values (e.g. phase difference per cell in the reflectarray model). In this workflow, the optical scattering response is initially investigated and, then, the solution is hill-climbed via an optimization loop relying on either field measurements or full-wave simulations. The approximate models are as follows.

The simplest model is the phased antenna-array analysis, where each single unit cell is treated as an independent antenna, excited by a single impinging ray, and emitting a single ray in response, with a local phase and amplitude alteration. Assuming a meta-material consisting of $M \times N$ unit cells, the scattered E-field complex amplitude pattern at a given frequency can be calculated by the envelope (coherent superposition) of all rays scattered from the meta-material [1]

$$E(\theta, \varphi) = \sum_{m=1}^{M}\sum_{n=1}^{N} A_{mn} e^{j\alpha_{mn}} f_{mn}(\theta_{mn}, \varphi_{mn}) \cdot \Gamma_{mn} e^{j\gamma_{mn}} f_{mn}(\theta, \varphi) e^{j\Phi_{mn}(\theta,\varphi)} \quad (4.2)$$

In (4.2), φ and θ are the azimuth and elevation angles in the scattering direction, $(\theta_{mn}, \varphi_{mn})$ denotes the direction of the wavefront 'ray' incident on the mnth cell, A_{mn} and α_{mn} are the amplitude and phase of the incident wavefront on the mnth cell, Γ_{mn} and γ_{mn} form the reflection coefficient (amplitude and phase) of the mnth cell, while f_{mn} defines the scattering pattern of the mnth cell, which, according to reciprocity, is identical for the incident and scattered direction, and, in this work, is assumed that $f_{mn}(\theta, \varphi) = \cos(\theta)$. Finally, $\Phi_{mn}(\theta, \varphi)$ is the phase shift in the mnth cell stemming from its geometrical placement, as

$$\Phi_{mn}(\theta, \varphi) = k \sin\theta \left[d_x m \cos\varphi + d_y n \sin\varphi\right] + \phi_0(\theta, \varphi) \quad (4.3)$$

where $d_{x,y}$ are the rectangular unit-cell lateral dimensions, $k = 2\pi/\lambda$ is the wavenumber in the medium enclosing the meta-material, and ϕ_0 is the reference phase denoting the spherical coordinate system center, typically in the middle of the meta-material aperture. Given a uniform, single-frequency impinging wave, any departing wavefront is essentially a Fourier composition of the individual meta-atom responses. Thus, we can, also, calculate the meta-atom amplitudes Γ_{mn} and phases γ_{mn} that yield a desired departing wavefront, by applying an inverse Fourier transform, as elaborately discussed in [1]. The calculated Γ_{mn} and γ_{mn} values must be mapped to the R_i and X_i values that generate them, since the latter are the actual tunable meta-material parameters. This process requires a set of simulations, yet it can be automated: existing model calibration techniques, such as the Regression and Goodness of Fit can be employed [63].

The shortcoming of the antenna-array approach is that the coupling between adjacent unit cells (e.g. compare against Figure 4.5) is not properly accounted for, which can result to model imprecision [1]. To this aim, the Meta-material Middleware user is presented with an alternative model. It utilizes the phased array and equivalent circuit model, which assumes not only the transmitting-responding antenna per meta-atom but also circuit elements that interconnect them and account for the cross-meta-atom meta-material interactions. The disadvantage of this approach is that an expert needs to define this circuit model, that is generally unique per meta-material design [6].

Once this model has been selected and provided in the proper format, the optimization workflow of Figure 4.9 continues, once again, with the calibration phase, which is identical as before. The key difference and merit is that the calibration is, now, extremely precise with regard to the full-wave simulations, while it takes much less time to complete, as detailed in the corresponding study of [1].

An intermediate solution, combining the precision of the circuit model and the automation of the antenna-array model, is an equivalent propagation model, mentioned here for the sake of completion. The main idea is to introduce a generic mechanism to capture the cross-interactions among meta-atoms (as opposed to the strict, physics-derived nature of the circuit model) and then proceed with automatic model calibration, avoiding the need for expert input. The equivalent ray model uses a neural network approach as the generic cross-talk descriptor [64]. A short summary is as follows. Each meta-atom is mapped to a neural network node, and the locally impinging wave amplitude and phase are its inputs. Then, we clone this layer (omitting the inputs) and form a number of intermediate, fully connected layers (usually 3–5), thereby emulating a recurrent network with a finite number of steps. We define links per node (shared among all node clones), which define an alteration of the local phase and amplitude, and its distribution to other neighbouring meta-atoms/nodes. Next, we proceed to calibrate the model via feed-forward/back-propagation, thereby obtaining a match between R, X, Γ_{mn}, and γ_{mn} values. Nonetheless, despite its automated nature, a major drawback of this model is the need for considerable computational resources, without which the model loses its value, since it becomes restricted only to very simple meta-material designs.

Since computational complexity is a concern regardless of the chosen model, the Meta-material Middleware workflow allows the user to define solution reduction across three directions. First, meta-atoms may be grouped into periodically repeated super-cells. Thus, the optimization workflow needs only to optimize the configuration parameters of a super-cell, as opposed to optimizing the complete meta-material. Second, the range of possible R and X values per meta-atom can be discretized into regular or irregular steps, reducing the solution space further.[1] Finally, some R and X values or ranges can be discarded due to the physical nature of the optimization request. For instance, if we seek to optimize a wave steering approach with an emphasis on minimal losses over the meta-material

---

1 Notably, contemporary optimization engines already incorporate equivalents to this direction, as they are able to detect strongly and loosely connected inputs–outputs [63].

(maximum reflection amplitude), the Ohmic resistance R needs to receive its boundary value. On a related track, machine learning-based approaches can quickly estimate the performance deriving from one set of R and X values, thereby discarding non-promising ones and accelerating convergence [65]. Subsequently, the Meta-material Middleware workflow moves to the optimization stage, where it attempts to hill-climb the initial solution detected via any of the described approximate models. At this point, the workflow is compatible to any modern optimization engine, which receives an input solution and outputs one or more proposed improvements upon it at each iteration. Herein, we stress the existence of engines that, also, incorporate machine learning mechanisms, to accelerate the optimization cycle [63]. The optimization can be based either on full-wave simulations or a real measurement test bed, described in Section 4.6. The optimization metric can be any reduction of the produced departing wavefront. Various metrics relevant to antenna and propagation theory may be extracted, namely: the number of main lobes (beam directions), the directivity of main lobes, the side (parasitic) lobes, and their levels, the beam widths, etc. Such metrics can be used to quantify the meta-material performance for the requested functionality, e.g. the main lobe directivity and beam width measures how 'well' a meta-material steers an incoming wavefront to a desired outgoing direction. Lastly, the hill-climbed meta-material configuration pertaining to the meta-material API callback is stored into a database for any future use by meta-material users.

Finally, we note that multiple simultaneous functionalities can be supported by interlacing different scattering profiles across the meta-material. In general, this is pre-formed by spatially mixing the profiles in phasor form:

$$A_{mn}e^{j\alpha_{mn}} = \sum_{c=1}^{N_c} A_{c,mn}e^{j\alpha_{c,mn}} \tag{4.4}$$

where c iterates over single, 'low-level' functionalities and n, m are the unit cell indices. Typically, low-level functionalities correspond to simple beam steering operations, which are produced exclusively by phase variations on the meta-material ($A_{c,mn} = 1$). In this case, a 'high-level' functionality will correspond to a multi-splitting operation with variable spatial distribution of A_{mn} amplitude, raising the hardware requirements for the meta-material. Therefore, a meta-material with no absorption capabilities (and thus no control over A_{mn}) will have limited access to high-level operations, unless a mathematical approximation is to be applied, skewing the scattering response from its ideal state. As discussed in Section 4.6, a method for minimizing amplitude variations has been successfully investigated by increasing the number of secondary parasitic lobes. Such a problem can be easily reformulated into an optimization task, where an optimal match to the ideal high-level operation can be pursued under specific constrains (e.g. $A_{mn} >$ const., $\forall m, n$).

4.5.2 The Meta-material Functionality Profiler

The optimization workflow of Figure 4.9 opts for the best meta-material configuration for a given, specific pair of impinging and departing wavefronts. However, in real deployments, it is not certain that a meta-material will always be illuminated by the intended wavefront [35]. For instance, user mobility can alter the impinging wavefront in a manner that has limited relation to the intended one and, consequently, to the running meta-material configuration. As such, there is a need for *fully profiling a meta-material*, i.e. calculate and cache its expected response for each intended meta-material configuration, but also for each possible (matching or not) impinging wavefront of interest. This profiling process is outlined in Figure 4.10.

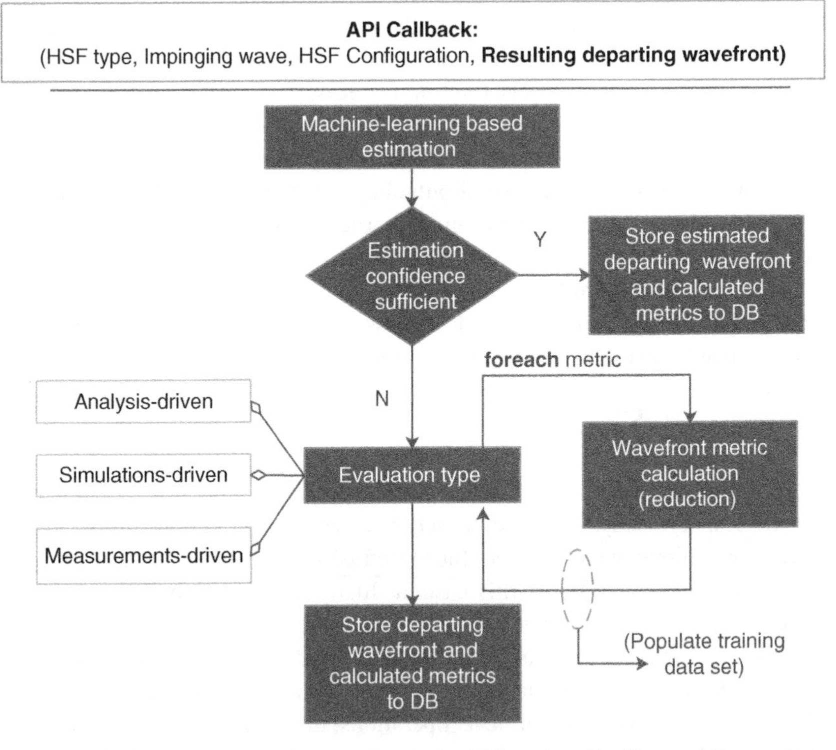

Figure 4.10 Workflow for profiling a meta-material functionality. The workflow seeks to produce a data set that describes the meta-material behaviour for any impinging wave type that does not match the one specified in the current meta-material configuration. An exhaustive evaluation takes place first for a wide set of possible impinging waves. For intermediate impinging wave cases, the workflow can rely on estimations produced by machine learning algorithms or simple extrapolation means, provided that it yields an acceptable degree of confidence.

The profiling process begins by querying the existing cache (part of the DB) or trained model for the given meta-material and an estimation (or existing calculated outcome) of the expected meta-material response for a given impinging wave. If it exists, this response is stored into a separate profile entry for the meta-material in the Meta-material Middleware DB.[2] If the response needs to be calculated anew, the process proceeds with either an analysis-, simulation-, or measurement-driven evaluation. Therefore, the choice is given as a means to facilitate the expert into reducing the required computational time, as allowed per case. Then, the profiler proceeds to also calculate all possible reductions of the departing wavefront, e.g. the number of main lobes (beam directions), the directivity of the main lobes, the side (parasitic) lobes and their levels, the beam widths, etc. Finally, once all required impinging wavefronts have been successfully processed, the profiling process is concluded.

It is clarified, that the middleware operations are one-time only, i.e. once the database containing the behaviour profile of a meta-surface is complete, it can be used in any application setting in the real world by any tile of the same type.

4.6 Software Implementation and Evaluation

Employing the concepts of the previous sections, we developed a complete Java implementation of the described software. The software is subdivided into two integral modules: (i) an implementation of the meta-material API that handles the communication and allocation of existing configurations and (ii) the Meta-material Middleware that populates the configuration DB with new data (new tiles, configurations, and functionalities). The Meta-material Middleware incorporates a full GUI environment, guiding the user through a step-by-step process to produce new configurations. It utilizes all available theoretical and computational tools for the accurate characterization of a meta-material tile. Furthermore, it offers direct access to the configuration DB, manually, via a custom-made Structured Query Language (SQL) manager or through the automated process following a successful meta-material characterization. Through this process, all the necessary data related to a newly produced configuration become explicitly available to the API.

A microwave meta-surface was selected to demonstrate the capabilities of the developed concepts and methods for software-tunable meta-materials. We adapted the design of [66], where a set of radio-frequency (RF) diodes can be

2 In case of an estimated response, the user has control over the process to filter out estimations with low confidence. However, the selected estimation engine must be able to provide a confidence degree for this automation.

employed to toggle the reflection-phase of each cell between 16 states. We numerically extracted the response of the meta-surface (i.e. its scattering pattern), and finally used the developed software to demonstrate how the meta-surface response can be controlled. For the practical demonstration of the developed software in the same measurement environment (anechoic chamber) with a simpler, 1-bit meta-surface hardware, we redirect the reader to [22, 45], since the hardware manufacturing topic is quite extensive and clearly beyond the software aspects that constitute the focus of this chapter.

In the following, we list and comprehensively describe the steps undertaken during a optimization process, as seen through the GUI environment of the Meta-material Middleware. In summary, this process involves:

- The definition of a new unit cell structure and tile array (if required).
- The parameterization of a new functionality.
- The analytical evaluation of the scattering profile on the meta-material for the selected functionality.
- The association of the meta-material profile to the set of element states, through the use of numerical simulations.
- The experimental evaluation of the exported configuration through physical measurement of a meta-material prototype. Notably, the presented software has been verified experimentally, and a full report can be found online [22].
- The final storing of all configuration parameters into the configuration DB to complete the function optimization process.

In this context, Figure 4.11a depicts all the individual steps in separate panels. If a new configuration is to be defined for a tile already present in the configuration DB, then, the first step can be skipped. Alternatively, the user must input all essential parameters of the unit cell structure, i.e. the number and type of all variables that correspond to the sum of reconfigurable meta-material elements. The definition of a new configuration begins with the parameterization of the desired functionality (Figure 4.11a.2). The current implementation supports plane wave or point source inputs (for far- and near-field energy sources) and a set of output options corresponding to all basic meta-material functionalities, discussed in Section 4.2.1. Here, we select a beam splitting operation and proceed to the first main step of the characterization process.

The analytical evaluation of the energy scattering profile is performed in the software locally and in real time. During this step, the Meta-material Middleware calculates the proper scattered field response of the impinging wave, for each unit cell at the $N \times M$ tile, via either the analytical methods of Section 4.5 or through an optimization process. The scattered fields are evaluated for each unit cell as a double complex variable ($A_1 e^{i\phi_1}$ and $A_2 e^{i\phi_2}$, magnitude and phase of the reflected transverse electric (TE) and transverse magnetic (TM) polarizations,

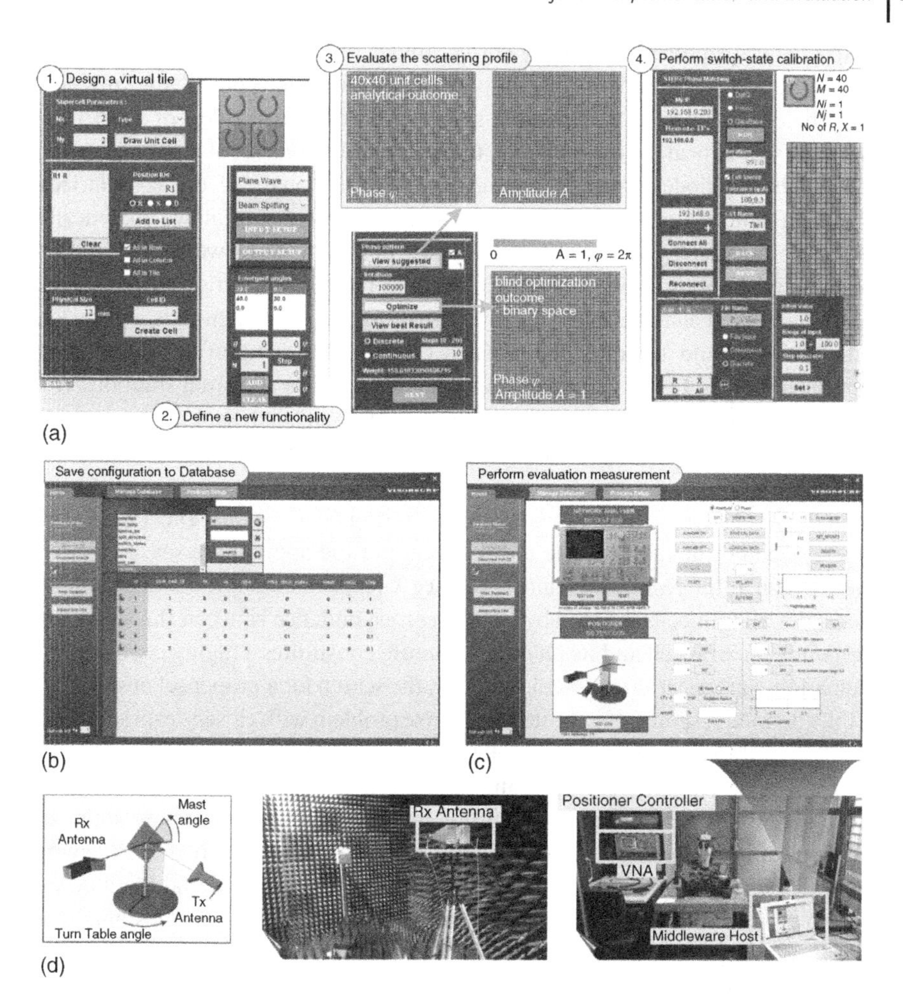

Figure 4.11 (a) Snapshots of the Meta-material Middleware GUI during the meta-material characterization process. (b) After a successful characterization process, the configuration is saved to the DB. Herein, as an instance, a snapshot of the DB manager that was created during development of the software is presented. (c) The Meta-material Middleware offers extensive experimental capabilities through a dedicated module. (d) Utilizing the hardware present in an anechoic chamber, the Meta-material Middleware was able to acquire full scattering diagrams of a meta-material tile, mounted on the available positioning equipment. The three-dimensional (3D) schematic, shown on the left, is updated live as the turntable or the positioner head rotates during a measurement or optimization process. On the right-hand side, the Meta-material Middleware host (white laptop) connects via Ethernet to the VNA (measuring the Rx/Tx antennas) and to the positioner controller.

respectively), a process physically correct under the condition that the unit cell is a subwavelength entity. This simply implies that the input of the 'physical size' field in Figure 4.11a.1 must comply with this specification or a warning message will appear. In our example, the selected tile consists of binary elements (D stands for diodes), which may only control the phase of the co-polarized scattered field (ϕ_1 term). By clicking the 'View suggested' button, we analytically calculate and display $\phi_0 - \phi_1$ (ϕ_0 is the phase of the incoming wavefront), which should give an indication of the diode states at the meta-material. Alternatively, a similar result can be extracted by launching the meta-heuristic optimizer, either via a blind optimization process (all-0 initial solution) or an assisted optimization, using the analytically evaluated profile as an initial solution. The latter practice leads to more refined results, over an analytical evaluation, by considering the finite size of the tile and the non-infinitesimal size of the unit cell.

In the final step, we seek to match the scattering field response (i.e. the $(A_1 e^{i\phi_1}, A_2 e^{i\phi_2})_{i,j}$ pairs calculated in the previous step) to the appropriate set of element states, such as the resistance R and reactance X (capacitance or inductance) of the loads, for all unit cells indexed by i, j. The correspondence between the scattering response of a unit cell and its physical structure constitutes a highly complex and demanding propagation problem. Actually, the search for a proper set of states for a fully defined response constitutes an inverse problem with closed-form solutions available only for very simple unit cell types. As such, it demands the use of highly efficient optimization algorithms and strong computation power to perform the necessary numerical simulations. In our implementation, we employ a cluster of interconnected computer clients (that serve as simulation nodes) receiving instructions through a TCP/IP network from the host Meta-material Middleware running on the main machine. In a lightweight scenario, a single-cell simulation assigned to one of the clients can be completed in less than a minute, but this may grow substantially when the complexity of the design is increased. An acceptable convergence is expected to be reached within a few thousands simulation trials. In perspective, the prototype studied in this work displayed a typical evaluation time of 8–12 hours for a full characterization of its configuration space and all possible impinging plane waves (θ, ϕ sets) at the operation frequency of the hardware. The evaluation is performed through Algorithm 1 (see next page) whose steps are described below. Specifically, the user:

- Selects one of the three options: (i) a gradient based optimizer, recommended for a small number of variables (less than four), (ii) a meta-heuristic optimizer, recommended for a higher count of variables (more than three), and (iii) a database-based optimization process that configures the proper switch-states, through simulation results already stored in the configuration DB.

- Suggests a convergence limit *Tol* to the optimizer in the Tolerance text box for all variable targets.
- Fills the IP list with all available simulation nodes and initiates a connection. All nodes will launch the installed simulation software, open the corresponding geometry model, and remain idle until further instructions are received.
- Modifies, if necessary, the value range, step and initial value for each variable in the unit cell.
- Begins the optimization process by clicking the 'RUN' button. The software will iterate over all unit cells (from top-left (0,0) to bottom-right (N, M)), seeking a set of optimal values for the variables of each cell. During this sequence, each unit cell (i,j) initiates an independent optimization process and a new series of simulations begin until a sufficiently close convergence to the targeted scattering field response $S_{ij} = (A_1 e^{i\phi_1}, A_2 e^{i\phi_2})_{i,j}$ is achieved. Prior to each simulation, the optimizer looks up all entries in the associated table of the configuration DB, in case a result (S_{db}) is already stored. If not, the simulation will start and the result will be stored afterwards. Over time, this process can populate the DB with enough results, making option (iii) of the initial step a highly efficient method for evaluating new functionalities for this particular tile.

By completing the previous steps, the software has successfully defined a new configuration for the chosen tile, which, now, remains to be stored in the configuration DB. Before doing so, we can apply an additional evaluation step by conducting an experimental measurement on a physical prototype (if available). Our current implementation is able to assess multi-splitting or absorption functionalities by measuring the full scattering diagram on the front meta-material hemisphere, which can then be compared to the scattering profile produced by the software. As presented in the corresponding technical report [22], several tests with a meta-material unit have already been successfully conducted in a fully equipped anechoic chamber (Figure 4.11). This test bed incorporates a variety of algorithms, meeting individual needs for accuracy and speed for various cases of scattering response measurements (e.g. a full 3D-pattern versus a 2D-slice might be required for arbitrary lobe scattering). A simple case is outlined in Algorithm 2, where the user has previously set the appropriate parameters in the GUI (Figure 4.11c). The GUI automates the process of measuring meta-material devices in the anechoic chamber by supporting several communication interfaces for the following equipment:

- **Vector network analyser (VNA)**: Produces the energy signal and receives the response (i.e. S_{21}-parameter) from the antenna setup.
- **Positioner**: Allows the mechanical support of the meta-material and antenna devices. Its controller can instruct the rotation of both heads (towards θ, ϕ), allowing a complete characterization of the scattering profile.

- **Meta-material controllers**: A meta-material hosting reconfigurable elements incorporates a communication network for the explicit control of its element states. The Meta-material Middleware implements the proper interface for the evaluation prototypes (serial port connection, Wi-Fi, and Bluetooth have been integrated). The same interfaces are, also, used for the meta-material API developed in Section 4.4.

A supplementary note is that the final switch-state configuration can be re-evaluated using the same meta-heuristic optimizer utilized in step 3 of Figure 4.12a via actual experimental results. The optimizer starts with the software-defined configuration as an initial solution and gradually adjusts the switch-state matrices to more optimally converge to the pursued functionality under true operational conditions. The implemented algorithm follows the template of Algorithm 2, where N correspond to the number or optimization variables (e.g. the number of scattering lobes) and a second *for* loop nests the existing loop, seeking to maximize the $Sum_i(E[i])$ metric.

For further evaluation purposes, we validate a number of indicative examples from the literature, based on previously measured and simulated results. Hence,

Algorithm 4.1 Physical Element Calibration. Highlighted commands run in a separate thread.

Set *index* to $i = 0, j = 0$
while index does not exceed max **do**
  Get S_{ij}; Get *Tol*
  if !(exists S_{db} where $|S_{db} - S_{ij}| < Tol$) **then**
    while searching **do**
      Select node where *state* == *free*
      Set to *busy*
      Initiate parallel *thread*;
      Optimizer: Fixes (R, L, C, D) variables;
      Optimizer → Requests new simulation;
      node ← Returns S_{sim}
      $S_{db} = S_{sim}$;
      if $\left|S_{sim} - S_{ij}\right| < Tol$ **then**
        searching = *false*
        Set (R, L, C, D) variables as *optimal*
      end if
      Set node *state* to *free*
    end while
    Iterate *index*
  end if
end while

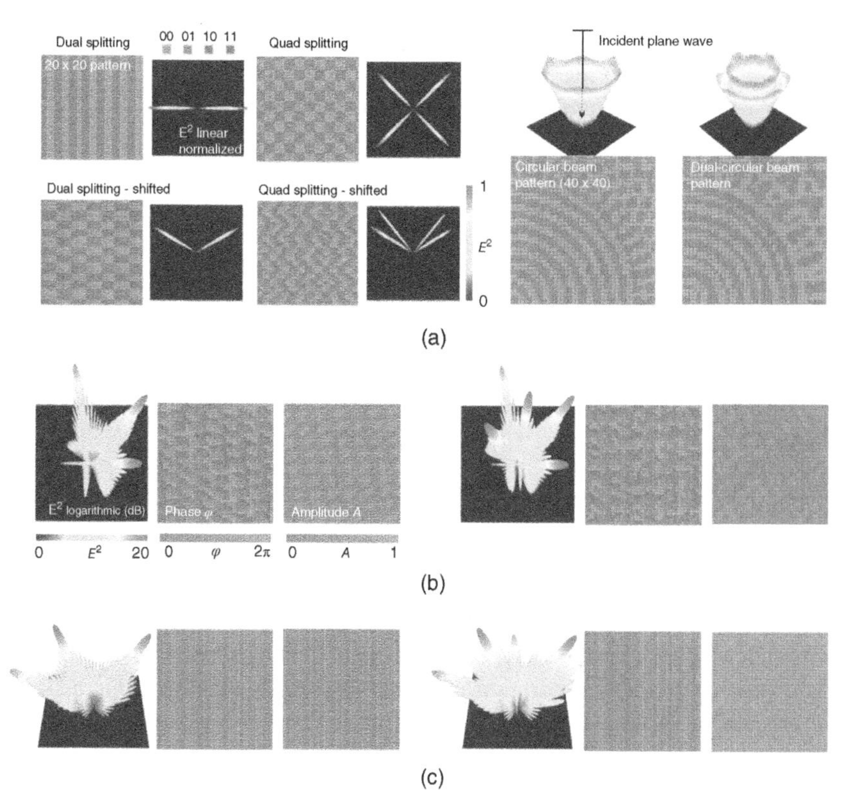

Figure 4.12 (a) Element states and far-field scattering diagram of a 4-bit meta-material array for six different cases. (b) Element states and scattering diagrams for triple- and quintuple-beam scattering, assuming continuously adjustable states with absorption capabilities. (c) Comparison between a tile with resistor elements (left) and a tile without (right) for an in-plane triple-beam splitting functionality. A non-uniform amplitude pattern (A_{mn}) can eliminate all side-lobes and provide increased security to the signal. The incident field in all cases is a vertically impinging plane wave, chosen for simplicity; yet any plane wave or point source input can be considered by a corresponding shift of the phase for each unit cell.

Figure 4.12a presents the optimization outcome for a 4-bit meta-material array, which can switch over four available states per unit cell with reflection phases $(-90, 0, 90, 180)$ and full reflection amplitude $(A = 1)$ [30, 67]. The outcome complies fully with the results provided by the corresponding authors, indicating that the Meta-material Middleware may cooperate with arbitrary hardware configurations and thus be compatible with any reasonable design in a future diverse meta-material market. Following these results, we, also, test four exclusive cases that highlight the additional capabilities of our software. In particular, Figure 4.12b displays the results for a triple- and a quintuple-beam splitting case,

Algorithm 4.2 Evaluate Scattering Response, 3D, slow rotation case. Highlighted commands refer to HW instructions.

Set $N \leftarrow$ number of equidistant points on the hemisphere
Set $P[N]$ $\leftarrow$Struct of (θ, ϕ) points
for all elements i in P **do**
  Send rotation command $\rightarrow$ Positioner
  (Mast - $P[i]$.theta, Head - $P[i]$.phi, Speed)
  Receive $\leftarrow$ Positioner feedback
  Refresh 3D Figure
  Send measurement command $\rightarrow$ VNA
  $E[i]$ = **Receive** $\leftarrow$ S21 power
  Refresh Plot graph
end for
Save ($P[i]$, $E[i]$) to DB

while Figure 4.12c shows the optimization outcome for an in-plane triple-beam splitting functionality. For the latter case, the integrated theoretical algorithms were able to eliminate all side lobes by suggesting a non-uniform pattern for the amplitude A of the co-polarized scattered field. This particular case demands a tile with controllable absorption elements (resistors). Finally, in Figure 4.13 we proceed to showcase the optimization of an arbitrary departing wavefront

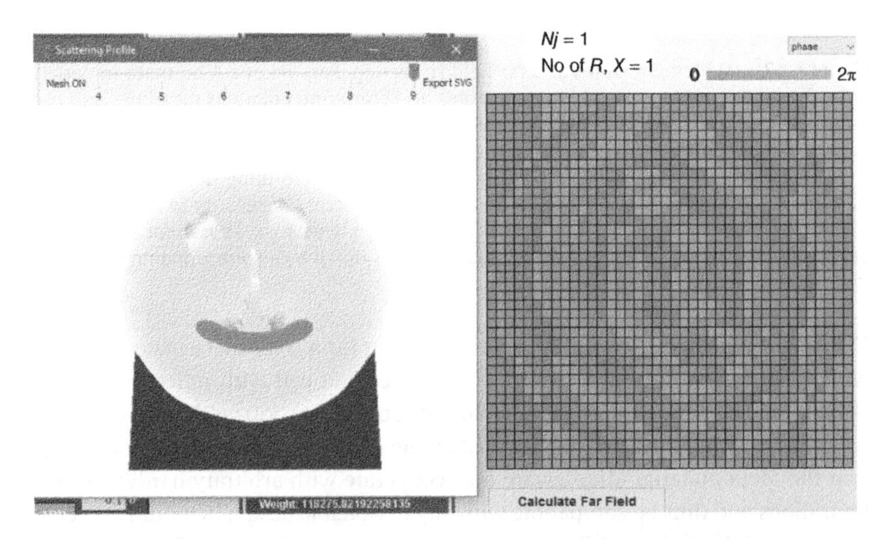

Figure 4.13 Arbitrary functionality optimization test. A smiley face-shaped scattering pattern is successfully produced (left). The corresponding meta-material element configuration (different meta-atom states expressed in colormap) is shown to the left.

formation. A smiley-face shaped scattering pattern is selected as the required energy wave response of the meta-material to a planar impinging wave. The optimization process successfully produces the required wavefront, and the corresponding meta-material element states are shown to the right of Figure 4.13.

4.7 Discussion: The Transformational Potential of the IoMMT and Future Directions

While the potential of the IoMMT paradigm alone may be worth the investigation [39–42], here we evaluate its practical opportunities affecting the industry, the end users and the environment, namely:

- How can the IoMMT prolong the life cycle of products across deployment scales?
- How can the IoMMT help maintain a high-speed product development pace, without sacrificing ecological concerns during the product design phase?

To these ends, we believe that the concept of Circular Economy (CE) and its associated performance indices is a fitting framework for the initial exploration and evaluation of the IoMMT paradigm [68, 69]. CE seeks to make technological products reusable, repairable, and recyclable across their lifetime (i.e. development, purchase, usage, and disposal) by introducing cross-product and cross-manufacturer interactions. Instead of the traditional, linear order of life cycles phases, i.e. (i) raw resource acquirement, (ii) processing, (iii) distribution, (iv) its use, and (v) disposal, the CE advocates to create links from disposal to all preceding phases, promoting (i) reprocessing or refurbishment, (ii) re-deployment and re-distribution, and (iii) multiple uses.

However, according to the literature [70], the CE introduces a paradoxical tension in the industry: While the industry is pressed for faster growth and, hence, a faster product development rate, CE can introduce a series of design considerations that make for a slower product development rate. In this view, the paradigm of IoMMT is by its nature impactful for the energy and ecological footprint of multiple products, across disciplines and scales.

The fact that it enables the tuning and optimization of the physical properties of matter allows not only for a tremendous impact both in terms of QoS per product and scale but also for energy savings in a horizontal manner. Moreover, the IoMMT can contribute a software-driven way for optimizing material properties. Using this new technology, the industrial players can maintain a fast-paced product design, where energy efficiency and sustainability can be upgraded programmatically via 'eco-firmware' during its use, thereby offloading the product design phase of such concerns.

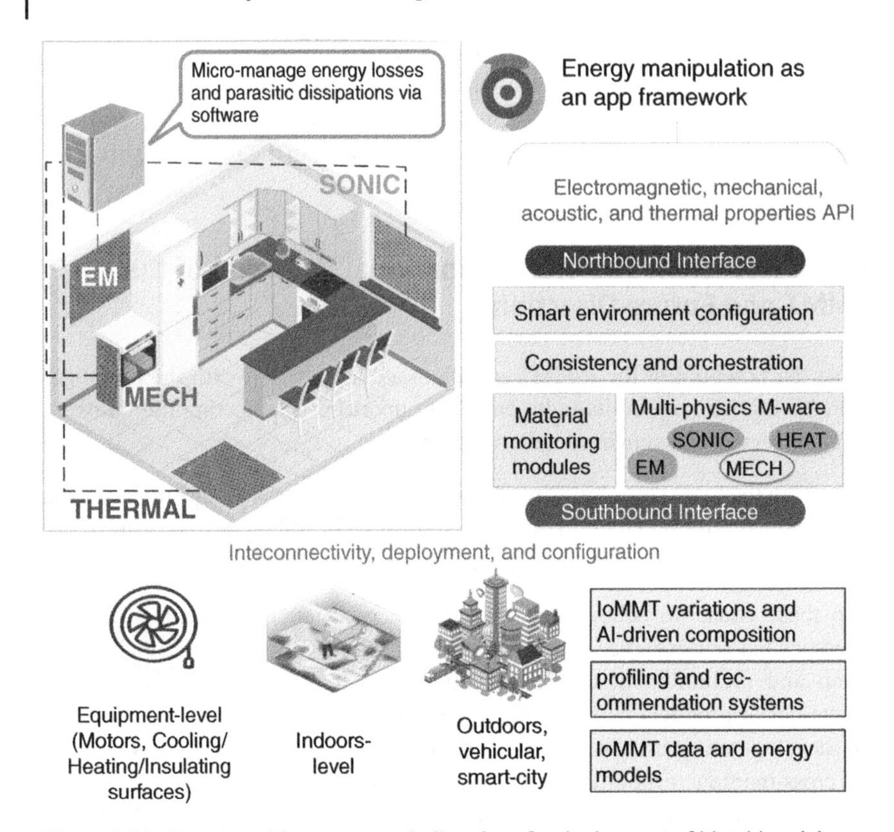

Figure 4.14 Envisioned future research directions for the Internet of MetaMaterials.

An important future research direction of the IoMMT is to quantify the financial savings stemming from adopting this technique. While the CE-derived paradoxical tensions have been around for long and are hard to eradicate, IoMMT can facilitate their resolution by quantifying them, potentially aligning fast-paced marketing with environmental sustainability.

Apart from the CE line of work, future work will seek to provide the theoretical and modelling foundations of the IoMMT. Our vision, overviewed in Figure 4.14, is for a full-stack study of this new concept, covering: the physical layer modelling, the internetworking and communications layer, and the application layer.

At the physical layer, future research needs to classify and model meta-materials in a functionality-centric way, and introduce fitting Key-Performance Indicators for each offered energy manipulation type. Meta-surface-internal control variations (technologies and ways of monitoring embedded active elements) can be taken into account and aspire to deduce models covering the aspects of control data traffic, energy expenditure and feature-based manufacturing cost estimation.

This will enable the creation of the first, cross-physical domain profiling and recommendation system for meta-materials, to match the requirements and specifications of any envisioned application.

Moving to the networking layer, research can follow the proposed northbound/southbound abstraction model inspired from the SDN paradigm. This will provide the necessary platform for: (i) Interconnecting the IoMMT to the vast array of existing networked devices and assorted standards, models and protocols. (ii) Provide the necessary software abstractions, to open the field of energy micromanagement to software developers (i.e. without specialty in Physics), enabling the energy-propagation-as-an-app paradigm. In this aspect, we also envision the need for algorithms optimizing IoMMT deployments for minimal-investment-maximal-control, and orchestrating IoMMT deployments for any set of generic energy micro-management objective. The control time granularity depends on the application scenario and the volatility of the factors affecting the energy propagation within an environment. In an indoors wireless communications setting, such as the one studied in [20], where the Intelligent Wireless Environment needs to continuously adapt to the position of user devices, the control granularity can be considered to be 10–25 ms (i.e. randomly walking users running a mobile application).

Finally, at the application layer, research can define key-applications of the IoMMT at multiple scales, starting from the internals of equipment, such as motors, heating, cooling, and insulating surfaces. This will provide the basic units for IoMMT incorporation to devices spanning home appliances (ovens, refrigerators, washers, heating, and cooling units), electronics (from interference cancellation, to smart cooling) and building materials (acoustic, thermal, and mechanical insulators). Various scales can be taken into consideration, from indoor (smart-house) and outdoor (city-level, vehicular networks) IoMMT deployments.

4.8 Conclusion

In this chapter, we show how prospective 6G computer networks stand to benefit from the new paradigm we introduce, the IoMMT, aimed to control the propagation of waves of any form, electromagnetic, acoustic or mechanical, within a space. Then, towards its realization, we pursue the systematic expansion of the concept of software-defined meta-materials by establishing the key software elements of a meta-material network. Finally, we demonstrate the IoMMT potential by harnessing electromagnetic waves in the microwave band, enabled by custom reconfigurable meta-surfaces, also called IRS.

With this goal in mind, we introduce two special categories of software, the Meta-material Middleware, which can methodically produce novel configurations for a single meta-material tile and the meta-material API, which is in charge of supervising the energy propagation within a meta-material-coated space. The two components were studied and developed autonomously, and then merged into a unified application via a common layer of abstraction.

For the API, we explored the means to interpret a meta-material, its configuration space, and the supported energy manipulation functionalities via a group of well-defined software objects. The key objective was to successfully conceal the physical layer of the meta-material, and only expose the essential parameters for configuring an operation. In light of this definition, the API is capable of instructing its environment to steer, focus, absorb, or split the incoming signals through a simple interface, without any reference to the underlying physics.

The Meta-material Middleware was developed by means of various theoretical and computational tools, including analytical algorithms that assess the scattering response, full-wave simulations that accurately evaluate a unit cell response, and experimental modules that can perform physical measurements of meta-material devices. Through a step-by-step process, we defined a robust characterization methodology, supporting a large set of meta-material operations with no restriction on the specifications for the meta-material. We also discuss the prospect of applying different optimization algorithms for each stage of the meta-material characterization process. Finally, the evaluation of the middleware outcomes demonstrated that the IoMMT paradigm can be employed successfully within a unified framework, covering any type of electromagnetic, acoustic, mechanical, and even thermal meta-material and meta-surface.

Acknowledgements

This work was supported by the European Union's Horizon 2020 research and innovation programme-project C4IIoT, GA EU833828. The authors also acknowledge FETOPEN-RIA project VISORSURF. All presented software modules were conceived, designed, and implemented in full by the Foundation for Research and Technology-Hellas (FORTH) and G. Pyrialakos served as the lead developer. The present chapter derives content first published at Liaskos, C., Pyrialakos, G. G., Pitilakis, A., Tsioliaridou, A., Christodoulou, M., Kantartzis, N., ... & Akyildiz, I. F. (2020). The Internet of MetaMaterial Things and their software enablers. Int. Telecommun. Union, Journal of Future Emerging Technologies, 1, 1-23. [Online: https://www.itu.int/dms_pub/itu-s/opb/itujnl/S-ITUJNL-JFETF.V1I1-2020-P05-PDF-E.pdf, DOI: 10.52953/PPWI5546]

References

1 Capolino, F. (2017). *Theory and Phenomena of Metamaterials*. CRC Press.

2 Liaskos, C., Tsioliaridou, A., Pitsillides, A. et al. (2015). Design and development of software defined metamaterials for nanonetworks. *IEEE Circ. Syst. Mag.* 15 (4): 12–25.

3 Pitilakis, A., Tasolamprou, A.C., Liaskos, C. et al. (2018). Software-defined metasurface paradigm: concept, challenges, prospects. *2018 12th International Congress on Artificial Materials for Novel Wave Phenomena (Metamaterials)*, pp. 483–485. IEEE.

4 Liaskos, C., Tsioliaridou, A., Pitilakis, A., et al. (2019). Joint compressed sensing and manipulation of wireless emissions with intelligent surfaces. In *2019 IEEE 15th International Conference on Distributed Computing Sensor Systems (DCOSS)*, pp. 318–325.

5 Pan, J., Jain, R., Paul, S. et al. (2015). An internet of things framework for smart energy in buildings: designs, prototype, and experiments. *IEEE Internet Things J.* 2 (6): 527–537. https://doi.org/10.1109/jiot.2015.2413397.

6 Li, A., Singh, S., and Sievenpiper, D. (2018). Metasurfaces and their applications. *Nanophotonics* 7 (6): 989–1011. https://doi.org/10.1515/nanoph-2017-0120.

7 Liu, F., Pitilakis, A., Mirmoosa, M.S. et al. (2018). Programmable metasurfaces: state of the art and prospects. *2018 IEEE International Symposium on Circuits and Systems (ISCAS)*, pp. 1–5. IEEE.

8 Mehrotra, R., Ansari, R.I., Pitilakis, A. et al. (2019). 3D channel modeling and characterization for hypersurface empowered indoor environment at 60 GHz millimeter-wave band. *2019 International Symposium on Performance Evaluation of Computer and Telecommunication Systems (SPECTS)*, pp. 1–6. IEEE.

9 Abadal, S., Liaskos, C., Pitsillides, A. et al. (2019). Terahertz programmable metasurfaces: networks inside networks. In: *Nanoscale Networking and Communications Handbook*, 47–67. CRC Press. John R. Vacca

10 Tasolamprou, A.C., Pitilakis, A., Tsilipakos, O. et al. (2019). The software-defined metasurfaces concept and electromagnetic aspects. *arXiv preprint arXiv:1908.01072*.

11 Dash, S., Liaskos, C., Akyildiz, I.F., and Pitsillides, A. (2019). Wideband perfect absorption polarization insensitive reconfigurable graphene metasurface for THz wireless environment. *2019 IEEE Microwave Theory and Techniques in Wireless Communications (MTTW)*, Volume 1, pp. 93–96. IEEE.

12 Manessis, D., Seckel, M., Fu, L. et al. (2019). High frequency substrate technologies for the realisation of software programmable metasurfaces on PCB hardware platforms with integrated controller nodes. *2019 22nd European Microelectronics and Packaging Conference & Exhibition (EMPC)*, pp. 1–7. IEEE.

13 Pitilakis, A., Tsilipakos, O., Liu, F. et al. (2020). A multi-functional reconfigurable metasurface: electromagnetic design accounting for fabrication aspects. *IEEE Trans. Antennas Propag.* 69 (3): 1440–1454.

14 Manessis, D., Seckel, M., Fu, L. et al. (2020). Manufacturing of high frequency substrates as software programmable metasurfaces on PCBs with integrated controller nodes. *2020 IEEE 8th Electronics System-Integration Technology Conference (ESTC)*, pp. 1–7. IEEE.

15 Taghvaee, H., Abadal, S., Alarcon, E. et al. (2020). The scaling laws of hypersurfaces. In *The Internet of Materials*, pp. 227–267. CRC Press. C. Liaskos.

16 Kadic, M., Bückmann, T., Schittny, R., and Wegener, M. (2013). Metamaterials beyond electromagnetism. *Rep. Prog. Phys.* 76 (12): 126501.

17 Surjadi, J.U., Gao, L., Du, H. et al. (2019). Mechanical metamaterials and their engineering applications. *Adv. Eng. Mater.* 21 (3): 1800864.

18 Liaskos, C., Tsioliaridou, A., Nie, S. et al. (2019). On the network-layer modeling and configuration of programmable wireless environments. *IEEE/ACM Trans. Netw.* 27 (4): 1696–1713.

19 Pishvar, M. and Harne, R.L. (2020). Foundations for soft, smart matter by active mechanical metamaterials. *Adv. Sci.* 7 (18): 2001384.

20 Liaskos, C., Tsioliaridou, A., Pitsillides, A. et al. (2018). Using any surface to realize a new paradigm for wireless communications. *Commun. ACM* 61: 30–33.

21 Akyildiz, I.F. and Jornet, J.M. (2010). The internet of nano-things. *IEEE Wireless Commun.* 17 (6): 58–63.

22 The VISORSURF Project Consortium (2019). A hypervisor for metasurface functionalities: progress report. *European Commission Project VISORSURF: Public Report, August 2019.* https://ec.europa.eu/research/participants/documents/downloadPublic?documentIds=080166e5c560b376&appId=PPGMS (accessed 24 June 2022).

23 Frenzel, T., Kadic, M., and Wegener, M. (2017). Three-dimensional mechanical metamaterials with a twist. *Science* 358 (6366): 1072–1074.

24 Haberman, M.R. and Guild, M.D. (2016). Acoustic metamaterials. *Phys. Today* 69 (6): 42–48. https://doi.org/10.1063/pt.3.3198.

25 Dresselhaus, M.S., Chen, G., Tang, M.Y. et al. (2007). New directions for low-dimensional thermoelectric materials. *Adv. Mater.* 19 (8): 1043–1053.

26 Glybovski, S.B., Tretyakov, S.A., Belov, P.A. et al. (2016). Metasurfaces: from microwaves to visible. *Phys. Rep.* 634: 1–72.

27 Chen, H.-T., Taylor, A.J., and Yu, N. (2016). A review of metasurfaces: physics and applications. *Rep. Prog. Phys.* 79 (7): 076401 (1–40).

28 Huang, C., Zhang, C., Yang, J. et al. (2017). Reconfigurable metasurface for multifunctional control of electromagnetic waves.*Adv. Opt. Mater.* 5 (22): 1700485 (1–6).

29 Özdogan, O., Björnson, E., and Larsson, E. (2019). Intelligent reflecting surfaces: physics, propagation, and pathloss modeling. *IEEE Wireless Commun. Lett.* 9 (5): 581–585.

30 Zhang, Q., Wan, X., Liu, S. et al. (2017). Shaping electromagnetic waves using software-automatically-designed metasurfaces. *Sci. Rep.* 7 (1): 3588 (1–11).

31 Mosk, A.P., Lagendijk, A., Lerosey, G., and Fink, M. (2012). Controlling waves in space and time for imaging and focusing in complex media. *Nat. Photonics* 6 (5): 283–292.

32 Tan, X., Sun, Z., Koutsonikolas, D., and Jornet, J.M. (2018). Enabling indoor mobile millimeter-wave networks based on smart reflect-arrays. *IEEE INFOCOM 2018-IEEE Conference on Computer Communications*, pp. 270–278. IEEE.

33 Liu, F., Pitilakis, A. Mirmoosa, M.S. et al. (2018). Programmable metasurfaces: state of the art and prospects. *2018 IEEE International Symposium on Circuits and Systems (ISCAS)*, Volume 2018, 8351817. IEEE, May 2018.

34 Saeed, T., Soteriou, V., Liaskos, C. et al. (2020). Toward fault-tolerant deadlock-free routing in hypersurface-embedded controller networks. *IEEE Netw. Lett.* 2 (3): 140–144.

35 Liaskos, C., Nie, S., Tsioliaridou, A. et al. (2020). Mobility-aware beam steering in metasurface-based programmable wireless environments. *ICASSP 2020-2020 IEEE International Conference on Acoustics, Speech and Signal Processing (ICASSP)*, pp. 9150–9154. IEEE.

36 Liaskos, C., Nie, S., Tsioliaridou, A. et al. (2020). End-to-end wireless path deployment with intelligent surfaces using interpretable neural networks. *IEEE Trans. Commun.* 68 (11): pp. 6792–6806.

37 Mathioudakis, F., Liaskos, C., Tsioliaridou, A. et al. (2020). Advanced physical-layer security as an app in programmable wireless environments. *2020 IEEE 21st International Workshop on Signal Processing Advances in Wireless Communications (SPAWC)*, pp. 1–5. IEEE.

38 Chen, Z., Guo, B., Yang, Y., and Cheng, C. (2014). Metamaterials-based enhanced energy harvesting: a review. *Physica B* 438: 1–8.

39 Liaskos, C., Tsioliaridou, A., Ioannidis, S. et al. (2020). Applications of the internet of materials: programmable wireless environments. In: *The Internet of Materials*, 269–317. CRC Press, C. Liaskos.

40 Liaskos, C., Tsioliaridou, A., Ioannidis, S. et al. (2021). Next generation connected materials for intelligent energy propagation in multiphysics systems. *IEEE Commun. Mag.* 59 (8): 100–106.

41 Liaskos, C., Pyrialakos, G., Pitilakis, A. et al. (2019). ABSense: sensing electromagnetic waves on metasurfaces via ambient compilation of full absorption. *Proceedings of the 6th Annual ACM International Conference on Nanoscale Computing and Communication*, pp. 1–6.

42 Liaskos, C., Tsioliaridou, A., and Ioannidis, S. (2019). Organizing network management logic with circular economy principles. *2019 15th International Conference on Distributed Computing in Sensor Systems (DCOSS)*, pp. 451–456. IEEE.

43 Florijn, B., Coulais, C., and van Hecke, M. (2014). Programmable mechanical metamaterials. *Phys. Rev. Lett.* 113 (17): 175503 (1–4).

44 Cummer, S.A., Christensen, J., and Alù, A. (2016). Controlling sound with acoustic metamaterials. *Nat. Rev. Mater.* 1 (3): 16001 (1–13).

45 Zhang, L., Chen, X.Q., Liu, S. et al. (2018). Space-time-coding digital metasurfaces. *Nat. Commun.* 9 (1): 4334 (1–11). https://doi.org/10.1038/s41467-018-06802-0.

46 Abadal, S., Liaskos, C., Tsioliaridou, A. et al. (2017). Computing and communications for the software-defined metamaterial paradigm: a context analysis. *IEEE Access* 5: 6225–6235.

47 Taghvaee, H., Abadal, S., Pitilakis, A. et al. (2020). Scalability analysis of programmable metasurfaces for beam steering. *IEEE Access* 8: 105320–105334.

48 Sergiou, C., Lestas, M., Antoniou, P. et al. (2020). Complex systems: a communication networks perspective towards 6G. *IEEE Access* 8: 89007–89030.

49 Tsilipakos, O., Tasolamprou, A.C., Pitilakis, A. et al. (2020). Toward intelligent metasurfaces: the progress from globally tunable metasurfaces to software-defined metasurfaces with an embedded network of controllers. *Adv. Opt. Mater.* 8 (17): 2000783.

50 Liaskos, C., Nie, S., Tsioliaridou, A. et al. (2018). Realizing wireless communication through software-defined hypersurface environments. *2018 IEEE 19th International Symposium on" A World of Wireless, Mobile and Multimedia Networks"(WoWMoM)*, pp. 14–15. IEEE.

51 Liaskos, C., Nie, S., Tsioliaridou, A. et al. (2018). A new wireless communication paradigm through software-controlled metasurfaces. *IEEE Commun. Mag.* 56 (9): 162–169.

52 Liaskos, C., Nie, S., Tsioliaridou, A. et al. (2019). A novel communication paradigm for high capacity and security via programmable indoor wireless environments in next generation wireless systems. *Ad Hoc Networks* 87: 1–16.

53 Liaskos, C., Tsioliaridou, A., Nie, S. et al. (2018). Modeling, simulating and configuring programmable wireless environments for multi-user multi-objective networking. *arXiv preprint arXiv:1812.11429*.

54 Oktian, Y.E., Lee, S.G., Lee, H.J., and Lam, J.H. (2017). Distributed SDN controller system: a survey on design choice. *Comput. Netw.* 121: 100–111.

55 Liaskos, C., Pitilakis, A., Tsioliaridou, A. et al. (2017). Initial UML definition of the hypersurface compiler middle-ware. *European Commission Project VISORSURF: Public Deliverable D2.2, 31 December 2017*. http://www.visorsurf.eu/m/VISORSURF-D2.2.pdf (accessed 24 June 2022).

56 Ashraf, N., Lestas, M., Saeed, T. et al. (2020). Extremum seeking control for beam steering using hypersurfaces. *2020 IEEE International Conference on Communications Workshops (ICC Workshops)*, pp. 1–6. IEEE.

57 Liaskos, C. (2020). Interim: drafting a stack. In: *The Internet of Materials*, pp. 225–226. CRC Press. C. Liaskos.

58 Tsioliaridou, A., Pyrialakos, G., Pitilakis, A. et al. (2020). Designing the internet-of-materials interaction software. In: *The Internet of Materials*, pp. 77–140. CRC Press. C. Liaskos.

59 Tsioliaridou, A., Liaskos, C., Pitsillides, A., and Ioannidis, S. (2017). A novel protocol for network-controlled metasurfaces. *Proceedings of the 4th ACM International Conference on Nanoscale Computing and Communication*, pp. 1–6.

60 Tsioliaridou, A., Liaskos, C., Ioannidis, S., and Pitsillides, A. (2015). CORONA: A coordinate and routing system for nanonetworks. *Proceedings of the 2nd Annual International Conference on Nanoscale Computing and Communication*, pp. 1–6.

61 Saeed, T., Abadal, S., Liaskos, C. et al. (2019). Workload characterization of programmable metasurfaces. *Proceedings of the 6th Annual ACM International Conference on Nanoscale Computing and Communication*, pp. 1–6.

62 Ashraf, N., Saeed, T., Ansari, R.I. et al. (2019). Feedback based beam steering for intelligent metasurfaces. *2019 2nd IEEE Middle East and North Africa COMMunications Conference (MENACOMM)*, pp. 1–6. IEEE.

63 Laguna, M. and Marti, R. (2003). The OptQuest callable library. In: *Optimization Software Class Libraries*, 193–218. Springer.

64 Liaskos, C., Tsioliaridou, A., Nie, S. et al. (2019). An interpretable neural network for configuring programmable wireless environments. *IEEE Interantional Workshop on Signal Processing Advances in Wireless Communications (SPAWC 2019)*, pp. 1–5.

65 Taghvaee, H., Jain, A., Timoneda, X. et al. (2021). Radiation pattern prediction for metasurfaces: a neural network-based approach. *Sensors* 21 (8): 2765.

66 Saifullah, Y., Waqas, A.B., Yang, G.-M. et al. (2019). 4-bit optimized coding metasurface for wideband RCS reduction. *IEEE Access* 7: 122378–122386.

67 Liu, S., Cui, T.J., Zhang, L. et al. (2016). Convolution operations on coding metasurface to reach flexible and continuous controls of terahertz beams. *Adv. Sci.* 3 (10): 1600156 (1–12).

68 Stahel, W.R. (2016). The circular economy. *Nature* 531 (7595): 435–438.

69 Liaskos, C., Tsioliaridou, A., and Ioannidis, S. (2018). Towards a circular economy via intelligent metamaterials. *2018 IEEE 23rd International Workshop on Computer Aided Modeling and Design of Communication Links and Networks (CAMAD)*, pp. 1–6. IEEE.

70 Daddi, T., Ceglia, D., Bianchi, G., and de Barcellos, M.D. (2019). Paradoxical tensions and corporate sustainability: a focus on circular economy business cases. *Corp. Soc. Responsib. Environ. Manage.* 26 (4): 770–780.

5

IRS Hardware Architectures

Jun Y. Dai[1,2,3], Qiang Cheng[1,2,3], and Tie Jun Cui[1,2,3]

[1]State Key Laboratory of Millimeter Waves, Southeast University, Nanjing, China
[2]Institute of Electromagnetic Space, Southeast University, Nanjing, China
[3]Frontiers Science Center for Mobile Information Communication and Security, Southeast University, Nanjing, China

5.1 Introduction

With the rapid development of wireless communication technology, intelligent reconfigurable surface (IRS) has been recognized as an innovative approach for the electromagnetic (EM) wave manipulation, signal modulation, and smart radio environment reconfiguration [1–5]. It is a two-dimensional (2D) ultra-thin artificial surface composed of periodically or aperiodically arranging sub-wavelength elements, which possess dynamic controllability of the EM waves with the superiority of low loss, ultra-thin thickness, and easy fabrication [6–10]. Such unique attributes are believed to bring in fundamental influence on the wireless communication framework due to the potential to construct controllable radio environments for EM signals, without relying on those complex signal processing techniques [11, 12].

In fact, the concept of IRS in the communication community was evolved from the programmable meta-surface in the EM community [8]. In the past 10 years, meta-surfaces have been well studied owing to their powerful capabilities in controlling the EM waves [13–15]. However, the passive meta-surfaces have fixed functions once they are fabricated. To dynamically control the EM waves, active meta-surfaces have been presented [16–19], which are usually classified into two types: tunnable meta-surfaces and reconfigurable meta-surfaces. In general, the tunnable meta-surfaces can reach continuous responses with similar functions by tuning the active device [18, 19]; while the reconfigurable meta-surface can realize

Intelligent Reconfigurable Surfaces (IRS) for Prospective 6G Wireless Networks, First Edition.
Edited by Muhammad Ali Imran, Lina Mohjazi, Lina Bariah, Sami Muhaidat,
Tie Jun Cui, and Qammer H. Abbasi.

different functions by switching the active device, but the number of realized functions is very limited [16, 17]. In order to obtain many different functionalities by a single platform, digital coding and programmable meta-surface has been proposed with the aid of field programmable gate array (FPGA), which can switch many different functions in ultra-fast speed by manipulating the digital coding patterns on the meta-surface [8]. This advantageous feature of the programmable meta-surface makes it a perfect platform of IRS. More importantly, the digital coding representation of meta-elements makes it possible to modulate the digital information directly on the meta-surface. Hence, a more general concept of information meta-surface was proposed to bridge the EM physical world and the digital world [20, 21]. In recent years, the information meta-surfaces have achieved significant developments in both space-domain coding and time-domain coding strategies [8, 9, 22–25].

Besides the advantages mentioned above, the digital coding meta-surface can also simplify the design of meta-element and biasing network, and hence is more suitable for IRS realization [8]. The basic idea underpinning this novel branch lies in the limited types of coding elements and the capability to manipulate the EM fields, which offer a simple and efficient way to design, optimize, and fabricate IRSs. In this way, every possible spatial distribution of the elements on IRS can be equivalently and univocally described in terms of digital coding sequences. In addition, a reprogrammable IRS can be further realized by independently controlling the digital elements via FPGA or micro-programmed control unit (MCU), which features real-time switching among different EM functions in a programmable manner [8, 22, 23]. Most importantly, the digital coding and programmable meta-surface evolves a more advanced version of IRS that is more familiar to the wireless communication community, with distinct features in the analysis and processing of information [20, 21]. In the premier researches, the coding process is performed only in the spatial domain, in which the digital coding sequence corresponds to the location of the elements to manipulate the spatial distribution of EM fields, including the far-field and near-field patterns [8, 9]. With the programmable characteristic, the coding process can be extended to the temporal domain, leading to the control of the spectral distribution of EM waves [24, 25]. Furthermore, the combination of spatial and temporal coding strategies offers the possibility of multi-domain joint control of EM waves, opening a new avenue for simultaneous manipulation of the wavefront and frequency spectrum [26–28].

In this chapter, we will detail the general hardware architecture of IRS, starting with the introduction to the concept, principle, and composition of IRS. Then we will elaborate on the operation mode of IRS when deployed in wireless communication systems. Several practical prototypes will be demonstrated for reference. Finally, the hardware configuration of IRS will be discussed.

5.2 Concept, Principle, and Composition of IRS

As mentioned in Section 5.1, an IRS is a 2D artificial surface composed of tens to hundreds of sub-wavelength elements, which can be employed like wallpaper to cover parts of walls, ceilings, buildings, and even traffic tools, and is capable of manipulating the parameters of the EM waves impinging upon it in a controllable and programmable manner. Therefore, it can be concluded that reconfigurability, programmability, and discretization are the prominent properties of IRS, among which reconfigurability dominates. Here we first take a 1-bit digital coding meta-surface as a simple IRS example [8]. As shown in Figure 5.1a, consider an IRS comprising $M \times N$ elements, each of which is encoded as '0' or '1' with a phase difference of π. With the integration of tunable devices like diodes, the states of elements can be further controlled in real time. The geometry and reflection phase spectra of the element embedded with PIN diode (Skyworks, SMP-1320)

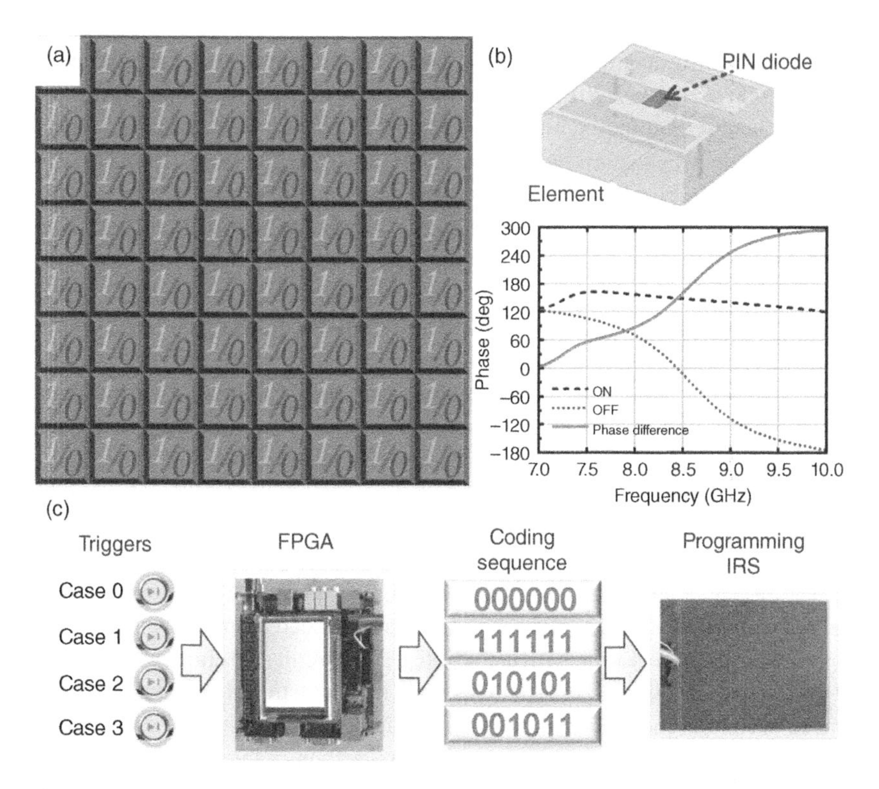

Figure 5.1 (a) Conceptional illustration of IRS (1-bit digital coding meta-surface) [8]. (b) Geometry and reflection phase spectra of the IRS element [8]. (c) Working flow of IRS with digital logic devices. Source: Cui et al. [8]/with permission of Springer Nature.

are illustrated in Figure 5.1b, in which the states '0' and '1' correspond to 'OFF' and 'ON' of the diode, respectively. A dielectric substrate (F4B, ϵ_r=2.65, δ=0.001) is sandwiched by two metal layers, on top of which a PIN diode is mounted and two electrodes are, respectively, connected to the bottom metal layer through metallic vias. The bit depth of element can certainly be extended by increasing the number of digital states. For example, a 2-bit digital coding meta-surface contains four digital states '00', '01', '10', and '11' that, respectively, represent four distinct phase responses of 0, $\pi/2$, π, and $3\pi/2$, and higher-bit also follows similar strategies. With the help of many digital logic devices, the exhibited EM functions of digital coding meta-surface can be switched in real time by pre-designed coding sequences, as demonstrated in Figure 5.1c.

Designing the coding sequence is critical to generate various functions for specific applications. The common approach is to perform the coding process in the spatial domain, which uses the coding sequence to determine the phase distribution of IRS directly, leading to the flexible control of scattering patterns [8, 9]. For instance, as respectively depicted in Figure 5.2a, the IRS can attain different scattering beams by varying the corresponding coding sequences of 0000 … /0000 …, 0101 … /0101 …, and 0101 … /1010 …, i.e. a single beam, two rabbit-ear beams, and four symmetrical beams. In this way, IRS can dynamically manipulate the propagation behaviour of EM waves, establishing a smart radio environment that is capable of improving the wireless communication adaptively. Another design approach of coding sequence focuses on modulating the IRS in the temporal domain using the dynamics and programmability of the element, which makes it a time-varying device. The spectrum of the modulated wave is that of the modulation signal shifted by the incident wave frequency, like a spatial mixing effect [24, 25]. For example, Figure 5.2b demonstrate the time-varying phase waveforms and the corresponding spectral amplitude distributions of two coding sequences, i.e. 1-bit sequence '… 0101 …' and 2-bit sequence '… − 11 − 10 − 01 − 00 − …', respectively. On this basis, it paves the way to spectrum control of EM waves by IRS, providing the possibility to realize information modulation without traditional costly radio frequency (RF) component.

It is worth noting that, most of the IRSs are regarded as passive and lossy devices because they only consume EM energy by ohmic losses, dielectric losses, structural resonances, etc. In fact, some new designs have been proposed to enable EM amplifications by integrating with active integrated circuit (IC) chips, such as voltage and power amplifiers [29, 30]. The basic mechanism lies in amplifying the induced current of the IRS element to realize signal enhancement. This branch of IRS could lead to distinctive applications in wireless communication frameworks, including wireless relay and signal repeater. Undoubtedly, these works expand the application scope of IRS to some extent at the cost of more power consumption, which, as a result, can certainly be classified to IRS in a broad sense.

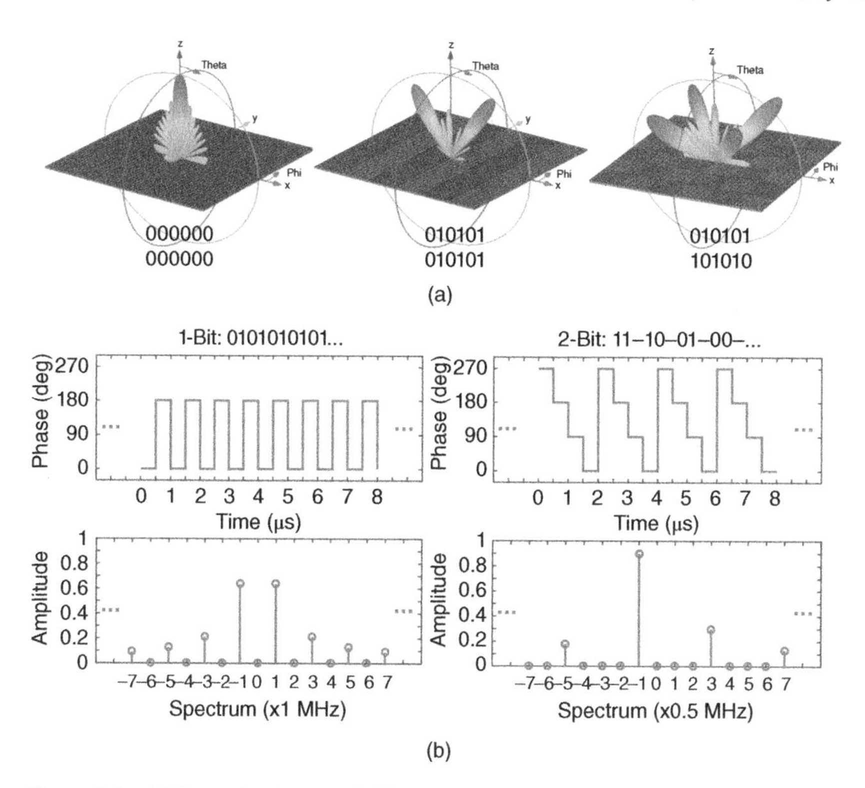

Figure 5.2 (a) Scattering beams of IRS with different spatial coding sequences. Source: Cui et al. [8]/with permission of Springer Nature. (b) Time-varying phase waveforms and corresponding spectral amplitude distributions of two temporal coding sequences. Source: Zhao et al. [24]/Oxford University Press/CC BY 4.0.

5.3 Operation Mode of IRS

Based on the above content, IRS shows the great potential to use different operation modes generate multifarious functions in wireless communications. Broadly speaking, the operation mode of IRS can be classified into two categories according to the role IRS plays in the wireless networks.

- **Wavefront manipulation**: In this mode, IRS focuses on manipulating the EM wavefront by synthesizing various spatial distributions of its element states. It requires the control chip and biasing network to have as many ports as possible to obtain precise control of the IRS, preferably for each sub-wavelength element, thereby realizing finer re-configurability of EM wavefront. As the straightforward application of IRS, this mode helps construct the smart radio environment that allows IRS to directly reconfigure the propagation behaviour of EM waves by software. Unlike uncontrollable natural propagation environments, the smart radio environment can optimize the scattering properties of the built-in IRSs to obtain adaptive programmable wireless channels, thereby maximizing the performance of the entire communication network. In addition, this mode of IRS can help the wireless signals bypass physical obstacles in the propagation path and establish virtual links among the base stations and the users, which shows great attractiveness for the coverage extension of wireless networks. Furthermore, it is also able to resist fast fading lead by the multi-path effects due to the promotion of multi-path mitigation, which finally improves the quality of the received signal.

- **Information modulation**: In this mode, IRS is responsible for modulating information into EM waves by generating corresponding temporal sequences of its element states. This operation mode raises high claims of modulation speed for the control board to achieve wider signal bandwidth, which is able to implement wireless communications with faster message rates. Owing to the unique capability of IRS for EM spectral regulation, it provides a promising method in constructing new wireless communication systems with greatly simplified architecture. Such an idea enables direct modulations and transmissions of information without using the traditional RF components, which is considered to be low-cost and power-saving.

In the following text, we will elaborate on different implementations of the two categories of operation mode as examples for better demonstrations.

5.3.1 Prototypes of Wavefront Manipulation Mode

In [31], a RFocus prototype is constructed by the researchers from the Massachusetts Institute of Technology (MIT), USA. They design a simple IRS element that utilized RF switches to either let the EM wave penetrate or be reflected. A software controller is used to set the state of each element with a majority-voting-based optimization algorithm to maximize the received signal strength, as shown in Figure 5.3a. It requires signal strength measurement only, which sidesteps the complexity and difficulty in measuring the phase without loss of reliability. Figure 5.3b illustrates the final built prototype that contains 3200 elements in total on a 6 m^2 surface, each of which occupies the size of $\lambda/4 \times \lambda/5$ and costs a few cents only. With the guidance of optimization algorithm in wavefront manipulation mode, the prototype adjusts the beam of impinging EM wave towards specified directions or focuses it at desired positions, thereby improving the signal quality of wireless communication. The MIT places the prototype next to a wall in the lab for the performance test, in which they fix the receiver position while move the transmitter to various positions and maximize the signal strength by the optimization algorithm. The results demonstrate that the RFocus

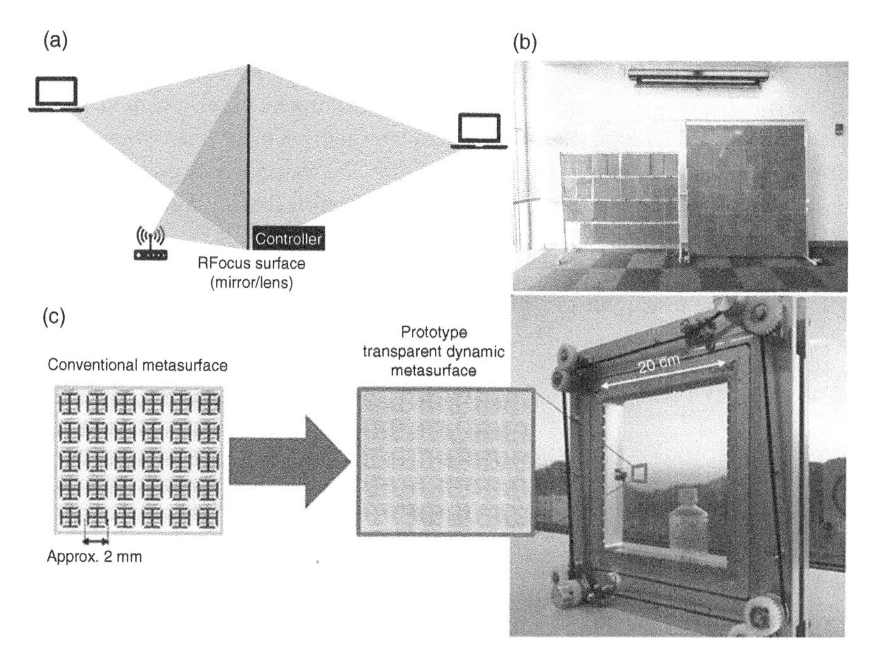

Figure 5.3 (a) Conceptional illustration of the RFocus prototype. Source: Arun and Balakrishnan [31]/USENIX/Public Domain. (b) Deploy environment of the RFocus prototype. Source: Arun and Balakrishnan [31]/The Usenix Association. (c) Prototype of optically transparent IRS. Source: Ref. [32]/NTT DOCOMO.

provides benefits throughout the entire floor, where the minimum, median, and maximum signal improvements across all locations are 3.8, 9.5, and 20.0 dB, respectively. Finally, the researchers announce that their prototype obtains remarkable improvements of the median signal strength (9.5×) and the median channel capacity (2.0×) in the practical scenario of an office building.

The researchers of NTT DOCOMO, Japan, have successfully fabricated a transparent IRS prototype operating at 28 GHz, as shown in Figure 5.3c [32]. The IRS uses a glass substrate to support a large number of sub-wavelength periodically arranged elements, thus achieving optically transparent property that is superior for unobtrusive use. With the movement of the another glass substrate, the IRS possesses three states of dynamical control for the incident EM waves in: the following: full penetration, partial reflection, and full reflection. The primary trial has successfully demonstrated two modes of the IRS at 28 GHz: the full penetration mode, where the meta-surface and movable glass substrate are attached to each other, and the full reflection mode, where the meta-surface and movable glass substrate are separated by more than 200 μm. Although the distance between the meta-surface and glass is manually controlled in the proof-of-concept test,

DOCOMO says that piezoelectric actuators will be used to switch between penetration and reflection modes at high speed in the future. On this basis, the prototype is quite suitable for some specific installation environment, especially in locations that are not suited for base station installation or high-security areas where reception needs to be blocked selectively. In addition, the transparency guarantees the mergence with the surrounding environment, indicating that it is a good candidate for IRS deployment in buildings, vehicles, and billboards.

In Figure 5.4a, an IRS-aided wireless communication prototype, developed and evaluated by the researchers in Huazhong University of Science and Technology,

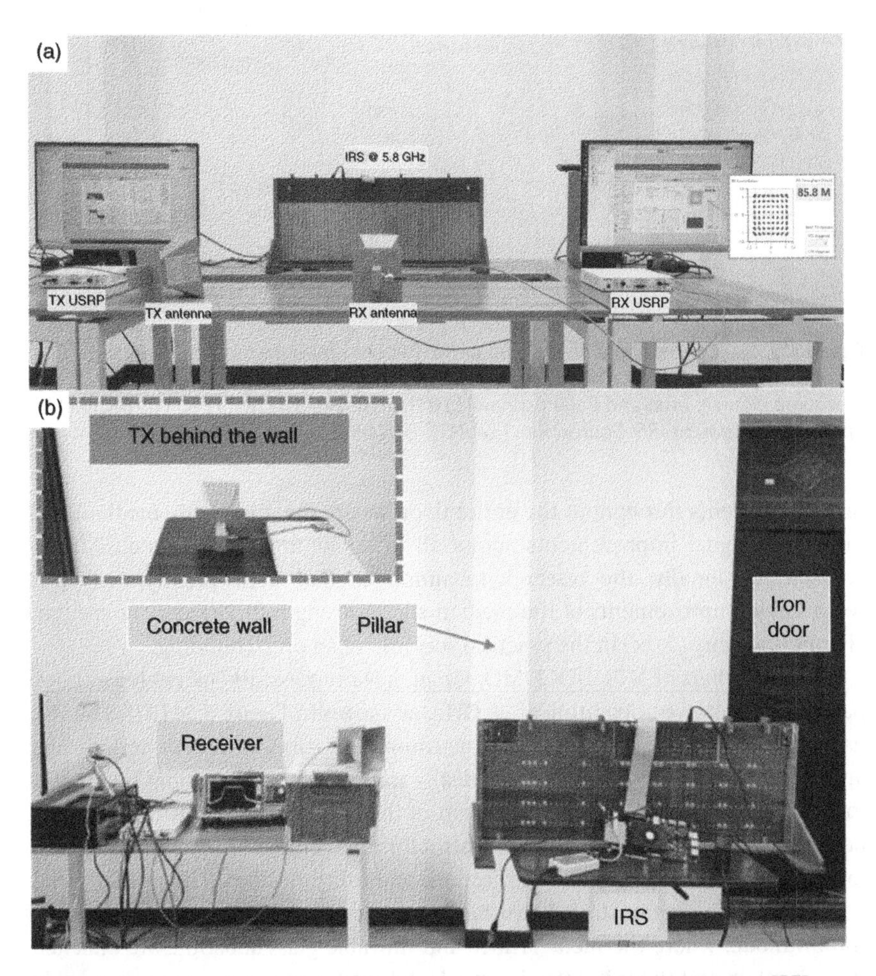

Figure 5.4 (a) Illustration of the IRS-aided wireless communication prototype [33]. (b) Experimental environments of the indoor test without LoS path. Source: Pei et al. [33]/Cornell University.

China, is depicted [33]. The authors use a traditional reflective element design based on varactor diode to realize their IRS, which is capable of reconfiguring the phase of reflected wave. The entire IRS is composed of 1100 controllable elements working at 5.8 GHz band. The wireless communication system is based on conventional setups, i.e. transmitter (base station) and receiver (user), with the introduction of IRS to optimize the performance of uplink and downlink channels. A central processor is used for IRS configuration and channel monitoring with the feedback-based adaptive reflection coefficient optimization algorithm. In the indoor test without Line-of-Sight (LoS) path, as shown in Figure 5.4b, the prototype achieves a 26 dB power gain compared to a same size copper plate. In addition, it also conduct long-range outdoor tests of 50 and 500 m, respectively, in which the prototype obtains up to 27 dB power gains. Finally, it implements a 1080p video live-streamed transmission with the data rate of 32 Mbps. The power consumption of the IRS is estimated around 1 W, demonstrating the outstanding property of IRS in power saving.

5.3.2 Prototypes of Information Modulation Mode

As mentioned in Section 5.2, IRS can act as a spatial mixer, in which the incident wave and time-varying parameters can, respectively, be regarded as the local oscillator (LO) and baseband signals in the traditional wireless communication systems, as show in Figure 5.5a [34]. In this way, as long as a mapping relationship between the baseband signal and the parameters is established properly, the digital information can be modulated on the scattered wave and propagated in free space. Figure 5.5b displays the block diagram of this kind of wireless communication system, in which IRS is the core component in the transmitter.

Based on this framework, the first new architecture of wireless communication system was proposed by the researchers in South-east University, China [24]. Drawing on the idea of harmonic generation, two 2-bit phase-encoding coding sequences are designed, namely '00 − 01 − 10 − 11' and '11 − 10 − 01 − 00', each of which will generate the -1st and $+1^{\text{st}}$ order harmonics to, respectively, represent the binary digits '0' and '1'. According to the mapping principle, it is reasonable to categorize this modulation scheme as the Binary Frequency Shift Keying (BFSK) modulation. Referring to [24] for details, a proof-of-concept prototype is constructed that successfully realizes the real-time digital information transmission (e.g. a photo picture) with a message rate of 78.125 kbps at the RF frequency of 3.6 GHz in the EM anechoic chamber, as shown in Figure 5.6a. To further improve the message rate, new schemes that use similar hardware settings but directly map the baseband signal to the reflection phase is proposed, namely Quadrature Phase Shift Keying (QPSK) [35] and 8 Phase Shift Keying (8PSK) [37] modulations. Owing to the shorter length of coding sequence required

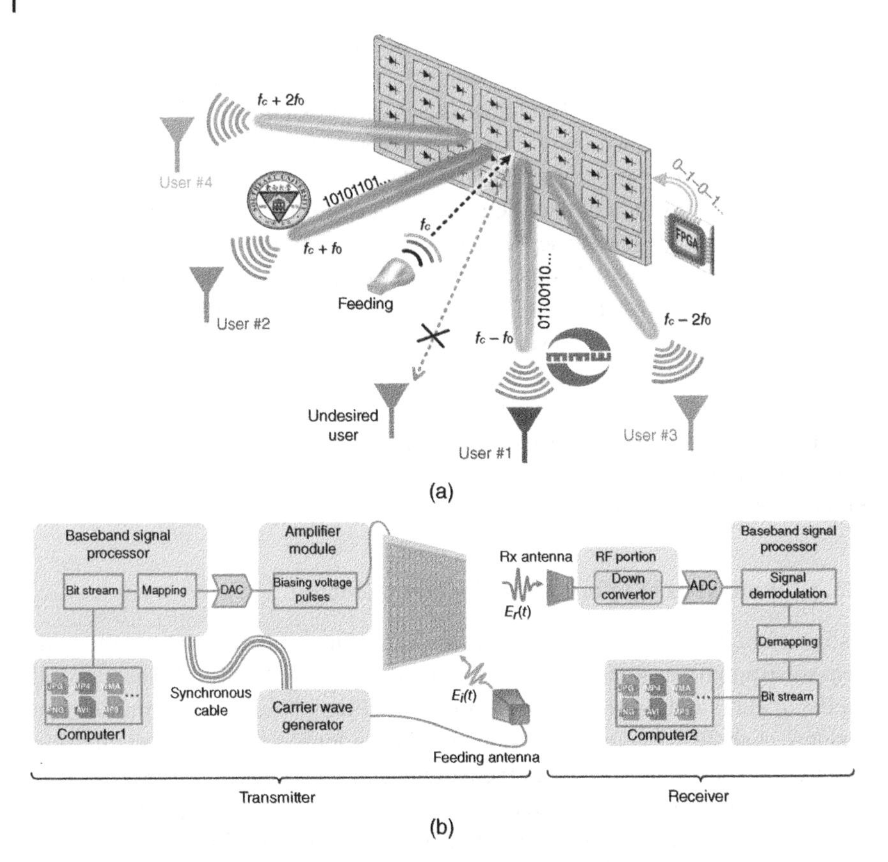

Figure 5.5 (a) Schematic of the wireless communication architecture based on IRS. Source: Zhang et al. [34]/with permission of Springer Nature. (b) Block diagram of the architecture. Source: Dai et al. [35]/with permission of John Wiley & Sons.

for each digital message, the phase shift keying (PSK) modulation schemes have higher message rates than the frequency shift keying (FSK) modulation scheme. It is worth noting that, since several traditional communication technologies have been introduced to improve the reliability and reduce the error rate of the system, the experimental verification of these works can be performed in the practical indoor scenarios, as shown in Figure 5.6b. Finally, these PSK systems achieve higher-message rates of 1.6 Mbps (QPSK) and 6 Mbps (8PSK) at the RF frequency of 4 and 4.25 GHz, respectively. Moreover, some advanced systems based on IRS are proposed with better performance, which modulate the baseband signal at harmonic frequencies using higher order modulation schemes, such as 16 Quadrature Amplitude Modulation (16QAM) [38]. With the employment of high-frequency PIN diode in the element design, 256QAM modulation scheme based on this architecture is also implemented in the millimetre-wave

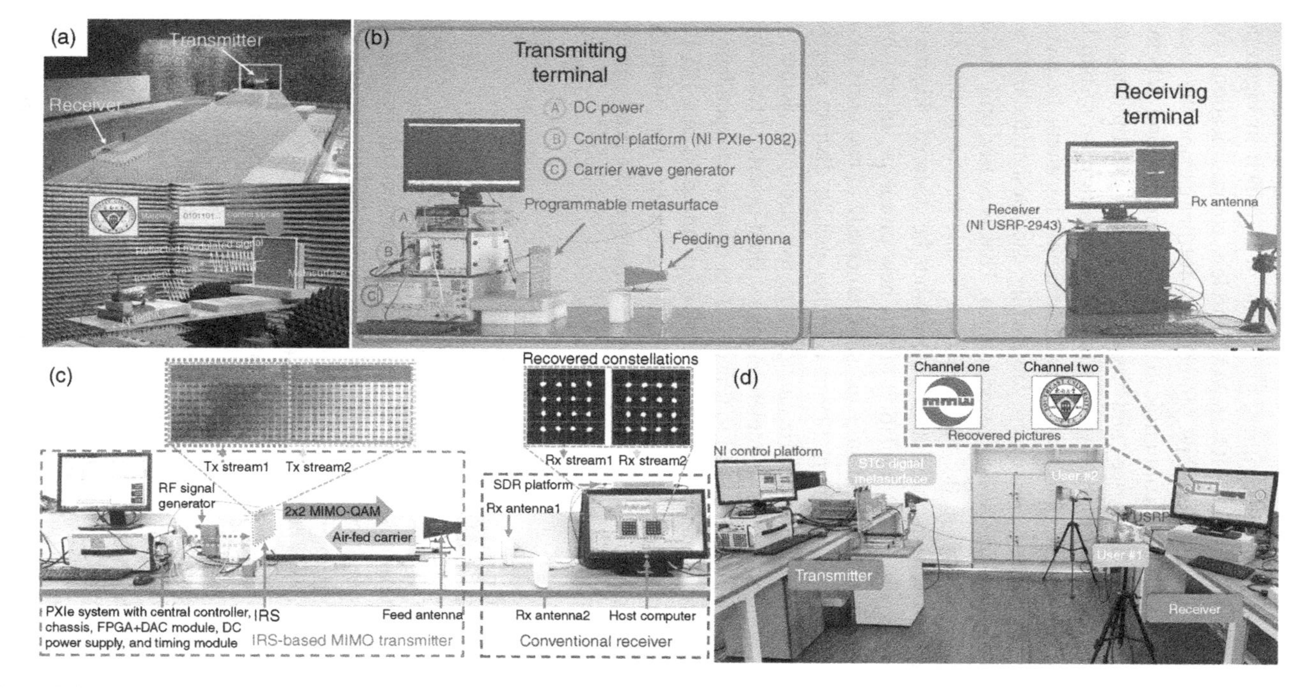

Figure 5.6 Experimental environments and results of the proof-of-concept prototypes of (a) BFSK system. Source: Zhao et al. [24]/Oxford University Press, (b) QPSK system. Source: Dai et al. [35]/John Wiley & Sons, (c) MIMO-16QAM system. Source: Tang et al. [36]/IEEE, and (d) space- and frequency-division multiplexing system. Source: Zhang et al. [34]/Springer Nature.

band [39]. In particular, the Multiple Input Multiple Output (MIMO) technology is introduced in [36] to realize simultaneous transmissions of multiple data streams. The corresponding experiment environment is shown in Figure 5.6c, which implements a 2 × 2 MIMO 16QAM wireless communication system with a message rate of up to 20 Mbps. Moreover, to further expand the capacity and improve the security of communications, a new modulation scheme is proposed to implement the space- and frequency-division multiplexing techniques simultaneously in a multichannel wireless communication system [34]. In this case, different data streams are assigned to different harmonic frequencies and routed to different spatial directions. Meanwhile, each designated user owns an independent channel in the frequency domain. That is to say, only the user that uses the right channel in the right direction is able to receive the correct information; otherwise, the transmitted date cannot be restored. Figure 5.6d demonstrates the experimental validation of this new prototype, in which two pictures are transmitted independently and simultaneously to two users located at different directions. In addition, IRS can further exhibit distinct responses to the EM waves with different polarizations as long as anisotropic elements are used. Therefore, besides the above implementations, dual-polarized transmitters based on IRS also can be constructed to transmit data streams independently via different polarization channels [40]. Based on the above introductions, IRS-based wireless communication systems exhibit excellent capability for information modulation and energy radiation, which is very promising in the next-generation communication technology.

5.4 Hardware Configuration of IRS

Recently highlighted investigations of IRS have demonstrated the huge potential in wireless communications. Those aforementioned prototypes indicate that multifarious methods and technologies can be utilized in IRS implementations. Despite the IRS designs may be poles apart, they share the same hardware configuration in a broad sense to enable their main operational principles, as shown in Figure 5.7 [2]. First, as the most prominent property of IRS, re-configurability makes the tunable element essential in the design process. Similarly, the unit cell of sub-wavelength scale is also important since it dominates the response to the EM waves at microscopic level. The configuration network can be regarded as the meridian vessel of the entire architecture that connects limbs of IRS, i.e. each unit cell or super-cell, to its brain, namely central controller. Therefore, it is easy to infer that the form of configuration network will affect the control accuracy and available functionalities to a large extent. In addition, for remote control and programming, the central controller should equip a gateway to establish a

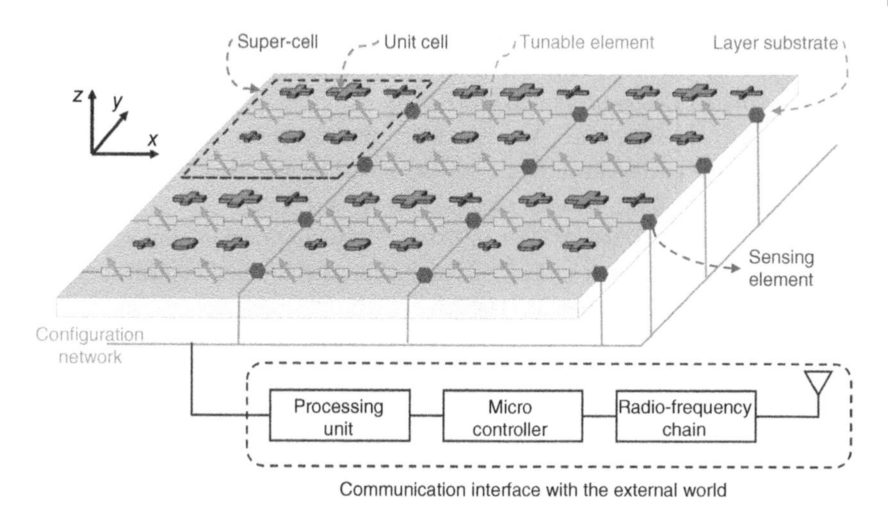

Figure 5.7 Hardware configuration of IRS. Source: Renzo et al. [2]/with permission of IEEE.

communication interface with the external world, and a processing unit to realize fast in situ computing. Finally, IRS can also integrate some sensing elements to estimate the desired environmental state information, which is preferable to enable self-adaptivity by constructing feedback links to the IRS. However, it is worth noting that the most of current researches realize IRSs without sensing elements owing to their inevitable cost and power consumption.

5.5 Conclusions

In this chapter, we have introduced the hardware architecture of IRS. The concept, principle, composition as well as the operation mode of IRS when it is deployed in wireless communication systems that are elaborated in detail. With several practical prototypes, the unique properties of IRSs in spatial and spectral control of EM waves have been fully demonstrated. Finally, the general hardware configuration of IRS is discussed.

References

1 Basar, E., Renzo, M.D., Rosny, J.D. et al. (2019). Wireless communications through reconfigurable intelligent surfaces. *IEEE Access* 7 (99): 116753–116773.

2 Renzo, M.D., Zappone, A., Debbah, M. et al. (2020). Smart radio environments empowered by reconfigurable intelligent surfaces: how it works, state of research, and road ahead. *IEEE J. Sel. Areas Commun.* 38(11): 1.

3 Huang, C., Zappone, A., Alexandropoulos, G.C. et al. (2019). Reconfigurable intelligent surfaces for energy efficiency in wireless communication. *IEEE Trans. Wireless Commun.* 18 (99): 4157–4170.

4 Tang, W., Chen, M.Z., Chen, X. et al. (2021). Wireless communications with reconfigurable intelligent surface: path loss modeling and experimental measurement. *IEEE Trans. Wireless Commun.* 20 (1): 421–439.

5 Tang, W., Dai, J.Y., Chen, M.Z. et al. (2020). MIMO transmission through reconfigurable intelligent surface: system design, analysis, and implementation. *IEEE J. Sel. Areas Commun.* 38 (11): 1.

6 Pfeiffer, C. and Grbic, A. (2013). Metamaterial Huygens' surfaces: tailoring wave fronts with reflection less sheets. *Phys. Rev. Lett.* 110 (19): 197401.

7 Kuester, E.F., Gordon, J.A., O'Hara, J. et al. (2012). An overview of the theory and applications of metasurfaces: the two-dimensional equivalents of metamaterials. *IEEE Antennas Propag. Mag.* 54 (2).

8 Cui, T.J., Qi, M.Q., Wan, X. et al. (2014). Coding metamaterials, digital metamaterials and programmable metamaterials. *Light: Sci. Appl.* 3 (10): e218.

9 Huang, C., Sun, B., Pan, W. et al. (2017). Dynamical beam manipulation based on 2-bit digitally-controlled coding metasurface. *Sci. Rep.* 7 (1): 1–8.

10 Kim, T.-T., Kim, H., Kenney, M. et al. (2018). Amplitude modulation of anomalously refracted terahertz waves with gated-graphene metasurfaces. *Adv. Opt. Mater.* 6 (1): 1700507.

11 Di Renzo, M., Zappone, A., Debbah, M. et al. (2020). Smart radio environments empowered by reconfigurable intelligent surfaces: how it works, state of research, and the road ahead. *IEEE J. Sel. Areas Commun.* 38 (11): 2450–2525.

12 Toumi, M. and Aijaz, A. (2021). System performance insights into design of RIS-assisted smart radio environments for 6G. *2021 IEEE Wireless Communications and Networking Conference (WCNC)*, pages 1–6. IEEE.

13 Holloway, C.L., Kuester, E.F., Gordon, J.A. et al. (2012). An overview of the theory and applications of metasurfaces: the two-dimensional equivalents of metamaterials. *IEEE Antennas Propag. Mag.* 54 (2): 10–35.

14 Yu, N., Genevet, P., Kats, M.A. et al. (2011). Light propagation with phase discontinuities: generalized laws of reflection and refraction. *Science* 334 (6054): 333–337.

15 Glybovski, S.B., Tretyakov, S.A., Belov, P.A. et al. (2016). Metasurfaces: from microwaves to visible. *Phys. Rep.* 634: 1–72.

16 Chen, K., Feng, Y., Monticone, F. et al. (2017). A reconfigurable active Huygens' metalens. *Adv. Mater.* 29 (17): 1606422.

17 Liu, M., Powell, D.A., Zarate, Y., and Shadrivov, I.V. (2018). Huygens' metadevices for parametric waves. *Phys. Rev. X* 8 (3): 031077.

18 Yao, Y., Shankar, R., Kats, M.A. et al. (2014). Electrically tunable metasurface perfect absorbers for ultrathin mid-infrared optical modulators. *Nano Lett.* 14 (11): 6526–6532.

19 Mao, R., Wang, G., Cai, T. et al. (2020). Tunable metasurface with controllable polarizations and reflection/transmission properties. *J. Phys. D: Appl. Phys.* 53 (15): 155102.

20 Cui, T.J., Liu, S., and Zhang, L. (2017). Information metamaterials and metasurfaces. *J. Mater. Chem. C* 5 (15): 3644–3668.

21 Cui, T.J. Li, L., Liu, S. et al. (2020). Information metamaterial systems. *Iscience* 23 (8): 101403.

22 Liu, C., Yu, W.M., Ma, Q. et al. (2021). Intelligent coding metasurface holograms by physics-assisted unsupervised generative adversarial network. *Photon. Res.* 9 (4): B159–B167.

23 Zeng, H., Lan, F., Zhang, Y. et al. (2020). Broadband terahertz reconfigurable metasurface based on 1-bit asymmetric coding metamaterial. *Opt. Commun.* 458: 124770.

24 Zhao, J., Yang, X., Dai, J.Y. et al. (2019). Programmable time-domain digital-coding metasurface for non-linear harmonic manipulation and new wireless communication systems. *Natl. Sci. Rev.* 6 (2): 231–238.

25 Dai, J.Y., Yang, L.X., Ke, J.C. et al. (2020). High-efficiency synthesizer for spatial waves based on space-time-coding digital metasurface. *Laser Photon. Rev.* 14 (6): 1900133.

26 Dai, J.Y., Zhao, J., Cheng, Q., and Cui, T.J. (2018). Independent control of harmonic amplitudes and phases via a time-domain digital coding metasurface. *Light: Sci. Appl.* 7 (1): 1–10.

27 Dai, J.Y., Yang, J., Tang, W. et al. (2020). Arbitrary manipulations of dual harmonics and their wave behaviors based on space-time-coding digital metasurface. *Appl. Phys. Rev.* 7 (4): 041408.

28 Zhang, L., Chen, X.Q., Liu, S. et al. (2018). Space-time-coding digital metasurfaces. *Nat. Commun.* 9 (1): 1–11.

29 Qiu, T., Jia, Y., Wang, J. et al. (2020). Controllable reflection-enhancement metasurfaces via amplification excitation of transistor circuit. *IEEE Trans. Antennas Propag.* 69 (3): 1477–1482.

30 Chen, L., Ma, Q., Jing, H.B. et al. (2019). Space-energy digital-coding metasurface based on an active amplifier. *Phys. Rev. Appl.* 11 (5): 054051.

31 Arun, V. and Balakrishnan, H. (2020). RFocus: Beamforming using thousands of passive antennas. *17th USENIX Symposium on Networked Systems Design and Implementation (NSDI 20)*, pp. 1047–1061, Santa Clara, CA, February 2020. USENIX Association.

32 NTT DOCOMO (2020). DOCOMO Conducts World's First Successful Trial of Transparent Dynamic Metasurface.

33 Pei, X., Yin, H., Tan, L. et al. (2021). RIS-aided wireless communications: prototyping, adaptive beamforming, and indoor/outdoor field trials. *arXiv preprint arXiv:2103.00534*.

34 Zhang, L., Chen, M.Z., Tang, W. et al. (2021). A wireless communication scheme based on space-and frequency-division multiplexing using digital metasurfaces. *Nat. Electron.* 4 (3): 218–227.

35 Dai, J.Y., Tang, W.K., Zhao, J. et al. (2019). Wireless communications through a simplified architecture based on time-domain digital coding metasurface. *Adv. Mater. Technol.* 4 (7): 1900044.

36 Tang, W., Dai, J.Y., Chen, M.Z. et al. (2020). MIMO transmission through reconfigurable intelligent surface: system design, analysis, and implementation. *IEEE J. Sel. Areas Commun.* 38 (11): 2683–2699.

37 Tang, W., Dai, J.Y., Chen, M. et al. (2019). Programmable metasurface-based RF chain-free 8PSK wireless transmitter. *Electron. Lett.* 55 (7): 417–420.

38 Dai, J.Y., Tang, W., Yang, L.X. et al. (2019). Realization of multi-modulation schemes for wireless communication by time-domain digital coding metasurface. *IEEE Trans. Antennas Propag.* 68 (3): 1618–1627.

39 Chen, M.Z., Tang, W., Dai, J.Y. et al. (2021). Accurate and broadband manipulations of harmonic amplitudes and phases to reach 256 QAM millimeter-wave wireless communications by time-domain digital coding metasurface. *Natl. Sci. Rev.* 9 (1): nwab134.

40 Huang, C.X., Zhang, J., Cheng, Q., and Cui, T.J. (2021). Polarization modulation for wireless communications based on metasurfaces. *Adv. Funct. Mater.* 31 (36): 2103379.

6

Practical Design Considerations for Reconfigurable Intelligent Surfaces

James Rains[1], Jalil ur Rehman Kazim[1], Anvar Tukmanov[2], Lei Zhang[1], Qammer H. Abbasi[1], and Muhammad Ali Imran[1]

[1]*James Watt School of Engineering, University of Glasgow, Glasgow, UK*
[2]*BT Labs, Adastral Park, Ipswich, UK*

6.1 Intelligent Reflecting Surface Architecture

There is an abundance of electromagnetic (EM) transformation and sensing capabilities that can be realized by intelligent reflecting surfaces (IRSs) and their related technologies. Achieving this level of functionality requires meticulous design of the IRS unit cells and associated processing, biasing, and control circuitry.

From a high-level perspective, IRS structures consist of three layers, as shown in Figure 6.1. These are the meta-surface layer, the configuration network layer, and the control layer.

This section provides the reader with a selection of past works regarding unit cell architecture for continuously and discretely tunable elements that can be utilized to achieve the elementary and advanced functions of RISs. In the literature, IRS meta-surface layers generally take two forms. One approach is that of reflectarray-type structures with array elements spaced in the region on $\lambda/2$, where λ is the guide wavelength on the surface, and another common approach is that of reconfigurable meta-surfaces, with elements of a periodicity much smaller than $\lambda/2$ [1].

Reconfigurable meta-surfaces employ tuning mechanisms that can be split into three categories, namely circuit tuning, geometric tuning, and material tuning. Circuit tuning is achieved through variation of impedances within the unit cell elements, such as with switches and variable capacitance. Geometric tuning involves altering the electrical shape of the elements, such as through

Intelligent Reconfigurable Surfaces (IRS) for Prospective 6G Wireless Networks, First Edition.
Edited by Muhammad Ali Imran, Lina Mohjazi, Lina Bariah, Sami Muhaidat,
Tie Jun Cui, and Qammer H. Abbasi.

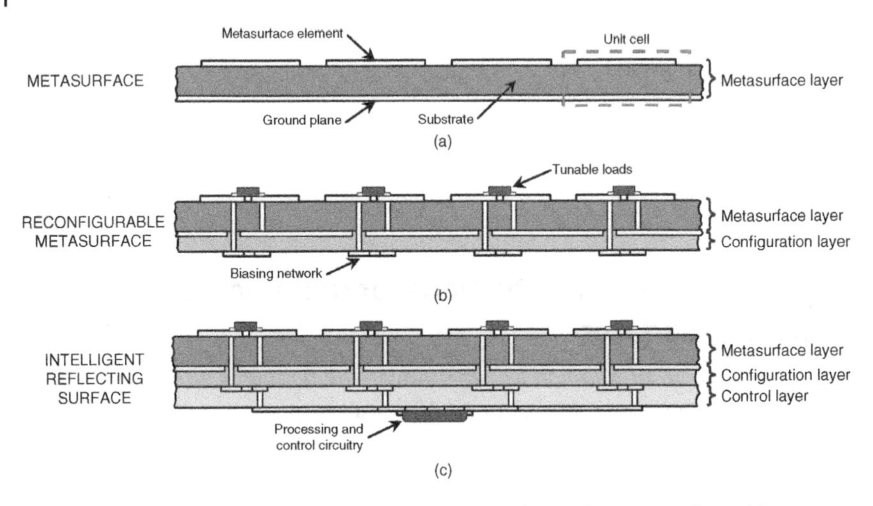

Figure 6.1 Comparison of the hardware structures of generic meta-surfaces (a), reconfigurable meta-surfaces (b), and intelligent reflecting surfaces (c).

rotation and elongation of resonators, and material tuning involves changes in the constitutive parameters and conductivity of the substrate [2].

Tuning mechanisms may utilize tunable materials such as liquid crystal, graphene, and ferroelectric thin-film [2]. Elements employing continuously tunable discrete components such as varactors and varistors are also common, as well as discretely reconfigurable components such as positive-intrinsic-negative (PIN) diodes and micro-electromechanical system (MEMS) switches [3]. Additionally, materials such as photo-induced plasma, carbon nanotubes, and thermally controlled materials such as vanadium dioxide (VO2) have also been explored for reconfigurable meta-surface operation [4–6]. In the microwave regime, control of the phase shifting properties of reflectarray and meta-surface unit cells is commonly achieved by embedding within them semi-conductor-based components, such as varactors and PIN diodes, as well as MEMS switches [2]. By bridging gaps made in unit cell resonators with varactor diodes, it is possible to adjust the capacitance and therefore the effective electrical length of the resonators, thereby altering the resonant frequency and consequently the reflection phase shift. Similarly, these switches can be utilized for connecting and isolating sections of a resonator to the same effect (Figure 6.2).

Reconfiguration of the unit cells requires a method of biasing the tuning mechanisms of each cell. The architecture of the biasing network depends on the tuning mechanism being employed and can take the form of electrical, magnetic, optical, mechanical, thermal, or EM stimuli. Due to the transversally electrically large surface and the large number of reconfigurable elements comprising it, there are

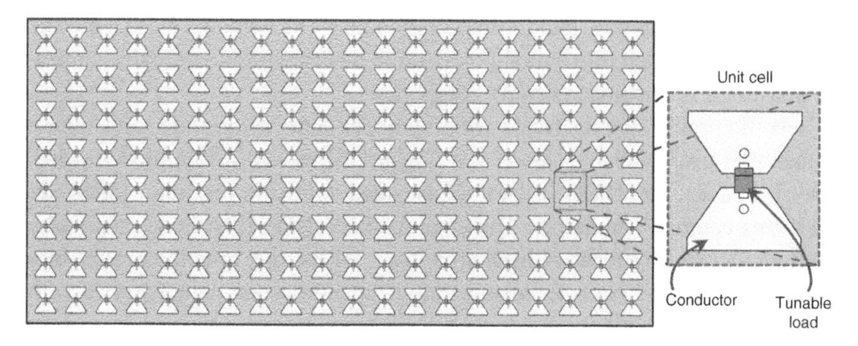

Figure 6.2 Front view of an example intelligent reflecting surface consisting of conducting unit cells with associated tunable loads (such as PIN diodes, varactors, or RF-MEMS for use in the microwave regime), mounted on a grounded dielectric substrate.

a number of complexities to addressing the individual unit cells and these will be set out in the *configuration networks* section.

6.1.1 Tunability of Unit-cell Elements

Literature on reconfigurable reflectarrays, being similar in architecture to transmission-based IRS systems, and reconfigurable meta-surfaces provides an abundant source of techniques for dynamic control of unit cell reflection characteristics [2, 7]. Two common methods of reflection response control in reconfigurable reflectarrays are that of tunable delay lines and tunable resonant elements. The former technique involves coupling a radiating element to a transmission line section, the electrical length of which is varied to provide differing phase shift characteristics, prior to coupling and re-radiation of the incident wave. The latter technique involves altering the resonant frequency of the radiating element itself through reactive loading or switching geometry [8].

Tsilipakos et al. showed that loading a pair of conducting patches with a variable resistive-capacitive (RC) element was shown to offer a 300° continuous phase range in [9]. The unit cell design offers minimal loss and, when the capacitance values across a super-cell were optimized, it was shown to offer anomalous reflection with a radiation efficiency of 98%. Alongside a resistance variation of the UC, the design may offer a tunable magnitude response for applications such as tunable perfect absorption and higher-order modulation schemes.

A tunable unit cell element employing a shorted substrate integrated waveguide (SIW) section, loaded with shunt varactors and slot-coupled to a resonant patch was demonstrated by J. Zang et al. [10]. The varactors provide continuous tuning of the effective length of the SIW section, whilst the biasing network is well shielded from the radiating and tuning components. The work in [11] extends the SIW slot-coupled patch into a dual-polarization device with individually controllable phases for the orthogonal wave components.

A reconfigurable meta-surface with independent magnitude and phase response was proposed by Ashoor and Gupta [12], where a PIN diode-loaded, dipole-ring resonator (DRR) and varactor-loaded split ring resonator (SRR) are operated as coupled resonators. The conductance of the PIN diode enables by magnitude control by forward-biasing voltage, whilst the varactor reverse-bias voltage enables phase control. Beamforming capabilities are demonstrated with a significant improvement of side-lobe level over uniform magnitude control at the expense of dissipative loss. Symmetric and asymmetric beam splitting with a Chebyshev magnitude distribution was demonstrated through full-wave simulation of a 48 unit cell structure.

While conventional meta-materials are described by effective medium parameters in analogue form, coding meta-surfaces are composed of digitally controlled unit cells with coarsely defined respective phase and/or magnitude responses. Coding meta-surfaces have the capability to be digitally controlled to perform elementary functions on EM waves such as anomalous reflection, scattering, beam splitting, and absorption [13]. These n-bit digital meta-surfaces consist of 2^n phase states spaced $360/2^{n\circ}$ apart.

Two state (1-bit) fully digital meta-surfaces were first introduced by Cui et al. [14] under the name of *coding meta-materials*. It was experimentally shown that diffuse scattering for radar cross section (RCS) reduction over a broad bandwidth was possible even with such a low-phase shift resolution. The authors utilized an field programmable gate array (FPGA) to drive sections of the surface composed of 1-bit reflection phase UCs. Each UC, composed of two grounded parallel rectangular loops are connected by a PIN diode, with 0° and 180° reflection phase conditions for PIN diode reverse-biased and forward-biased states, respectively. Digital meta-surfaces bring together the disciplines of information theory with EM theory to provide a powerful framework for future communications and sensing applications.

The authors of [15] created a multi-functional FPGA-controlled reconfigurable meta-surface capable of varying reflection phase of 1600 individually controllable, 1-bit, single-polarization unit-cells (UCs). The anisoptropic property of this surface is utilized for linear-to-linear polarization conversion. Beam steering, beam shaping, and agile scattering are demonstrated, the performance of which was optimized via a genetic algorithm and a 2D-inverse fast Fourier transfer (2D-IFFT) technique to substantially improve optimization efficiency.

The aforementioned 1-bit unit cell works involved a similar unit cell design of resonant elements connected to or isolated from ground, the resulting phase difference over the frequency band being of a convex shape with its peak at the artificial magnetic conductor (AMC) frequency of the UC, as is the case in similar designs [16, 17]. However, it is possible to achieve a broadband linear curve for reflection phase through the current-reversal mechanism [18]. Montori et al. [19] presented a micro-strip patch coupled to a polarization-shifting slotline network which is switched, through two MEMS switches, between alternate edges of the patch in the orthogonal polarization to provide a precise 180° phase difference. This was utilized for a 1-bit reflectarray for imaging applications at W-band.

An improved bandwidth of 1-bit meta-surface behaviour can be achieved through compensation of the non-linearity of the phase-frequency curve. Huang et al. [20] presented a unit-cell design for 1-bit operation extended over a broad bandwidth through a combination of coarse tuning with a PIN diode and fine tuning with a varactor to compensate for the dispersion of the scatterers. These unit cells were utilized to provide multiple elementary functions, realizing beam splitting, beam steering, polarization conversion, and RCS reduction, demonstrating significant improvement in RCS reduction bandwidth over previous works.

Although the non-idealities of such coarse phase partitions are revealed in the form of directivity reduction and increased sidelobe levels (SLLs) when compared to continuous phase tuning, the merits of the 1-bit architecture include the simplicity of its control network and reduced computational complexity. It has been shown that a multitude of elementary functions are possible with only two reflection phase states per unit cell [14].

In order to mitigate the effects of poor quantization-lobe level performance in 1-bit reflecting element designs, Kashyap et al. [21] proposed a phase-delay randomization scheme where each unit cell is subject to a uniformly distributed random phase offset such that the periodicity of the phase-rounding quantization error is broken. It was shown that a significant SSL improvement was possible by introducing random line lengths in delay-line loaded reflecting patches.

Through 2-bit IRS unit cells, reflection phase states of $\phi_r \in [0, \pi/2, \pi, 3\pi/2]$ are available. Several examples of 2-bit coding meta-surface unit cells based on segmented resonant elements are available in the literature [22–25]. The authors in [23] developed a 2-bit digital unit cell element composed of an irregular hexagonal patch with PIN diodes connected to biasing lines at its top and bottom edges. In [22], the unit cells are composed of two nested capacitively loaded square loops, each embedded with PIN diodes, to provide a narrowband phase partition of 90°. Similar to the polarization-switching 1-bit element detailed in [19], the authors in [26] developed an extension of this element to provide 2-bit operation. A slotline coupling network facilitating the current reversal mechanism is given switchable paths that provide an additional 90° phase shift to each of the original 0° and 180° states. This unit cell was then utilized to create an IRS-based communication system with beam steering capability [27].

A 2-bit meta-surface element based on an antenna-filter-antenna type structure has been documented in [28] which exhibits a bandpass filter type response. A crossed slot is coupled to stripline resonators to form switchable 3- and 4-pole magnitude and phases responses. The filter responses at the centre frequency exhibit switchable 180° and 90°, respective, phase shifts and, when combined with the current reversal mechanism, produce the desired 360° and 270° phase shifts as well. Like the above-mentioned current reversal mechanism unit cells, the reflected wave for this design is polarized in the orthogonal mode to the incident wave. While the increased complexity may be undesirable, the nature of this design produces a flat magnitude response over the passband and an improved phase-frequency error performance over a broader bandwidth compared to segmented resonant elements.

In order to enable greater flexibility with required phase shifts and oblique incidence behaviour, higher-resolution IRS unit cells are required. However, with increased operational capacity comes greater complexity, volume, power, and fabrication cost. The costs to performance of a higher resolution should not outweigh the benefits. For instance, the increased number of lossy components results in a marked absorption of incident energy. Where the directivity improvement for a shift from 2 bits to 3 bits of resolution might result in a directivity improvement at oblique angles of 1–1.5 dB [29], the additional losses through component resistances and bias network parasitic reactances may in fact result in a net directivity reduction.

Higher-resolution discrete-phase unit cells are less common in the literature. While 3-bit segmented resonator-type unit cells can be implemented through connecting four optimized patches with three PIN diodes or MEMS switches, it is difficult to realize sufficient phase alignment over a useful bandwidth. While 2-bit unit cell designs only require a phase range of 270°, 3-bit unit cell designs require 315°. This additional phase range is typically achieved through implementing a higher-unit cell reflective Q-factor, leading to a steeper phase-frequency gradient [30]. The higher Q of the resonators results in a larger current flow through the discrete elements and therefore greater losses, as well as increased phase-frequency error due to the increased dispersion associated with narrower resonance. Saifullah et al. [31] recently proposed a 3-bit unit cell design employing four PIN diodes driven by three control lines to facilitate eight phase states. The design is addressable column-wise for azimuthal reflection control and maintains losses of less than 1.3 dB at X-band.

Multi-bit reflection phase behaviour is also possible through the method of vector synthesis, where the unit cell reflection phases are subject to temporal variation. This has been demonstrated by Zhang et al. [32], where a 2-bit meta-surface is driven with 16 sets of time-coding sequences to realise 4-bit operation with 360° phase coverage at a harmonic frequency. An FPGA-controlled prototype was constructed and anomalous reflection measurements revealed a significant SSL improvement over the 2-bit design.

6.1.2 Configuration Networks

The biasing network of an IRS provides a means of control over the individual unit cell reflection characteristics. The structure of this network is dependent on the tuning mechanism and the desired degrees of freedom of EM control. For example, for a 1-bit IRS structure utilizing an array of PIN diode-loaded unit cells with control in the azimuthal-plane, a biasing network might consist of columns driven by voltages such that the respective PIN diodes in a single column are reverse or forward-biased for 0° and 180°, respectively. These types of networks are commonly designed to be driven by FPGAs. For control in both ϕ and θ, IRSs require a more complex biasing network that supports individually addressable UCs.

The complexity of biasing circuitry for IRSs is dependent on factors such as the tuning scheme, the number of unit cells, and the periodicity of the UCs. Continuous tuning is typically dependent on digital-to-analogue converters (DACs) where, for individual control of unit cell elements with analogue signals, an analog-to-digital converter (ADC) channel is required for each UC. The DACs subsequently require digital control signals which may be in the form of well-known serial communication protocols such as SPI and I2C, with the associated circuitry and clock signals.

Electrical tuning of varactor diodes, requiring a variable DC voltage typically between 0 and 30 V, is achieved through employing DACs. DACs are a mature technology with the capability for a single DAC IC to provide multiple channels, thereby serving multiple unit cells each. The required volume of DACs for a large surface of individually controllable unit cells may be costly and the output voltage range can be a limiting factor, thereby requiring additional amplification circuitry. Biasing of PIN diodes is readily achieved through providing a forward bias voltage usually between 0.7 and 1.5 V with a typical operating current around 1–10 mA. These requirements can be readily achieved by digital circuitry, such as through microcontroller IO ports, thereby reducing bias network volume and cost at the expense of coarser tuning.

An effective method of mitigating the parasitic effects of biasing circuitry is to utilize light sources in a similar fashion to optoisolators, allowing the bias routing circuitry to be electrically isolated from the radiating components. An optically controlled reconfigurable meta-surface was developed by Zhang et al. [33], where a set of PIN photodiodes provide biasing voltage for unit cell sub-arrays loaded with varactor diodes. The work introduces the concept of an optically integrated digital platform (OIDP) meta-surface, where photodiodes, connected to a set of varactor-controlled unit cells, are illuminated with varying intensity light, allowing remote tuning of the meta-surface and therefore eliminating the bulky biasing circuitry associated with conventional reconfigurable meta-surfaces.

A reduction in biasing network complexity in reconfigurable reflectarrays through row-column control was recently proposed by Artiga [34]. In this scheme, element-phase shifts are controlled row-wise and column-wise such that, where rows and columns overlap, an additive phase shift results. For an $N \times M$ meta-surface, this scheme only requires $N + M$ analogue control lines as opposed to $N \times M$. The work also proposes a 1-bit row-column control scheme in which an XOR function is applied to digital UCs. The work demonstrates that gains close to those of per-element control of 1- and 2-bit unit cells with a row-column control scheme utilizing an iterative random search algorithm, at the expense of increased SSLs.

A potential enabling technology of programmable propagation environments is the concept of hyper-surfaces, where unit cells are separated into tiles and each tile is connected to a control node offering switching control alongside intra-tile networking capability. Kossifos et al. [35] proposed a hyper-surface consisting of an array of unit cells connected to application-specific integrated circuits (ASICs), each providing a tuneable impedance and grid communication node. These embedded controllers are arranged in a grid and form a network where each controller can communicate with its neighbours, enabling information such as switching instructions to be routed to the respective UCs. The appeal of ICs is not only in their simplicity from an all-in-one tuning and control solution but also in the ability to produce ICs of very small size, thereby rendering the prospect of implementing the required unit cell periodicity of approximately 1 mm×1 mm at 60 GHz a viable endeavour [36]. The hyper-surface concept is being explored in great depth by the VISORSURF project [37].

While there are techniques to reduce the volume of routing circuitry between tuning mechanisms at the expense of some functionality, such as employing a distributed configuration network, it is possible to eliminate the need for physical connections between unit cells through wireless inter-cell connectivity, enabling high-speed individual unit cell control [38, 39]. The scheme proposed in [38] involves a hyper-surface structure with ASICs operating as wireless transceiver-capable unit cell controllers. Two communication channels are considered, with the first scenario employing a blind via as a radiating element within the meta-surface substrate above the ground plane. The second scenario considers a dedicated layer in which a monopole is embedded in a dielectric between two conducting planes in the form of a parallel-plate waveguide. For the first scenario, full-wave simulations are carried out and the structure optimized to maximize unit cell to unit cell communication links whilst minimizing perturbations to the unit cell behaviour and path loss through leakage from the substrate. The second scenario is considered qualitatively, given the reduced complexity of propagation within the parallel-plate structure, demonstrating an improved transmission efficiency over first scenario at the expense of greater fabrication complexity.

The aforementioned techniques rely on constantly available bias signals to maintain the tuning mechanisms in their required states, resulting in control circuitry requirements that scale up vastly with desired degrees of control. To overcome the need for an isolated connection to each unit cell in element-wise control, it would be ideal to have a means of maintaining the unit cell state when the bias signal is removed, thereby exhibiting a form of memory. The unit cell tuning elements with memory could enable a grid arrangement of biasing lines, where each unit cell can be individually addressed by a row-column pair. Chua Mem-components [40] may offer this functionality in the form of mem-resistors, mem-inductors, and mem-capacitors, that is components whose respective resistance, inductance, and capacitance exhibit hysteresis.

Driscoll et al. demonstrated that electrically controlled persistent frequency tuning of a meta-surface in a work entitled *Memory Meta-materials* [41]. The authors utilize the highly hysteretic insulator-to-metal (IMT) phase transition of VO2 to generate a form of mem-capacitance. A meta surface consisting of gold split-ring resonators on a VO2 film was shown to respond an increase in temperature with a decrease in resonant frequency. The effect is due to the increase in the material permittivity of the VO2 specimen when exposed to the current-induced local heating. The shift in resonant frequency remains present after the VO2 has thermalized back to its original temperature due to the strong hysteresis in the IMT, enabling the electrical characteristics of the meta-surface to be controlled by electrical pulses.

Georgiou et al. [42] proposed a type of reactive mem-component–based on a poly disperse red 1 acrylate (PDR1A) substrate. When the PDR1A specimen is exposed to circularly polarized (CP) light, a photomechanical response ensues which results in an expansion of the substrate of up to 25%. On removal of the CP light, the substrate gradually contracts to its original state, retaining a type of memory. The thermal relaxation time can be reduced by a factor of 8 through exposing the substrate to linearly polarized (LP) light. The work proposes employing the PDR1A substrate as the dielectric in a capacitor, with one of the capacitor plates consisting of a transparent conductor, indium tin oxide (ITO), such that illumination of the PDR1A may occur, resulting in a tunable capacitance range of 20%. unit cells with mem-capacitors as tuning mechanisms may each include LP and CP LED pairs arranged in an anti-parallel fashion, driven individually through respective forward and reverse currents. A means of creating a mem-inductor is also proposed and can be achieved through modification of the aspect ratio of the capacitor and connecting lines, tuned in a similar fashion. This technique was subsequently utilized for an optically programmable meta-surface absorber [43] in which unit cells consisting of square patches are connected at the corners through mem-capacitors, enabling a tunable impedance to maintain a 95% absorbance over a 2.7% bandwidth.

6.1.3 IRS Control Layer

The IRS control layer facilitates the intelligence of the IRS through interfacing with the network at large to receive and apply configuration commands, applying configuration signals to the configuration network, and processing any on-board sensor information. This oversight could be achieved by an off-the-shelf micro-controller or FPGA, as set out in Figure 6.3.

In order to facilitate information transfer between the IRS and network control equipment, it has been proposed that a gateway interface be employed [44, 45] with its main responsibility that of providing unit cell configuration information to the configuration network. For incorporation into the network, the gateway might employ a wireless transceiver which is interfaced with a microcontroller. The IRS can utilize existing technologies employing IoT communication protocols, which are easily miniaturizable, are inexpensive to produce and exhibit low-power consumption. The microcontroller provides configuration signals to the biasing network according to the commands received from network equipment and can also process data received from any on-board fault detection or sensory hardware prior to relaying this information to the network controller.

While there are techniques to reduce the volume of routing circuitry between tuning mechanisms at the expense of some functionality, such as employing a distributed configuration network, it is possible to eliminate the need for physical

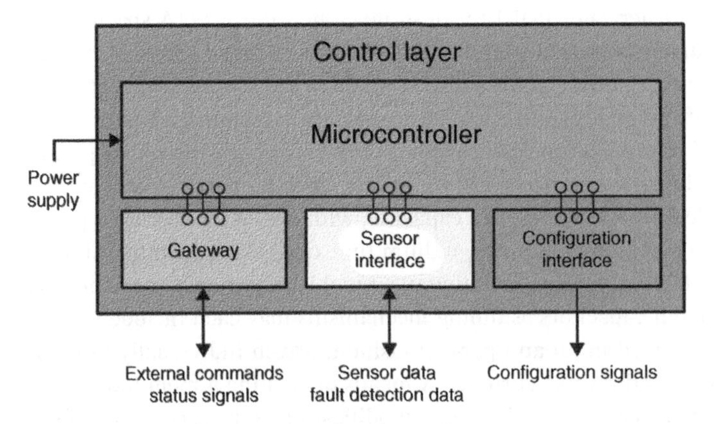

Figure 6.3 An example outline of a control layer for intelligent reflecting surfaces.

connections between unit cells through wireless inter-cell connectivity, enabling high-speed individual unit cell control [38, 39]. The scheme proposed by Tasolamprou et al. [38] involves a hyper-surface structure with ASICs operating as wireless transceiver-capable unit cell controllers. Two communication channels are considered, with the first scenario employing a blind via as a radiating element within the meta-surface substrate above the ground plane. The second scenario considers a dedicated layer in which a monopole is embedded in a dielectric between two conducting planes in the form of a parallel-plate waveguide. For the first scenario, full-wave simulations are carried out and the structure optimized to maximize unit cell to unit cell communication links whilst minimizing perturbations to the unit cell behaviour and path loss through leakage from the substrate. The second scenario is considered qualitatively, given the reduced complexity of propagation within the parallel-plate structure, demonstrating an improved transmission efficiency over first scenario at the expense of greater fabrication complexity.

On-board sensors can extract information from incident EM energy in order for the IRS to make informed decisions regarding the nature the required EM transformations. This information can also be sent to other network equipment to facilitate optimization at other communication nodes as well as for utilization by a wider smart environment with utility that is not necessarily communication-related. The literature in meta-surfaces with embedded sensors is not well established, though IRS-based sensing techniques such as those presented in the *IRS-Based Imaging and Wave Sensing* section show that it may not be necessary to significantly alter existing reconfigurable meta-surface technologies to extract environmental information.

A polarization sensing meta-surface was recently developed by Ma et al. [46] for the purpose of providing control information to a dual-polarization unit cell 1-bit meta-surface, the first of its kind to combine sensing and actuation of this type. If the polarization of incident energy is determined, the meta-surface can accordingly switch the meta-surface polarization for correct beamforming behaviour. Single- polarization-sensing units, consisting of two pairs of orthogonally polarized square patches, were embedded in a meta-surface structure consisting of 100 elements. These sensing units were connected to RF power detectors which were used to determine the dominant polarization of impinging EM energy, followed by an executing unit which performs the desired EM transformation at the sensed polarization. This simple-sensing regime employs minimal additional hardware that only requires an analogue to digital converter for each sensor, thereby easily interfacing with off-the-shelf microcontroller units.

IRSs might also employ sensors commonly found in embedded systems in order to adapt to the environment in which they are placed. This concept was also explored by Ma et al. [47]. In this work, the authors embedded a reconfigurable meta-surface with a gyroscope for orientation detection to aid in beamstaring (i.e.

the act of directing a beam in a fixed direction regardless of the device orientation). The system is capable of real-time, self-adaptive behaviour when exposed to variations in spatial orientation, achieving single- and multi-beam spatial modulation. A light sensor was embedded into a 2-bit reflecting meta-surface to demonstrate switching between two meta-surface functions from optical stimulus, where the meta-surface operated as a dual-beam reflecting meta-surface when exposed to light and as a diffusion meta-surface in the dark. The ability for an IRS to adapt to its orientation and environment in this manner would make embedding IRSs into moving and moveable objects such as vehicles and furniture, respectively, viable through self-adaptation with minimal additional components.

6.2 Physical Limitations of IRSs

The vast majority of the literature on IRS hardware implementations in the microwave regime employ a similar meta-surface structure of shaped conducting elements on a grounded dielectric substrate. The physical implications of these types of arrangements are well documented and theory available in the literature can be used to predict the resulting phase range, bandwidth, reflection loss, polarization performance, and oblique incidence performance [13, 30, 48–50]. However, there are still gaps in the understanding as to why and when certain meta-surface functionalities manifest themselves [30]. This section provides a brief overview of physical characteristics that have been considered in the design and modelling of IRS behaviour.

6.2.1 Bandwidth versus Phase Resolution

Continuously and independently tunable unit cell reflection magnitude and phase would enable IRSs precise control over EM transformations. However, phase-dependent magnitude is inevitable due to the resonant nature of the elements. Variations in field intensity within the unit cells depend on the tuned state, with the highest field strength typically at the AMC frequency, where in-phase (i.e. 0°) reflection occurs. The intensity of the magnitude reduction brought about by this behaviour depends on the loss mechanisms present in the materials, such as dielectric loss, and tuning component resistance.

In order to demonstrate these non-ideal reconfigurable meta-surface characteristics, results from a LP PIN diode-loaded reflecting meta-surface UC, as shown in Figure 6.4, shall be discussed [51]. The unit cell consists of three PIN diodes, D_1 to D_3 to provide eight discrete phase states, with capacitor C_c providing DC

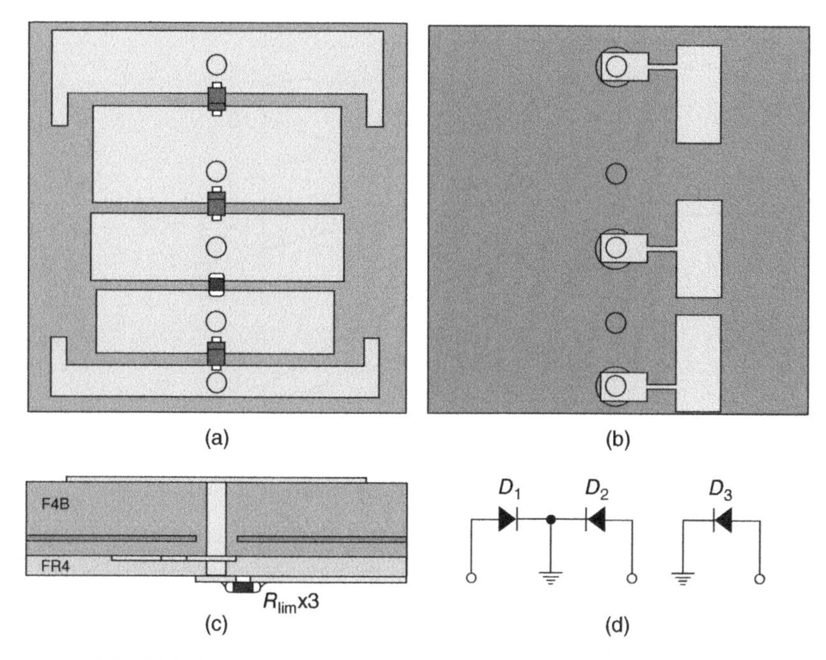

Figure 6.4 Digital meta-surface unit cell employing three PIN diodes as active switching components. Source: Rains et al. [51]/with permission of IEEE. Top view (a) with copper patches connected by PIN diodes and a capacitor, ground layer (b) with open circuit stepped-impedance stubs to isolate the DC and RF sections, cut view (c) with biasing lines connected to current-limiting resistors, and DC equivalent circuit of the diode arrangement (d).

isolation for the required diode biasing arrangement seen in Figure 6.4d. Full-wave simulation of the unit cell with periodic boundary conditions was carried out using a commercial solver.

The resulting reflection characteristics for the eight unit cell states are shown in Figure 6.6. The centre frequency for the device is 3.75 GHz, where it can be seen that the device offers eight distinct reflection phase values. Diverging from this centre frequency, it can be observed that the dispersive nature of the device causes the phase separation behaviour to become less desirable, with reflection phase values beginning to group together (Figure 6.5).

Three measures of the phase versus frequency performance that can be found in the literature are the *phase frequency error*, the *phase standard deviation*, and the *equivalent bit number*. The phase frequency error for reflectarray elements defined by Perruisseau-Carrier and Skrivervik [53] is a measure of the uniformity of phase partitioning and can be used to quantify the average imperfection of the phase response of a unit cell over a given bandwidth.

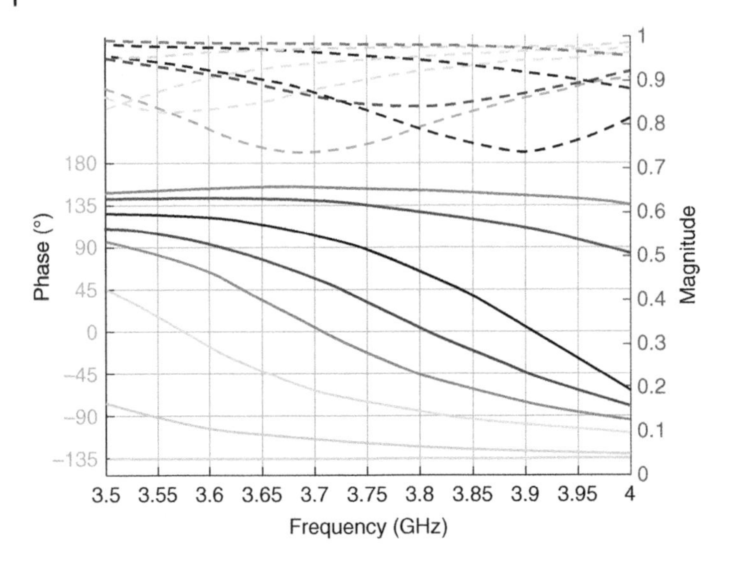

Figure 6.5 Reflection phase versus magnitude for the three PIN diode-based reflecting meta-surface at 3.75 GHz. Ideal relationship shown by crosses, whilst simulated phase versus magnitude shown by circles. The dashed line is calculated using an approximation introduced by Abeywickrama et al. [52] to aid in incorporating the phase-dependent magnitude into channel models. Source: Adapted from Abeywickrama et al. [52].

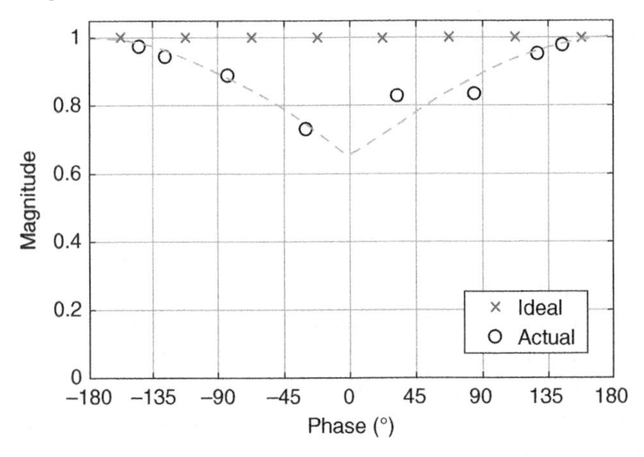

Figure 6.6 Reflection phase (solid curves) and magnitude (dashes curves) for normal incidence upon unit cell with three PIN diodes from Figure 6.4.

The phase standard deviation for an M-state (2^M-bit) unit cell is defined by Pereira et al. [54] as in (6.1):

$$\sigma_p = \sqrt{\frac{\sum_{m=1}^{M} (\Delta\phi_m)^3}{12 \times 360}} \tag{6.1}$$

where $\Delta\phi_m$ are the phase differences, in degrees, between adjacent unit cell states. For example, the phase standard deviations for 2- and 3-bit unit cells with perfect phase alignment are 26° and 13°, respectively.

The equivalent bit number takes into account the phase standard deviation of the discrete unit cell states to provide a figure of merit of the phase resolution. The equivalent bit number is defined as follows:

$$N_{\text{bit}} = \log_2\left(\frac{360}{\sqrt{12}\sigma_p}\right) \tag{6.2}$$

For instance, several works [54–56] have defined a working bandwidth for a 2-bit reflectarray element as that over which the equivalent bit number is 1.7 or higher, corresponding to a phase standard deviation of 32°. The equivalent bit number aids in visualizing the operating point of a given unit cell design and is particularly useful for calculating phase state alignment during optimization [51].

The phase range of IRS unit cells is typically limited by the tolerance for losses in magnitude and bandwidth reduction. Qu et al. recently introduced a model for the functionalities of metal-insulator-metal (MIM) meta-surfaces by considering meta-surface unit cells as single-port, single-mode resonators whose reflection coefficients are comprised of a combination of absorptive and radiative Q-factors with associated resonance lifetimes [30]. In order to realize a sufficiently large phase range over a useful bandwidth, it is necessary to operate the unit cells in an under-damped resonance. In this region, the absorptive Q-factor is greater than the radiative Q-factor, resulting in magnetic resonance and a full 360° phase range is theoretically achievable. Cong et al. utilized this Q-factor based model to explore the phase range behaviour of a multi-functional MEMS-based meta-surface [57].

For single resonance meta-surface unit cells, two extremes of phase shifts at $\pm\pi$ correspond to an under-damped resonance occurring at frequencies above and below the frequency range of operation, respectively. The extent to which these frequencies must be separated in order to maximize the phase range is dependent on the narrowness of the resonance. The extent to which it is possible to separate these resonances is dependent on the tuning ratio of the employed technology alongside the sensitivity of the unit cells to changes in the tuning mechanism.

One of the obstacles to achieving a large phase range in single resonance devices is that the sensitivity in the phase response to a tuned impedance asymptotically decreases on approaching $\pm\pi$ [58]. Several works [59–61] have introduced two or more tunable resonances in close proximity in order to increase the available phase range, at the expense of higher complexity and losses. In the context of IRSs, this increase in unit cell complexity, thus increased biasing network complexity and power consumption, should be weighed up against any benefits from a higher-phase range. The phase range consideration plays a role in the trade-offs between 2- and 3-bit unit cell designs, with corresponding phase ranges of 270° and 315°.

Zhu et al. [59] discussed the phase range limitation of single-resonance unit cells and introduced a dual-resonant unit cell employing two varactor diodes. The respective in-phase reflection resonant states were engineered such that they exist at the ±180° sides of the phase versus voltage curves. Instead of exhibiting a minimal magnitude at 0°, the magnitude is instead a minimum at ±180°. This configuration enables a large phase range without a wide tuning range which can be beneficial to keep the required DAC voltages low and/or varactor capacitance values high.

Rodrigo et al. [8] utilized a combination of capacitive loading and geometry switching of a segmented resonant dog-bone type element is employed to achieve greater operational frequency range at the expense of instantaneous bandwidth. PIN diodes provide a means of connecting segments of the resonator for coarse tuning, whilst a varactor is used for fine tuning. To achieve uniform phase shift characteristics over the tuning range, resonant elements in this work are shaped such that the Q factor shift caused by the effective electrical thickness of the substrate across the frequency range, alongside the variation in current density, are compensated for.

Tsilipakos et al. [62] recently introduced a method of achieving arbitrarily broadband linear reflection through anti-matched EM meta-surfaces. Electric and magnetic resonant surface admittivities are interlaced in an admittivity anti-matching condition which zeroes out transmission across the band, resulting in a linear phase response. This design is particularly suited to wideband systems that require a highly linear phase response to minimize dispersion.

6.2.2 Incidence Angle Response

The magnitude and phase shift behaviour of a unit cell design can vary profusely with the angle of incidence, particularly for oblique incidence angles above about 40° from broadside [58]. Figure 6.7 shows the effect of incidence angle for a TM-polarized wave on the equivalent bit number for the unit cell of Figure 6.4, where it is seen that reduced phase resolution occurs with increasing incidence angle. Figure 6.8 shows the effects incidence angle has on the magnitude of the device, with higher losses on approaching normal incidence about the frequency of operation due to the deeper resonance in this region. As could be expected, as the available phase range drops, the reflection magnitude increases.

The oblique incidence behaviour of unit cells can be derived by full-wave simulation tools, such as HFSS and CST, by placing a Floquet port above a single unit cell and employing periodic boundary conditions or their equivalent in the transverse plane. For the zeroth order mode, the resulting simulation is equivalent

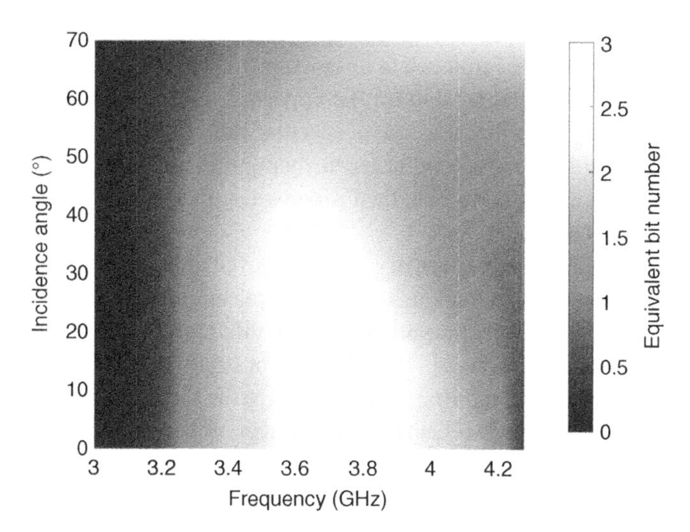

Figure 6.7 Plot showing equivalent bit number for the digital meta-surface unit cell of Figure 6.4 versus frequency and incidence angle for TM polarization. The operating region can be seen to be centred about 3.75 GHz, with reducing resolution for increasing incidence angle.

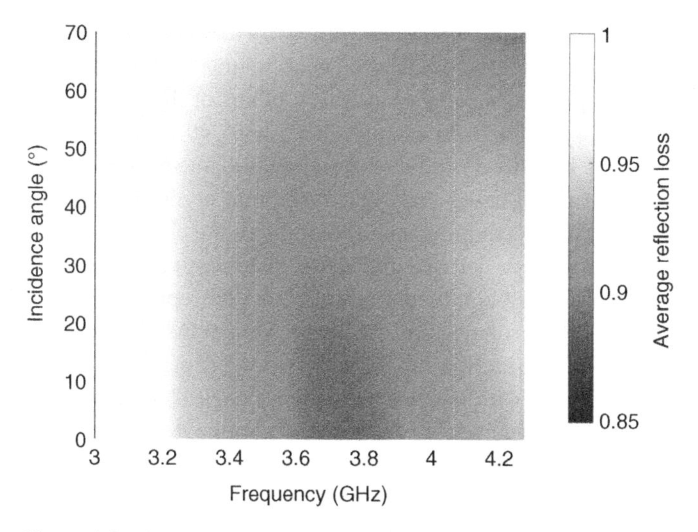

Figure 6.8 Plot showing average reflection magnitude for the digital meta-surface unit cell of Figure 6.4 versus frequency and incidence angle for TM polarization. A drop in magnitude of about 1 dB exists in the operating region due to losses in the diodes and substrate about resonance. The magnitude performance improves with incidence angle, coinciding with a reduction in phase range.

to measuring the specular reflection at a given angle from an infinite surface composed of translated copies of the unit cell of interest, with the reflection being sampled at the same angle of incidence but on the opposing side of the surface normal. This technique only describes the case of specular reflection in a purely periodic structure but provides a starting point for further optimization as part of a quasi-periodic structure, where there may be a progressive phase distribution between adjacent elements [63].

Synthesis of a surface impedance profile can only be achieved granularly due to the discrete nature of the UCs. In anomalous reflection applications with normal incidence SC periodicity are inversely proportional to the difference in the angle between the incident and reflected waves [64]. Although not strictly accurate, this is analogous to the phase profile resembling a mirror tilted to the same gradient as the phase profile in order to specularly reflect incoming waves. With more extreme angles, the periodicity becomes very small and any effects due to the granularity in the meta-surface unit cell arrangement are magnified, resulting in increased parasitic reflections [65]. Therefore, it is advantageous to keep unit cell periodicity as small as is practicable.

Taghvaee et al. [66] explored a method for mitigating the effects of oblique incidence in 2-bit unit cells by introducing redundant unit cell states in the design. The unit cells could be optimized with additional states, such as 16 as opposed to 4 in the 2-bit design, such that the unit cells can be tuned to exhibit 2-bit behaviour over a broader incidence angle range. Optimization goals in unit cell design could be set such that at differing incidence angles at least 4 of any of the 16 unit cell states align with 90° phase partitions – for example at $\theta_{\text{inc}} = 0°, 40°$, and $70°$.

Costa and Borgese [67] recently introduced an analytical model describing the behaviour of resonant patch-based IRSs that takes into account effects of incidence angle, element mutual coupling, and proximity of the elements to the ground plane. The model consists of surface impedance representing the periodic elements in parallel with a short-circuited transmission line representing the grounded dielectric. The effects of oblique incidence from transverse electric (TE) waves are modelled by considering the reduction in the capacitance of the gaps between adjacent unit cells with increasing incidence angle. The equivalent impedance of the grounded dielectric is modified according to the normal component of the propagation constant and TE/TM-dependent characteristic impedance, with the resulting model closely resembling the oblique incidence behaviour of full-wave simulations of a varactor-loaded patch UC. A link budget formulation is modelled as a non-LoS classical backscattering communication system, where it is shown that a received power gain improvement of 3.9 dB is achieved by computing varactor voltage values for a desired phase profile according to the oblique incidence angle of interest, rather than using the values computed for normal incidence.

6.2.3 Quantization Effects: How Many Bits?

A study by Wu et al. on the effects of phase quantization on the directivity of reflectarrays at varying angles of incidence revealed, when averaging over oblique angles of incidence, a distinct loss in directivity. When compared to continuous phase resolution, a directivity loss of 3.8, 0.8, and 0.2 dB was calculated for 1-bit, 2-bit, and 3-bit designs, respectively [29]. The authors concluded a 3-bit design to be an adequate choice of balancing gain improvement with unit cell complexity.

Several beamsteering performance metrics of programmable meta-surfaces with quantized phase shifts were parametrically studied in [66]. The authors compare the directivity, target deviation, SLL, and half-power beamwidth (HPBW) of beamsteering meta-surfaces of 1-, 2-, and 3-bits phase resolution subject to normally incident waves and subsequent reflection in the direction of $\theta_r = \phi_r = 45°$, with a fixed surface size and variation in unit cell periodicity. It was shown that when changing from 1- to 2-bits phase resolution, an increase in directivity of 3 dB is achieved, along with an appreciable improvement in target deviation and SSL. For unit cell periodicities below a half wavelength, the effects of an increase in phase resolution from 2 to 3 bits are less pronounced, with the biggest effect being a slight improvement in SSLs, which are mostly affected by quantization errors in phase and unit cell size. However, the authors note that by increasing the number of available unit cell states, the shift in operating point with increasing incidence angle could be compensated for, enabling a more robust programmability at the expense of increased losses and circuit complexity.

References

1. ElMossallamy, M.A., Zhang, H., Song, L. et al. (2020). Reconfigurable intelligent surfaces for wireless communications: principles, challenges, and opportunities. *IEEE Trans. Cognit. Commun. Network.* 6: 990–1002.
2. Turpin, J.P., Bossard, J.A., Morgan, K.L. et al. (2014). Reconfigurable and tunable metamaterials: a review of the theory and applications. *Int. J. Antennas Propag.* 2014: 1–18.
3. Debogovic, T. and Perruisseau-Carrier, J. (2015). Dual-polarized low loss reflectarray cells with MEMS-based dynamic phase control. *2015 9th European Conference on Antennas and Propagation (EuCAP)*, pp. 1–5.
4. Hashemi, M.R.M., Yang, S.-H., Wang, T. et al. (2016). Fully-integrated and electronically-controlled millimeter-wave beam-scanning. *2016 IEEE MTT-S International Microwave Symposium (IMS)*, IEEE, May 2016.

5 Zhang, Y., Wang, T., Wang, X. et al. (2019). Thermally switchable terahertz metasurface devices. *2019 44th International Conference on Infrared, Millimeter, and Terahertz Waves (IRMMW-THz)*IEEE, September 2019.

6 Wu, X., Lan, F., Shi, Z., and Yang, Z. (2017). Switchable terahertz polarization conversion via phase-change metasurface. *2017 Progress In Electromagnetics Research Symposium - Spring (PIERS)*, IEEE, May 2017.

7 Hum, S.V. and Perruisseau-Carrier, J. (2014). Reconfigurable reflectarrays and array lenses for dynamic antenna beam control: a review. *IEEE Trans. Antennas Propag.* 62: 183–198.

8 Rodrigo, D., Jofre, L., and Perruisseau-Carrier, J. (2013). Unit cell for frequency-tunable beamscanning reflectarrays. *IEEE Trans. Antennas Propag.* 61: 5992–5999.

9 Tsilipakos, O., Liu, F., Pitilakis, A. et al. (2018). Tunable perfect anomalous reflection in metasurfaces with capacitive lumped elements. *2018 12th International Congress on Artificial Materials for Novel Wave Phenomena (Metamaterials)*, IEEE, August 2018.

10 Zang, J., Correas-Serrano, D., Do, J. et al. (2019). Nonreciprocal wavefront engineering with time-modulated gradient metasurfaces. *Phys. Rev. Appl.* 11: 054054.

11 Spatola, S., Gomez-Diaz, J.S., and Carrasco, E. (2020). Time modulated reflectarray unit-cells with nonreciprocal polarization control. *2020 14th European Conference on Antennas and Propagation (EuCAP)*, IEEE, March 2020.

12 Ashoor, A.Z. and Gupta, S. (2020). Metasurface reflector with real-time independent magnitude and phase control.

13 Renzo, M.D., Zappone, A., Debbah, M. et al. (2020). Smart radio environments empowered by reconfigurable intelligent surfaces: how it works, state of research, and the road ahead. *IEEE J. Sel. Areas Commun.* 38: 2450–2525.

14 Cui, T.J., Qi, M.Q., Wan, X. et al. (2014) Coding metamaterials, digital metamaterials and programming metamaterials. *Light: Sci. Appl.* 3 (10): e218.

15 Yang, H., Cao, X., Yang, F. et al. (2016). A programmable metasurface with dynamic polarization, scattering and focusing control. *Sci. Rep.* 6.

16 ur Rehman Kazim, J., Ur-Rehman, M., Al-Hasan, M. et al. (2020). Design of 1-bit digital subwavelength metasurface element for sub-6 GHz applications. *2020 International Conference on UK-China Emerging Technologies (UCET)*, IEEE, August 2020.

17 Wang, D., Yin, L.-Z., Huang, T.-J. et al. (2020). Design of a 1 bit broadband space-time-coding digital metasurface element. *IEEE Antennas Wireless Propag. Lett.* 19: 611–615.

18 Yang, F., Xu, S., Pan, X. et al. (2018). Reconfigurable reflectarrays and transmitarrays: From antenna designs to system applications. *12th European*

Conference on Antennas and Propagation (EuCAP 2018), Institution of Engineering and Technology.

19 Montori, S., Chiuppesi, E., Marcaccioli, L. et al. (2010). 1-bit RF-MEMS-reconfigurable elementary cell for very large reflectarray. *2010 10th Mediterranean Microwave Symposium*, IEEE, August 2010.

20 Huang, C., Zhang, C., Yang, J. et al. (2017). Reconfigurable metasurface for multifunctional control of electromagnetic waves. *Adv. Opt. Mater.* 5: 1700485.

21 Kashyap, B.G., Theofanopoulos, P.C., Cui, Y., and Trichopoulos, G.C. (2020). Mitigating quantization lobes in mmWave low-bit reconfigurable reflective surfaces. *IEEE Open J. Antennas Propag.* 1 604–614.

22 Zhang, X., Zhang, H., Su, J., and Li, Z. (2017). 2-bit programmable digital metasurface for controlling electromagnetic wave. *2017 Sixth Asia-Pacific Conference on Antennas and Propagation (APCAP)*, IEEE, October 2017.

23 Zhang, L. and Cui, T.J. (2019). Angle-insensitive 2-bit programmable coding metasurface with wide incident angles. *2019 IEEE Asia-Pacific Microwave Conference (APMC)*, IEEE, December 2019.

24 Shuang, Y., Zhao, H., Ji, W. et al. (2020). Programmable high-order OAM-carrying beams for direct-modulation wireless communications. *IEEE J. Emerging Sel. Top. Circuits Syst.* 10: 29–37.

25 Zhang, F., Saifullah, Y., Yang, G.-M., and Jin, Y.-Q. (2018). 1-bit, 2-bit polarization insensitive reflection programable metasurface. *2018 IEEE International Symposium on Antennas and Propagation & USNC/URSI National Radio Science Meeting*, IEEE, July 2018.

26 Yang, X., Xu, S., Yang, F., and Li, M. (2017). A novel 2-bit reconfigurable reflectarray element for both linear and circular polarizations. *2017 IEEE International Symposium on Antennas and Propagation & USNC/URSI National Radio Science Meeting*, IEEE, July 2017.

27 Dai, L., Wang, B., Wang, M. et al. (2020). Reconfigurable intelligent surface-based wireless communications: antenna design, prototyping, and experimental results. *IEEE Access* 8: 45913–45923.

28 Cheng, C.-C. and Abbaspour-Tamijani, A. (2009). Design and experimental verification of steerable reflect-arrays based on two-bit antenna-filter-antenna elements. *2009 IEEE MTT-S International Microwave Symposium Digest*, IEEE, June 2009.

29 Wu, B., Sutinjo, A., Potter, M., and Okoniewski, M. (2008). On the selection of the number of bits to control a dynamic digital MEMS reflectarray. *IEEE Antennas Wireless Propag. Lett.* 7: 183–186.

30 Qu, C., Ma, S., Hao, J. et al. (2015). Tailor the functionalities of metasurfaces based on a complete phase diagram. *Phys. Rev. Lett.* 115: 235503.

31 Saifullah, Y., Zhang, F., Yang, G.-M., and Xu, F. (2018). 3-bit programmable reflective metasurface. *2018 12th International Symposium on Antennas, Propagation and EM Theory (ISAPE)*, IEEE, December 2018.

32 Zhang, L., Wang, Z.X., Shao, R.W. et al. (2020). Dynamically realizing arbitrary multi-bit programmable phases using a 2-bit time-domain coding metasurface. *IEEE Trans. Antennas Propag.* 68: 2984–2992.

33 Zhang, X.G., Jiang, W.X., Jiang, H.L. et al. (2020). An optically driven digital metasurface for programming electromagnetic functions. *Nat. Electron.* 3: 165–171.

34 Artiga, X. (2018). Row–column beam steering control of reflectarray antennas: benefits and drawbacks. *IEEE Antennas Wireless Propag. Lett.* 17: 271–274.

35 Kossifos, K.M., Petrou, L., Varnava, G. et al. (2020). Toward the realization of a programmable metasurface absorber enabled by custom integrated circuit technology. *IEEE Access* 8: 92986–92998.

36 Petrou, L., Karousios, P., and Georgiou, J. (2018). Asynchronous circuits as an enabler of scalable and programmable metasurfaces. *2018 IEEE International Symposium on Circuits and Systems (ISCAS)*, IEEE.

37 VISORSURF Project (2017). Visorsurf project.

38 Tasolamprou, A.C., Mirmoosa, M.S., Tsilipakos, O. et al. (2018). Intercell wireless communication in software-defined metasurfaces. *IEEE International Symposium on Circuits and Systems (ISCAS)*: 1–5.

39 Tasolamprou, A.C., Pitilakis, A., Abadal, S. et al. (2019). Exploration of intercell wireless millimeter-wave communication in the landscape of intelligent metasurfaces. *IEEE Access* 7: 122931–122948.

40 Yin, Z., Tian, H., Chen, G., and Chua, L.O. (2015). What are memristor, memcapacitor, and meminductor? *IEEE Trans. Circuits Syst. II: Express Briefs* 62: 402–406.

41 Driscoll, T., Kim, H.-T., Chae, B.-G. et al. (2009). Memory metamaterials. *Science* 325: 1518–1521.

42 Georgiou, J., Kossifos, K.M., Antoniades, M.A. et al. (2018). Chua mem-components for adaptive RF metamaterials. *2018 IEEE International Symposium on Circuits and Systems (ISCAS)*, IEEE, May 2018.

43 Kossifos, K.M., Antoniades, M.A., Georgiou, J. et al. (2018). An optically-programmable absorbing metasurface. *2018 IEEE International Symposium on Circuits and Systems (ISCAS)*, IEEE, May 2018.

44 Liaskos, C., Tsioliaridou, A., Pitsillides, A. et al. (2018). Using any surface to realize a new paradigm for wireless communications. *Commun. ACM* 61: 30–33.

45 Yuan, X., Zhang, Y.-J.A., Shi, Y. et al. (2021) Reconfigurable-intelligent-surface empowered wireless communications: challenges and opportunities. *IEEE Wireless Communications* 28 (2): 136–143.

46 Ma, Q., Hong, Q.R., Gao, X.X. et al. (2020). Smart sensing metasurface with self-defined functions in dual polarizations. *Nanophotonics* 9: 3271–3278.

47 Ma, Q., Bai, G.D., Jing, H.B. et al. (2019). Smart metasurface with self-adaptively reprogrammable functions. *Light: Sci. Appl.* 8.

48 Kuester, E., Mohamed, M., Piket-May, M., and Holloway, C. (2003). Averaged transition conditions for electromagnetic fields at a metafilm. *IEEE Trans. Antennas Propag.* 51: 2641–2651.

49 Zhang, Y., von Hagen, J., Younis, M. et al. (2003). Planar artificial magnetic conductors and patch antennas. *IEEE Trans. Antennas Propag.* 51: 2704–2712.

50 Tretyakov, S. (2003). *Analytical Modeling in Applied Electromagnetics*. Artech House Inc.

51 Rains, J., ur Rehman Kazim, J., Zhang, L. et al. (2021). 2.75-bit reflecting unit cell design for reconfigurable intelligent surfaces, IEEE International Symposium on Antennas and Propagation and USNC-URSI Radio Science Meeting (APS/URSI).

52 Abeywickrama, S., Zhang, R., Wu, Q., and Yuen, C. (2020). Intelligent reflecting surface: practical phase shift model and beamforming optimization. *IEEE Trans. Commun.* 68: 5849–5863.

53 Perruisseau-Carrier, J. and Skrivervik, A. (2007). Requirements and challenges in the design of high-performance MEMS reconfigurable reflectarray (RRA) cells. *2nd European Conference on Antennas and Propagation (EuCAP 2007)*, Institution of Engineering and Technology.

54 Pereira, R., Gillard, R., Sauleau, R. et al. (2012). Dual linearly-polarized unit-cells with nearly 2-bit resolution for reflectarray applications in X-band. *IEEE Trans. Antennas Propag.* 60: 6042–6048.

55 Nguyen, B.D., Tran, V.-S., Mai, L., and Dinh-Hoang, P. (2016). A two-bit reflectarray element using cut-ring patch coupled to delay lines. *REV J. Electron. Commun.* 68 (12): 7937–7946.

56 Luyen, H., Booske, J.H., and Behdad, N. (2020). 2-bit phase quantization using mixed polarization-rotation/non-polarization-rotation reflection modes for beam-steerable reflectarrays. *IEEE Trans. Antennas Propag.* 5 (2): 1600716.

57 Cong, L., Pitchappa, P., Wu, Y. et al. (2016). Active multifunctional microelectromechanical system metadevices: applications in polarization control, wavefront deflection, and holograms. *Adv. Opt. Mater.* 5: 1600716.

58 Hum, S.V., Okoniewski, M., and Davies, R.J. (2007). Modeling and design of electronically tunable reflectarrays. *IEEE Trans. Antennas Propag.* 55: 2200–2210.

59 Zhu, B.O., Zhao, J., and Feng, Y. (2013). Active impedance metasurface with full 360∘ reflection phase tuning. *Sci. Rep.* 3.

60 Zhu, L., Gao, X., and Hou, H. (2020). Design of a programmable meta-atom in C band. *2020 IEEE 3rd International Conference on Electronic Information and Communication Technology (ICEICT), IEEE, November 2020.*

61 Sun, Z., Huang, F., and Fu, Y. (2021). Graphene-based active metasurface with more than 330∘ phase tunability operating at mid-infrared spectrum. *Carbon* 173: 512–520.

62 Tsilipakos, O., Koschny, T., and Soukoulis, C.M. (2018). Antimatched electromagnetic metasurfaces for broadband arbitrary phase manipulation in reflection. *ACS Photonics* 5: 1101–1107.

63 Chen, Z. (2016). *Handbook of Antenna Technologies*. Singapore: Springer Reference.

64 Liu, F., Tsilipakos, O., Pitilakis, A. et al. (2019). Intelligent metasurfaces with continuously tunable local surface impedance for multiple reconfigurable functions. *Phys. Rev. Appl.* 11 044024.

65 Estakhri, N.M. and Alù, A. (2016). Wave-front transformation with gradient metasurfaces. *Phys. Rev. X* 6 041008.

66 Taghvaee, H., Abadal, S., Pitilakis, A. et al. (2020). Scalability analysis of programmable metasurfaces for beam steering. *IEEE Access* 8: 105320–105334.

67 Costa, F. and Borgese, M. (2021). Electromagnetic model of reflective intelligent surfaces. *IEEE Open J. Commun. Soc.*, 2: 1577–1589.

7

Channel Modelling in RIS-Empowered Wireless Communications

Ibrahim Yildirim[1,2] and Ertugrul Basar[1]

[1]*Department of Electrical and Electronics Engineering, Koç University, Sariyer, Istanbul, Turkey*
[2]*Faculty of Electrical and Electronics Engineering, Istanbul Technical University, Sariyer, Istanbul, Turkey*

7.1 Introduction

Despite the vast drastic expectations, fifth-generation (5G) wireless communication networks can be considered as an evolution of fourth-generation (4G) wireless networks by providing a more adaptive and flexible structure in terms of adapted physical-layer technologies. In sixth-generation (6G) wireless networks, which are expected to be rolled out in 2030s, it is an undeniable fact that there should be paradigm-shifting developments, especially in the physical layer, in order to meet the increasing massive demand in data consumption [1]. Within this context, millimetre wave (mmWave) and Terahertz (THz) communications, massive multiple-input multiple-output (MIMO) systems, cell-free networks, high-altitude platform stations, integrated space and terrestrial networks, and reconfigurable intelligent surfaces (RISs) can be considered as promising candidate technologies for 6G systems in order to satisfy broadband connectivity by keeping new use-cases with very high mobility and extreme capacity requirements [2].

RIS-empowered communication can be regarded as a paradigm-shifting revolution that provides operators software-based and dynamic control capability over the wireless propagation channel [3]. An RIS, which consists of large number of nearly passive reflecting elements, can intelligently direct the beams to the desired users and ensure reliable communication by creating virtual line-of-sight (LOS) links even when the direct link between terminals is blocked. Due to these unprecedented contributions of RISs, researchers have shown remarkable interest in RIS-empowered communication systems, and plentiful applications that are candidates for use in next-generation communication systems have been

Intelligent Reconfigurable Surfaces (IRS) for Prospective 6G Wireless Networks, First Edition.
Edited by Muhammad Ali Imran, Lina Mohjazi, Lina Bariah, Sami Muhaidat,
Tie Jun Cui, and Qammer H. Abbasi.

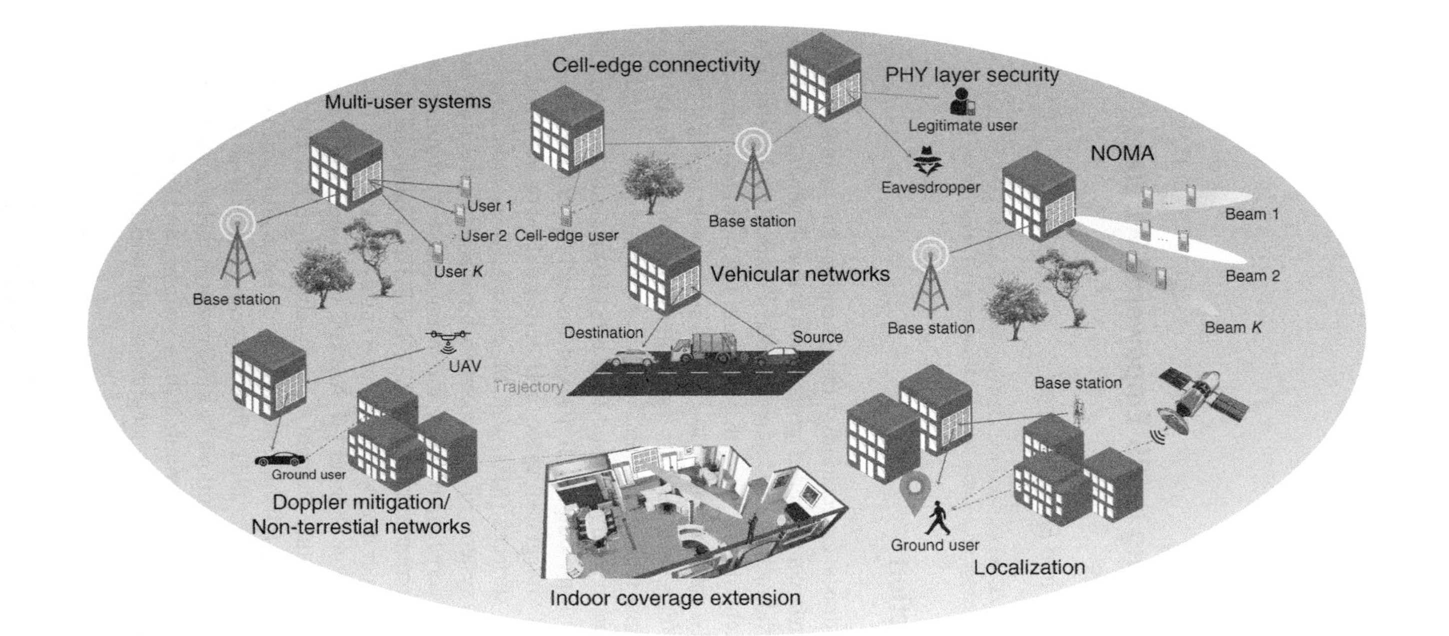

Figure 7.1 Miscellaneous RIS-assisted transmission scenarios and applications that are likely to emerge in future wireless communication networks. Source: Basar and Yildirim [9]/with permission of IEEE.

explored [4–8]. The various use-cases of RISs, including multi-user systems, physical (PHY) layer security, indoor-coverage extension, Doppler mitigation, vehicular and non-terrestrial networks, localization and sensing, and non-orthogonal multiple access (NOMA), are illustrated in Figure 7.1. Despite these rich RIS use-cases, there is no strong consensus to identify the killer applications that effectively exploit the potential of RISs to control the transmission medium with intelligent reflections for enhanced end-to-end system performance [9]. The first step in clearing this ambiguity is to form a unified and physical RIS-assisted channel model that can be adapted to different use-cases and operating frequencies, taking into account the physical characteristics of RISs. Since intelligent reflection is the art of manipulating the channel characteristics in a brilliant and dynamic way, modelling the RIS-assisted transmission link is essential to provide detailed insights into the practical use-cases. This chapter mainly aims to present a vision on channel modelling strategies for the RIS-empowered communications systems considering the state-of-the-art channel and propagation modelling efforts in the literature. Another objective of the chapter is to draw attention to open-source and standard-compliant physical channel modelling efforts to provide comprehensive insights regarding the practical use-cases of RISs in future wireless networks.

The rest of the chapter is organized as follows: Section 1.2 presents a general framework on channel modelling strategies for the RIS-empowered communications systems considering the state-of-the-art channel modelling efforts in the literature. Sections 1.3 and 1.4 provide a framework on the cluster-based statistical channel model for sub-6 GHz and mmWave bands, respectively. Section 1.5 introduces the open-source *SimRIS Channel Simulator* package that considers a proposed narrowband channel model for RIS-empowered communication systems. Section 1.6 presents numerical results obtained by using *SimRIS Channel Simulator* and the chapter is concluded in Section 1.7.

7.2 A General Perspective on RIS Channel Modelling

One of the most important aspects of enabling next-generation wireless technologies is the development of an accurate and consistent channel model to be validated effectively with the help of real-world measurements. From this point of view, remarkable research has recently been conducted to model propagation channels involving the modification of the wireless propagation environment through the inclusion of RISs. While many studies in the literature have modelled the path loss for RIS-assisted communication systems, modelling the end-to-end channel has also recently attracted significant attention. The near-field and far-field effects of RISs should also be considered when modelling the total path loss of RIS-empowered systems. Another critical point is the methodology followed when modelling the end-to-end channel. From this perspective, while many researchers follow physics-based channel modelling strategies by examining the physical properties of reflection and scattering effects, statistical channel modelling strategies have also been followed by considering the statistical properties of scatterers and clusters in the environment. Further, electromagnetic (EM) compatible channel models have also been proposed that take into account EM properties, such as mutual coupling among the sub-wavelength unit cells of the RIS. By considering the existing RIS-assisted channel modelling studies, the channel and propagation modelling approaches can be mainly classified as in Figure 7.2. Since there is no obvious distinction between certain approaches in this classification, channel modelling studies in which more than one approach is used can provide a broader perspective.

Initial studies on RIS propagation modelling have predominantly focused on the path-loss and scattered power characterization by an RIS in RIS-assisted communication systems [10–16]. In [10], the authors characterize the path-loss in near-field and far-field of RISs by using the general scalar theory of diffraction and the Huygens–Fresnel principle. Moreover, the power reflected from an RIS is obtained as a function of the distance between the transmitter (Tx)/receiver (Rx) and the RIS, the size of the RIS, and the phase shifts induced by the RIS. By modelling RISs as a sheet of EM material of negligible thickness, the conditions under which an RIS acts as an anomalous mirror are identified. In [11], the

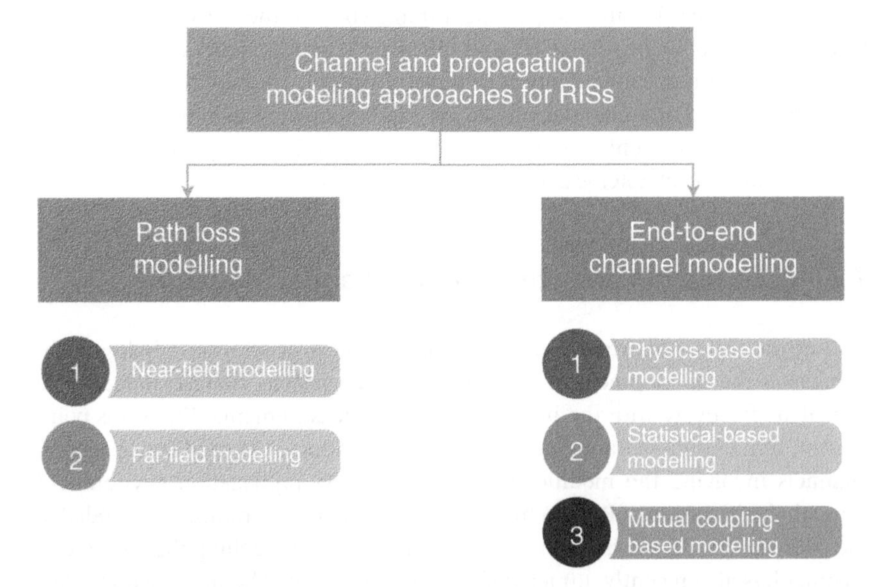

Figure 7.2 Classification of the channel and propagation modelling approaches based on the existing RIS literature.

authors characterize the path-loss using physics-based approaches and exploited the antenna theory on the calculation of the electric field both in the far-field and near-field of a finite-size RIS. While it is shown that an RIS can act as an anomalous mirror in the near field of the array, the received power expression is not formulated analytically for varying distances. The author in [12] calculates the path-loss as a function of RIS size, link geometry, and practical gain of RIS elements by considering a passive reflectarray-type RIS design under the far-field and near-field cases. The practical power scaling laws are introduced in [13] for the far-field region of array using the physical optic techniques. The authors also clarify why the surface comprises a large number of reflecting elements that individually act as diffuse scatterers but can jointly beamform the signal in the desired direction with a specific beamwidth. In [14], the free-space path-loss models for RIS-enabled communication systems are characterized with physics and EM-based approaches by considering the distances from the Tx/Rx to the RIS, the size of the RIS, the near-field/far-field effects of the RIS, and the radiation patterns of antennas and unit cells. The proposed models are validated through extensive computer simulation results and experimental measurements using a specifically manufactured RIS. In [15], the authors leverage the vector generalization of Green's theorem and characterize the free space path-loss of the RIS-assisted communications by using physical optic methods in order to overcome the limitation of geometric optics. The analyses are conducted for two-dimensional (2D) homogenized metasurfaces that can operate in reflection or refraction mode under both far-field and near-field cases. Furthermore, the power scaling laws for asymptotically large RISs are studied in [16] using a deterministic propagation model and observed that the asymptotic limit could only be achieved in the near-field case.

While afore-mentioned studies characterize the path loss by using physics and EM-based approaches, in [17], a physics-based, end-to-end channel model for RIS-empowered communication systems is introduced and a scalable optimization framework for large RISs is presented. This model includes the impact of all RIS tiles, the transmission modes of all tiles, and the incident, reflection, and polarization angles. Furthermore, each tile is modelled as an anomalous reflector, and physical optics-based analyses are conducted by adopting the concepts from the radar literature under the far-field conditions.

In [18], the authors propose a physics and EM-compliant end-to-end channel model for the RIS-assisted communication systems by considering the mutual coupling among the sub-wavelength unit cells of the RIS. This mutual coupling and unit cell aware model can be applied to an RIS consisting of closely spaced scattering elements controlled by tunable impedances. Inspiring from the impedance-based channel model in [18], an analytical framework and a iterative algorithm are presented to obtain the end-to-end received power in [19].

In [20], a multi-user MIMO interference network is investigated in the presence of multiple RISs by using a circuit-based model for the transmitters, receivers, and RISs. The mutual coupling between impedance-controlled thin dipoles and the impact of the tuning elements are also investigated. Additionally, a provably convergent optimization algorithm is proposed to maximize the sum-rate of RIS-assisted MIMO interference channels by assessing the mutual coupling among closely spaced scattering elements.

Recently, there have been a rapid and growing interest on stochastic non-stationary channel modelling activities for RIS-empowered wireless systems [21–25]. Within this context, in [21], a three-dimensional (3D) geometry-based stochastic channel model (GBSM) is introduced for a massive MIMO communication system in the presence of an RIS. The authors split the end-to-end channel into the sub-channels and characterize the large- and small-scale fading, respectively. In [22], practical phase shifts are considered, and the reflection phases of the RIS are represented using 2-bit quantization set in addition to the work in [21]. However, these two studies do not consider the practical deployment and physical properties of RISs. The study in [23] proposes a geometric RIS-assisted MIMO channel model for fixed-to-mobile communications based on a 3D cylinder model. While this study examines the non-stationary channels in which the receiver moves, the characteristic features of the scatterers in the environment are ignored, and analyses are merely conducted under sub-6 GHz frequency bands. Recently, the authors in [24] propose a general wideband non-stationary channel model for RIS-empowered communication systems operating at sub-6 GHz bands. By splitting the RIS-assisted MIMO channel model into subchannels, equivalent end-to-end channel models based on these subchannels are characterized under time-varying characteristics of MIMO systems in the presence of an RIS. Although the authors consider the physical characteristics of the wideband subchannel model between the mobile Tx and RIS, and RIS and mobile Rx for different propagation delays, LOS links between the Tx and Rx are ignored, and frequency bands above 6 GHz are not considered. More recently, a non-stationary 3D wideband GBSM for RIS-empowered MIMO communication systems are introduced in [25] by including propagation characteristics of RISs, such as unit number and size, relative terminal locations, and RIS configuration. Nevertheless, this study does not consider a direct link between the Tx and the mobile Rx due to the blockage, and the applicability of this model is limited to sub-6 GHz frequency bands in an outdoor environment.

Against this background, due to the spectrum shortage of sub-6 GHz frequency bands, it is unavoidable to migrate to the mmWave and THz bands in future wireless networks. Since transmission at higher frequencies makes the signals more vulnerable to blockage and interference, the signal attenuation and blockage prevent mmWaves from reaching long distances. Especially in mmWave frequencies,

Table 7.1 An overview of RIS channel modelling studies and their main contributions.

Work	Contributions
[10]	• Path-loss in near-field and far-field of RISs is characterized by using Huygens–Fresnel principle and physical optics-based methods • RISs are modelled as a sheet of EM material of negligible thickness
[11]	• Path-loss is characterized with physics-based approach • Electric field of a finite-size RIS is computed both in its far-field and near-field
[12]	• Path-loss is calculated for a passive reflectarray-type RIS • The path-loss is characterized as a function of RIS size, link geometry, and practical gain of RIS elements
[13]	• Practical power scaling laws are introduced for the far-field region of array using the physical optic principles
[14]	• Free-space path-loss with physic and EM-based approaches is characterized by considering the near-field/far-field effects of the RIS • It is validated through experimental measurements using manufactured RIS
[15]	• Path-loss is characterized as a function of the transmission distance and the RIS size by using physical optic methods and Huygens–Fresnel principle • 2D homogenized metasurfaces that can operate in reflection or refraction mode are considered under both far-field and near-field case
[16]	• Power scaling laws for asymptotically large RISs are analysed using a deterministic propagation model • It is shown that the asymptotic limit can only be achieved in the near-field
[17]	• End-to-end channel model is characterized including the impact of the physical parameters of an RIS • Physical optics-based analysis are conducted by adopting the concepts from the radar literature • Each tile is modelled as an anomalous reflector
[18–20]	• EM-compliant communication model is characterized by considering the mutual coupling among the RIS elements • The entire system is represented with a circuit-based model for all terminals • The mutual coupling between impedance-controlled thin dipoles and the impact of the tuning elements are investigated
[21–25]	• A statistical geometry-based non-stationary channel models are derived for the RIS-empowered MIMO systems • The various physical properties of RIS are considered, such as unit number and size, relative locations among Tx, RIS, and Rx
[9, 26–28]	• Comprehensive channel modelling efforts are introduced by considering 3D channel models for practical deployments and physical RIS characteristics • Open-source and widely applicable *SimRIS Channel Simulator* is introduced • Technical specifications on sub-6 GHz and mmWave bands are considered

using an RIS in the transmission can enable additional transmission paths when the direct link between the Tx and Rx is blocked or not sufficiently robust. Therefore, modelling a physical open-source and widely applicable mmWave channel for the RIS-assisted systems in indoor and outdoor environments is of great importance to shed light on realistic use-cases of RISs in future wireless networks. From this point of view, a unified narrowband channel model for RIS-assisted systems both in indoor and outdoor environments is introduced in [9, 26, 27] by including the 5G mmWave channel model with a random number of clusters/scatterers and the characteristics of the RIS. Moreover, a physical channel model for RIS-assisted systems is proposed in [28] by considering the currently used technical specifications on sub-6 GHz bands.

An overview of the aforementioned channel modelling studies and brief explanations of their main contributions and characteristics can be seen in Table 7.1 [9–27].

7.3 Physical Channel Modelling for RIS-Empowered Systems at mmWave Bands

This section introduces a unified signal/channel model for RIS-assisted 6G communication systems operating under mmWave frequencies. This model is generic and can be applied to indoor and outdoor environments and operating frequencies. Considering the promising potential of RIS-assisted systems in future wireless networks at mmWave frequencies, a framework on the clustered statistical MIMO model is proposed by considering 3GPP standardization [29], while a generalization is possible. Because of their unique functionality, channel modelling methodology for RISs is different from other counterpart technologies, such as relaying. Although the major steps for generation RIS-assisted channel modelling are included in this chapter, interested readers are referred to [26, 27] for the comprehensive view of technical details and statistical characteristics of the channel parameters.

In the following, our channel modelling methodology is explained to generate Tx–RIS, RIS–Rx, and Tx–Rx subchannels for indoors and outdoors. It is assumed that the Tx lies on the yz plane, while the RIS lies either on the xz plane (Scenario 1 – side wall) or yz plane (Scenario 2 – opposite wall) for indoor and outdoor environments. In Figure 7.3, the considered 3D geometry for a typical large indoor office is given as a reference for Scenario 1 (side-wall). For outdoor environments, this 3D geometry can be easily extended by modifying certain system parameters.

The existing interacting objects (scatterers) are assumed to be grouped under C clusters, each having S_c sub-rays for $c = 1, \dots, C$, that is $M = \sum_{c=1}^{C} S_c$. Therefore,

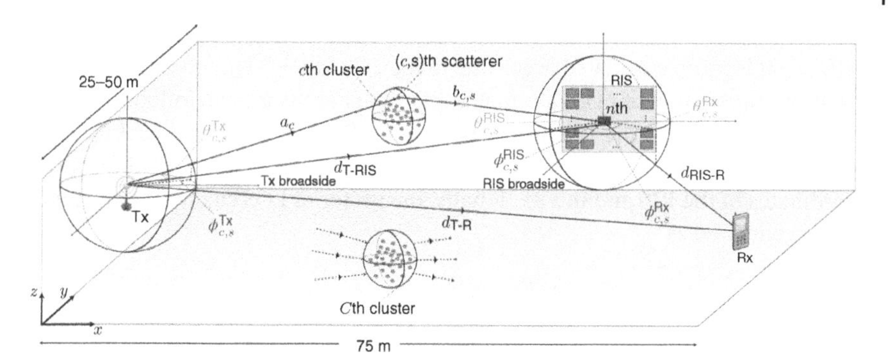

Figure 7.3 Generic InH Indoor Office environment with *C* clusters between Tx–RIS and an RIS mounted in the *xz* plane (side wall). Source: Basar et al. [27]/with permission of IEEE.

the vector of Tx–RIS channel coefficients $\mathbf{h} \in \mathbb{C}^{N\times1}$ can be obtained for a clustered model by considering array responses and path attenuations:

$$\mathbf{h} = \gamma \sum_{c=1}^{C} \sum_{s=1}^{S_c} \beta_{c,s} \sqrt{G_e(\theta_{c,s}^{\text{RIS}}) L_{c,s}^{\text{RIS}}}\ \mathbf{a}(\phi_{c,s}^{\text{RIS}}, \theta_{c,s}^{\text{RIS}}) + \mathbf{h}_{\text{LOS}} \tag{7.1}$$

where $\gamma = \sqrt{\frac{1}{\sum_{c=1}^{C} S_c}}$ is a normalization factor, $\mathbf{h}_{\text{LOS}}$ is the LOS component, $\beta_{c,s} \sim \mathcal{CN}(0,1)$ and $L_{c,s}^{\text{RIS}}$, respectively, stand for the complex path gain and attenuation associated with the (c,s)th propagation path, and $G_e(\theta_{c,s}^{\text{RIS}})$ is the RIS element pattern [30] in the direction of the (c,s)th scatterer. Here, $\mathcal{CN}(0,\sigma^2)$ denotes complex Gaussian distribution with zero mean and σ^2 variance, and $\mathbf{a}(\phi_{c,s}^{\text{RIS}}, \theta_{c,s}^{\text{RIS}}) \in \mathbb{C}^{N\times1}$ is the array response vector of the RIS for the considered azimuth ($\phi_{c,s}^{\text{RIS}}$) and elevation ($\theta_{c,s}^{\text{RIS}}$) arrival angles (with respect to the RIS broadside) and carefully calculated for our system due to the fixed orientation of the RIS. Here, the number of clusters, number of sub-rays per cluster, and the locations of the clusters can be determined for a given environment and frequency.

For the attenuation of the (c,s)th path, we adopt the 5G path loss model (the close-in free space reference distance model with frequency-dependent path loss exponent, in dB), which is applicable to various environments including Urban Microcellular (UMi) and Indoor Hotspot (InH) [31]. The LOS component of $\mathbf{h}$ is calculated by

$$\mathbf{h}_{\text{LOS}} = I_{\mathbf{h}}(d_{\text{T-RIS}}) \sqrt{G_e(\theta_{\text{LOS}}^{\text{RIS}}) L_{\text{LOS}}^{\text{T-RIS}}} e^{j\eta} \mathbf{a}(\phi_{\text{LOS}}^{\text{RIS}}, \theta_{\text{LOS}}^{\text{RIS}}) \tag{7.2}$$

where $L_{\text{LOS}}^{\text{T-RIS}}$ is the attenuation of the LOS link, $G_e(\theta_{\text{LOS}}^{\text{RIS}})$ is the RIS element gain in the LOS direction, $\mathbf{a}(\phi_{\text{LOS}}^{\text{RIS}}, \theta_{\text{LOS}}^{\text{RIS}})$ is the array response of the RIS in the direction of the Tx, and $\eta \sim \mathcal{U}[0, 2\pi]$. Here, $\mathcal{U}[a,b]$ is a random variable uniformly distributed in $[a,b]$, and $I_{\mathbf{h}}(d_{\text{T-RIS}})$ is a Bernoulli random variable taking values from the set $\{0, 1\}$ and characterizes the existence of a LOS link for a Tx–RIS separation of $d_{\text{T-RIS}}$. It is again calculated according to the 5G model [31].

For the calculation of LOS-dominated RIS–Rx channel $\mathbf{g}$ in an indoor environment, we re-calculate the RIS array response in the direction of the Rx by calculating azimuth and elevation departure angles $\phi_{\text{Rx}}^{\text{RIS}}$ and $\theta_{\text{Rx}}^{\text{RIS}}$ for the RIS from the coordinates of the RIS and the Rx. Finally, the vector of LOS channel coefficients can be generated as

$$\mathbf{g} = \sqrt{G_e(\theta_{\text{Rx}}^{\text{RIS}}) L_{\text{LOS}}^{\text{RIS-R}}} e^{j\eta} \mathbf{a}(\phi_{\text{Rx}}^{\text{RIS}}, \theta_{\text{Rx}}^{\text{RIS}}) \tag{7.3}$$

where $G_e(\theta_{\text{Rx}}^{\text{RIS}})$ is the gain of RIS element in the direction of the Rx, $L_{\text{LOS}}^{\text{RIS-R}}$ is the attenuation of LOS RIS–Rx channel, $\eta \sim \mathcal{U}[0, 2\pi]$ is the random phase term and $\mathbf{a}(\phi_{\text{Rx}}^{\text{RIS}}, \theta_{\text{Rx}}^{\text{RIS}})$ is the RIS array response in the direction of the Rx.

For outdoor channel modelling, the major change will be in the channel between the RIS and the Rx, which might be subject to small-scale fading as well with a random number of unique clusters. For this case, the RIS–Rx channel can be expressed as follows:

$$\mathbf{g} = \bar{\gamma} \sum_{c=1}^{\bar{C}} \sum_{s=1}^{\bar{S}_c} \bar{\beta}_{c,s} \sqrt{G_e(\theta_{c,s}^{\text{Rx}}) L_{c,s}^{\text{Rx}}} \; \mathbf{a}(\phi_{c,s}^{\text{Rx}}, \theta_{c,s}^{\text{Rx}}) + \mathbf{g}_{\text{LOS}} \tag{7.4}$$

where, similar to (7.1), $\bar{\gamma}$ is a normalization term, $\bar{C}$ and $\bar{S}_c$ stand for number of clusters and sub-rays per cluster for the RIS–Rx link, $\bar{\beta}_{c,s}$ is the complex path gain, $L_{c,s}^{\text{Rx}}$ is the path attenuation, $G_e(\theta_{c,s}^{\text{Rx}})$ is the RIS element radiation pattern in the direction of the (c, s)th scatterer, $\mathbf{a}(\phi_{c,s}^{\text{Rx}}, \theta_{c,s}^{\text{Rx}})$ is the array response vector of the RIS for the given azimuth and elevation angles, and $\mathbf{g}_{\text{LOS}}$ is the LOS component.

The RIS-assisted channel has a double-scattering nature, as a result, the single-scattering link between the Tx and Rx has to be taken into account in the proposed channel model. Even if the RIS is placed near the Rx, the Tx–Rx channel is relatively stronger than the RIS-assisted path, and cannot be ignored in the channel model.

For indoors, using single-input single-output (SISO) mmWave channel modelling, the channel between these two terminals can be easily obtained (by ignoring arrival and departure angles) as follows:

$$h_{\text{SISO}} = \gamma \sum_{c=1}^{C} \sum_{s=1}^{S_c} \beta_{c,s} e^{j\eta_c} \sqrt{L_{c,s}^{\text{SISO}}} + h_{\text{LOS}} \tag{7.5}$$

where γ, C, S_c, and $\beta_{c,s}$ are as defined in (7.1) and remain the same for the Tx–Rx channel under the assumption of shared clusters with the Tx–RIS channel, while h_{LOS} is the LOS component. Here, $L_{c,s}^{\text{SISO}}$ stands for the path attenuation for the corresponding link and η_e is the excess phase caused by different travel distances of Tx–RIS and Tx–Rx links over the same scatterers.

For outdoor environments, we assume that the RIS and the Rx are not too close to ensure that they have independent clusters (small-scale parameters) as in the 3GPP 3D channel model [29]. Using SISO mmWave channel modelling, the Tx–Rx channel can be easily obtained as follows:

$$h_{\mathrm{SISO}} = \tilde{\gamma} \sum_{c=1}^{\tilde{C}} \sum_{s=1}^{\tilde{S}_c} \tilde{\beta}_{c,s} \sqrt{L_{c,s}^{\mathrm{SISO}}} + h_{\mathrm{LOS}} \tag{7.6}$$

where the number of clusters $\tilde{C}$, sub-rays per cluster $\tilde{S}_c$, complex path gain $\tilde{\beta}_{c,s}$, and path attenuation $L_{c,s}^{\mathrm{SISO}}$ are determined as discussed earlier for the Tx–RIS path and $\tilde{\gamma}$ is the normalization term.

In Figure 7.4, an exemplary 3D geometry for the UMi Street Canyon outdoor environment is illustrated for Scenario 1, where Tx is mounted at 20 m height and Rx is a ground-level user. In this particular 3D geometry, each path has a single cluster with a different number of scatterers, while the number of clusters for each path can vary randomly in general.

The major steps of RIS-assisted physical channel modelling for indoor and outdoor environments can be summarized as in Figure 7.5. In general, Steps 1–5

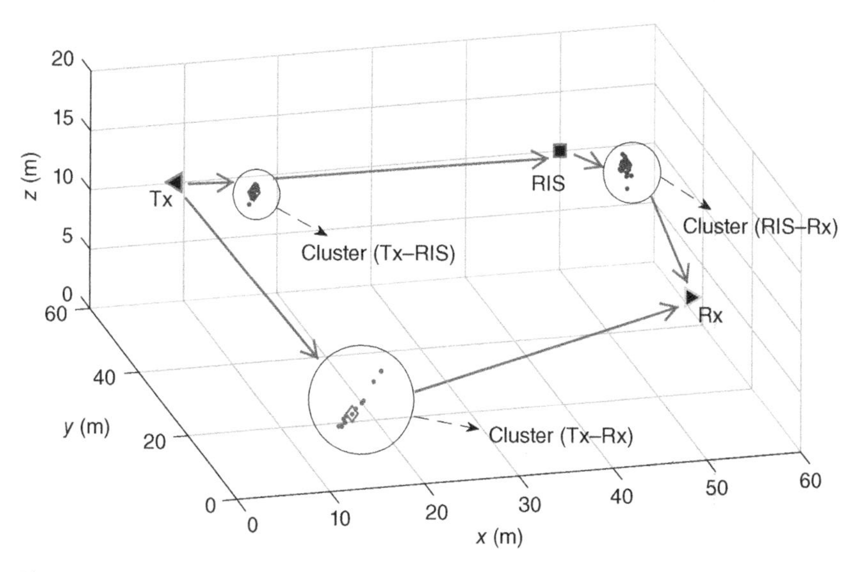

Figure 7.4 The considered UMi Street Canyon outdoor environment with random number of clusters/scatterers and an RIS on the *xz* plane. Source: Basar et al. [27]/with permission of IEEE.

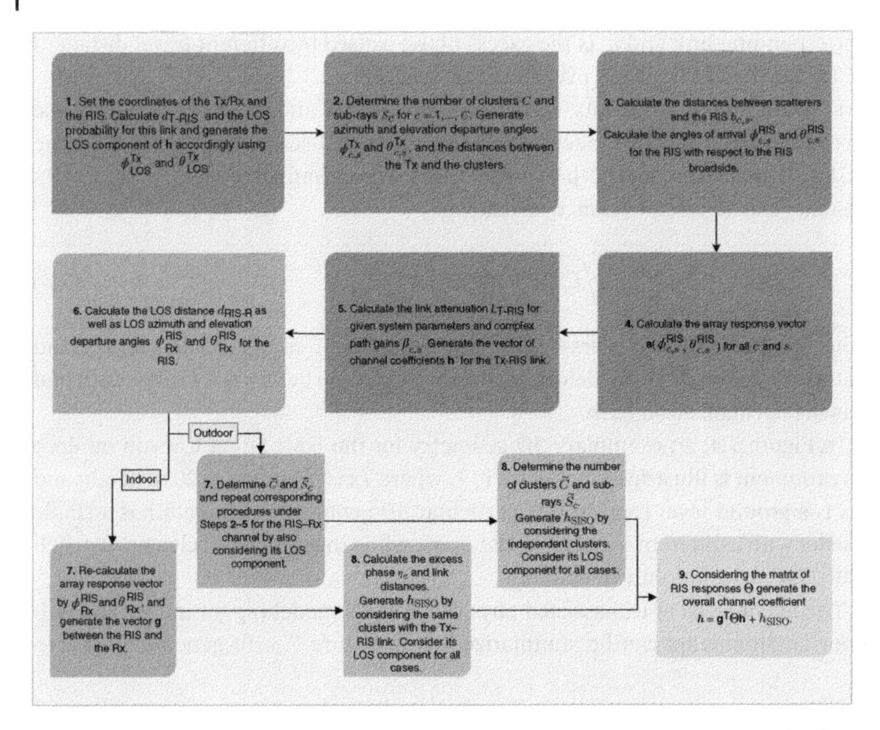

Figure 7.5 Summary of the major steps of RIS-assisted physical channel modelling for indoor and outdoor environments.

focus on the generation of **h**, while Steps 6, 7, and 8, respectively, deal with **g** and h_{SISO}.

Furthermore, in [9], the proposed channel modelling strategy is extended to a MIMO system with N_t transmit, and N_r receive antennas operating in the presence of an RIS with N reflecting elements. Consequently, the end-to-end channel matrix $\mathbf{C} \in \mathbb{C}^{N_r \times N_t}$ of an RIS-assisted MIMO system can be obtained as follows:

$$\mathbf{C} = \mathbf{G}\boldsymbol{\Phi}\mathbf{H} + \mathbf{D} \tag{7.7}$$

In this model, $\mathbf{H} \in \mathbb{C}^{N \times N_t}$ is the matrix of channel coefficients between the Tx and the RIS, $\mathbf{G} \in \mathbb{C}^{N_r \times N}$ is the matrix of channel coefficients between the RIS and the Rx, and $\mathbf{D} \in \mathbb{C}^{N_r \times N_t}$ stands for the direct channel (not necessarily a LOS-dominated one and is likely to be blocked for mmWave bands due to obstacles in the environment) between the Tx and the Rx. The considered system configurations and general expressions of the channel matrices for the MIMO case are given in Table 7.2.

Table 7.2 System configurations RIS-assisted MIMO channel modelling.

Channel matrices[a)]	$\mathbf{H} = \gamma \sum_{c=1}^{C} \sum_{s=1}^{S_c} \beta_{c,s} \sqrt{G_e(\theta_{c,s}^{\text{T-RIS}}) L_{c,s}^{\text{T-RIS}}}\ \mathbf{a}(\phi_{c,s}^{\text{T-RIS}}, \theta_{c,s}^{\text{T-RIS}}) \mathbf{a}^T(\phi_{c,s}^{\text{Tx}}, \theta_{c,s}^{\text{Tx}}) + \mathbf{H}_{\text{LOS}}$ $\mathbf{G} = \bar{\gamma} \sum_{c=1}^{\bar{C}} \sum_{s=1}^{\bar{S}_c} \bar{\beta}_{c,s} \sqrt{G_e(\theta_{c,s}^{\text{RIS-R}}) L_{c,s}^{\text{RIS-R}}}\ \mathbf{a}(\phi_{c,s}^{\text{Rx}}, \theta_{c,s}^{\text{Rx}}) \mathbf{a}^T(\phi_{c,s}^{\text{RIS-R}}, \theta_{c,s}^{\text{RIS-R}}) + \mathbf{G}_{\text{LOS}}$ $\mathbf{D} = \tilde{\gamma} \sum_{c=1}^{\tilde{C}} \sum_{s=1}^{\tilde{S}_c} \tilde{\beta}_{c,s} \sqrt{L_{c,s}^{\text{T-R}}} \mathbf{a}(\phi_{c,s}^{\text{Rx}'}, \theta_{c,s}^{\text{Rx}'}) \mathbf{a}^T(\phi_{c,s}^{\text{Tx}'}, \theta_{c,s}^{\text{Tx}'}) + \mathbf{D}_{\text{LOS}}$
Environments	Indoor: InH Indoor Office and Outdoor: UMi Street Canyon
Frequencies	28 and 73 GHz
Array Type	Uniform linear array (ULA) and uniform planar array (UPA)

a) $\gamma/\bar{\gamma}/\tilde{\gamma}$: normalization factor [26], $\beta_{c,s}/\bar{\beta}_{c,s}/\tilde{\beta}_{c,s}$: complex Gaussian distributed gain of the (c, s)th propagation path, $G_e(.)$: RIS element gain, $L_{c,s}^{i}$: attenuation of the (c, s)th propagation path [29], $\mathbf{a}\left(\phi_{c,s}^{i}, \theta_{c,s}^{i}\right)$: array response vectors for the considered azimuth $(\phi_{c,s}^{i})$ and elevation angles $(\theta_{c,s}^{i})$, $\mathbf{H}_{\text{LOS}}/\ \mathbf{G}_{\text{LOS}}/\ \mathbf{D}_{\text{LOS}}$: LOS component of sub-channels (i: indicator for the corresponding path/terminal as $i \in$ {T-RIS, RIS-R, T-R, Tx, Rx})

7.4 Physical Channel Modelling for RIS-Empowered Systems at Sub-6 GHz Bands

This section introduces a channel modelling strategy for RIS-assisted wireless networks in sub-6 GHz bands by investigating far-field and near-field behaviours in transmission. In addition to mmWave bands, RIS-assisted communication systems draw attention as an effective solution in sub-6 GHz frequencies, which are widely used in wireless communications. According to the technical specifications on sub-6 GHz bands [32, 33], an end-to-end channel model is derived for SISO systems employing an RIS by following the 3D channel modelling approach. The generic system model for the considered RIS-assisted wireless communication system is demonstrated in Figure 7.6, where $d_{\text{3D}}^{\text{T-RIS}}$, $d_{\text{3D}}^{\text{T-R}}$ and $d_{\text{3D}}^{\text{RIS-R}}$ denotes the 3D distances between the Tx–RIS, Tx–Rx, and RIS–Rx, respectively. It is assumed that the RIS with N number of elements is located on the xz-plane. There exist C clusters in the environment, each containing S rays. The channels between the Tx–RIS, RIS–Rx, and Tx–Rx are represented by $\mathbf{h} \in \mathbb{C}^{N\times1}$, $\mathbf{g} \in \mathbb{C}^{N\times1}$, and $h_{\text{SISO}} \in \mathbb{C}^{1\times1}$, respectively, where N is the number of RIS elements. It is assumed that the Tx and Rx are equipped with unity gain isotropic antennas. The positions of the Tx, Rx, and RIS are given in the cartesian coordinate system as $\mathbf{r}^{\text{Tx}} = (x^{\text{Tx}}, y^{\text{Tx}}, z^{\text{Tx}})$, $\mathbf{r}^{\text{Rx}} = (x^{\text{Rx}}, y^{\text{Rx}}, z^{\text{Rx}})$ and $\mathbf{r}^{\text{RIS}} = (x^{\text{RIS}}, y^{\text{RIS}}, z^{\text{RIS}})$, respectively.

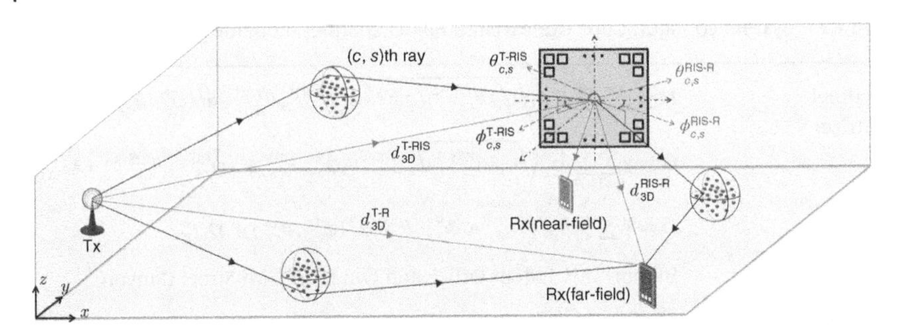

Figure 7.6 Generic system model for an RIS-assisted network with C number of clusters and S number of rays. Source: Kilinc et al. [28]/arXiv/CC BY 4.0.

Table 7.3 System parameters of physical RIS-assisted channel modelling for sub-6 GHz.

P_c	Power of the nth cluster
P_L	Path loss component
$G_e(\theta_{c,s}^{\text{T-RIS}})$	Radiation pattern of an RIS element in the direction of (c, s)th path
$\Phi_{c,s}$	Random initial phase
$\mathbf{a}(\theta_{c,s}^{\text{T-RIS}}, \phi_{c,s}^{\text{T-RIS}})$	Array response vector of the RIS in the direction of the Tx
$\theta_{c,s}^{\text{T-RIS}}$	Zenith angle of arrival (ZoA)
$\phi_{c,s}^{\text{T-RIS}}$	Azimuth angle of arrival (AoA) of the RIS for the (c, s)th path
$G_e(\theta_{c,s}^{\text{RIS-R}})$	RIS element radiation pattern in the direction of (c, s)th path
$\phi_{c,s}^{\text{RIS-R}}$	Azimuth of departure (AoD) and from the RIS
$\theta_{c,s}^{\text{RIS-R}}$	Zenith of departure (ZoD) angles from the RIS
$\mathbf{a}(\theta_{c,s}^{\text{RIS-R}}, \phi_{c,s}^{\text{RIS-R}})$	Array response vector of the RIS in the direction of the Rx

The definitions of the system parameters are given in Table 7.3 for the RIS-assisted channels. The major generation steps of the Tx–RIS channel in RIS-assisted physical channel modelling are also summarized in Figure 7.7. Interested readers are referred to [28] for the details channel generation procedure and statistical background of the considered channel parameters. By considering the certain system parameters and following the channel generation steps in Figure 7.7, the channel between the Tx and RIS is expressed as follows:

$$\mathbf{h} = \sum_{c=1}^{C}\sum_{s=1}^{S}\sqrt{\frac{P_c}{S}}\sqrt{\frac{G_e(\theta_{c,s}^{\text{T-RIS}})}{P_L}}e^{j\Phi_{c,s}}\mathbf{a}(\theta_{c,s}^{\text{T-RIS}}, \phi_{c,s}^{\text{T-RIS}}) \tag{7.8}$$

Furthermore, the direct link between the Tx and Rx can be obtained by following the Steps 1, 2, 3, 4, and 6 in the channel generation procedure

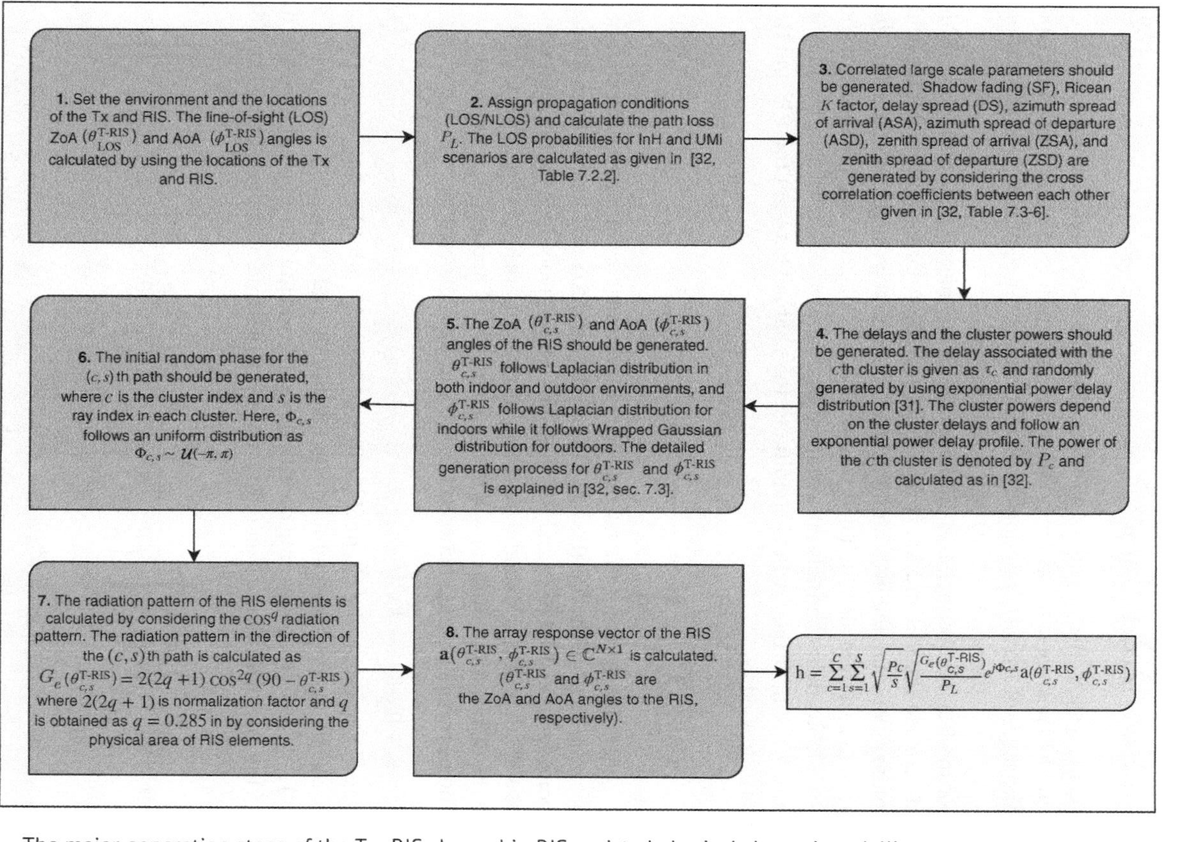

Figure 7.7 The major generation steps of the Tx–RIS channel in RIS-assisted physical channel modelling.

described in Figure 7.7. Therefore, the direct link between the Tx and Rx is also given by

$$h_{\text{SISO}} = \sum_{c=1}^{C} \sum_{s=1}^{S} \sqrt{\frac{P_c}{S}} \sqrt{\frac{1}{P_L}} e^{j\Phi_{c,s}} \tag{7.9}$$

Here, in all steps, the Tx and Rx positions should be taken into account instead of Tx and RIS positions, respectively. Moreover, Azimuth angle of arrival (AoA) and Zenith angle of arrival (ZoA) angles calculations are not required in the generation of h_{SISO}, since the SISO (direct) link is established.

For the RIS–Rx channel, two different channel scenarios are separately considered between the RIS and Rx. If the size of the RIS is large enough, the far-field boundary between the RIS and Rx will be extremely large. Accordingly, the near-field cases are also included when the RIS is placed close proximity of the Rx, while far-field channels are obtained by following the similar procedure in Figure 7.7. Under the far-field conditions, the channel between the RIS and the Rx is expressed as

$$\mathbf{g} = \sum_{c=1}^{C} \sum_{s=1}^{S} \sqrt{\frac{P_c}{S}} \sqrt{\frac{G_e(\theta_{c,s}^{\text{RIS-R}})}{P_L}} e^{j\Phi_{c,s}} \mathbf{a}(\theta_{c,s}^{\text{RIS-R}}, \phi_{c,s}^{\text{RIS-R}}) \tag{7.10}$$

The generation steps for the far-field RIS–Rx channel are similar to the channel generation steps of the Tx–RIS link. Differently, the RIS and Rx positions should be considered instead of Tx and RIS positions, respectively. In Steps 5, 7, and 8, the angles $\theta_{c,s}^{\text{RIS-R}}$ and $\phi_{c,s}^{\text{RIS-R}}$ should be taken into account instead of $\theta_{c,s}^{\text{T-RIS}}$ and $\phi_{c,s}^{\text{T-RIS}}$, and the distributions of them will be same with $\theta_{c,s}^{\text{T-RIS}}$ and $\phi_{c,s}^{\text{T-RIS}}$, respectively.

The near-field channel between the RIS and Rx is denoted by

$$\mathbf{g} = [g_1, g_2, \dots, g_N]^{\text{T}} \tag{7.11}$$

where g_n represents the channel coefficient from the nth RIS element to the Rx, and can be expressed in terms of its magnitude and phase as $g_n = |g_n| e^{-j\gamma}$ for $n = 1, 2, \dots, N$. By considering the RIS geometry and the RIS element locations in [28], the near-field channel gain from the nth RIS element is approximated as [16]

$$|g_n|^2 \approx \frac{1}{4\pi} \sum_{x \in \mathbb{X}} \sum_{z \in \mathbb{Z}} \left(\frac{\frac{xz}{y^2}}{3\left(\frac{z^2}{y^2}+1\right)\sqrt{\frac{x^2}{z^2}+\frac{z^2}{y^2}+1}} + \frac{2}{3}\tan^{-1}\left(\frac{\frac{xz}{y^2}}{\sqrt{\frac{x^2}{y^2}+\frac{z^2}{y^2}+1}} \right) \right) \tag{7.12}$$

where $\mathbb{X} = \{d/2 + x^n - x^{\text{Rx}}, d/2 + x^{\text{Rx}} - x^n\}$, $\mathbb{Z} = \{d/2 + z^n - z^{\text{Rx}}, d/2 + z^{\text{Rx}} - z^n\}$, and $y = |y^n - y^{\text{Rx}}|$. Moreover, the phase of g_n can be calculated as follows:

$$\gamma = 2\pi \bmod \left(\frac{\|\mathbf{r}^n - \mathbf{r}^{\text{Rx}}\|}{\lambda}, 1 \right) \tag{7.13}$$

7.5 SimRIS Channel Simulator

This section introduces the open-source, user-friendly, and widely applicable *SimRIS Channel Simulator v2.0* [9, 26]. *SimRIS Channel Simulator* aims to open a new line of research for physical channel modelling of RIS-empowered networks by including the LOS probabilities between terminals, array responses of RISs and Tx/Rx units, RIS element gains, realistic path loss and shadowing models, and environmental characteristics in different propagating environments and operating frequencies. More specifically, InH – Indoor Office and Urban Microcellar (UMi) – Street Canyon environments are considered for popular mmWave operating frequencies of 28 and 73 GHz from the 5G channel model [29]. Compared to its earlier version (*SimRIS Channel Simulator v1.0*) with only SISO Tx/Rx terminals, its v2.0 considers MIMO terminals with different types of arrays [9]. The simulator also supports terminals in the far-field of the RIS only, which is a reasonable assumption for mmWaves with shorter wavelengths and smaller RIS sizes, while the near-field RIS models will be easily included by extending the existing modelling strategy in its future releases.

The considered generic 3D geometry is given in Figure 7.3 for the representation of physical channel characteristics. In this setup, the RIS is mounted on the *xz*-plane while a generalization is possible in the simulator. The proposed model considers various indoor and outdoor wireless propagation environments in terms of physical aspects of mmWave frequencies while numerous practical 5G channel model issues are adopted to our channel model [29]. For a considered operating frequency and environment, the number of clusters (C), number of sub-rays per cluster (S_c), and the positions of the clusters can be determined by following the detailed steps and procedures in [27]. More specifically, according to the 5G channel model, the number of clusters and scatterers are determined using the Poisson and uniform distributions with certain parameters, respectively. Although the clusters between Tx–RIS and Tx–Rx can be modelled independently, it can be assumed that the Rx and the RIS might share the same clusters when the Rx is located relatively closer to the RIS. Due to the fixed orientation of the Tx and the RIS, the array response vectors of the Tx and the RIS are easily calculated for given azimuth and elevation departure/arrival angles. However, it is worth noting that if azimuth and elevation angles are generated randomly for the Tx, due to fixed orientation of the RIS, they will not be random anymore at the RIS and should be calculated from the 3D geometry using trigonometric identities. Nevertheless, the array response vector of the Rx can be calculated with randomly distributed azimuth and elevation angles of arrival due to the random orientation.

Using the SimRIS Channel Simulator, the wireless channels of RIS-aided communication systems can be generated with tunable operating frequencies, number of RIS elements, number of transmit/receive antennas, Tx/Rx array types, terminal locations, and environments. As discussed earlier, in SimRIS Channel Simulator v2.0, MIMO-aided Tx and Rx terminals are incorporated into its earlier version, and the array response vectors and receiver orientation are reconstructed by considering this MIMO system model. This new version also offers two different types of antenna array configurations: Uniform linear array (ULA) and uniform planar array (UPA), and the corresponding array response vectors are calculated according to the selected antenna array type. The graphical user interface (GUI) of our SimRIS Channel Simulator v2.0 is given in Figure 7.8. In this GUI, the selected scenario specifies the RIS position for both indoor and outdoor environments as xz-plane (Scenario 1) and yz-plane (Scenario 2). Considering the 3D geometry illustrated in Figure 7.3, Tx, Rx, and RIS positions can be manually entered into the SimRIS Channel Simulator. Furthermore, N_t (the number of Tx antennas),

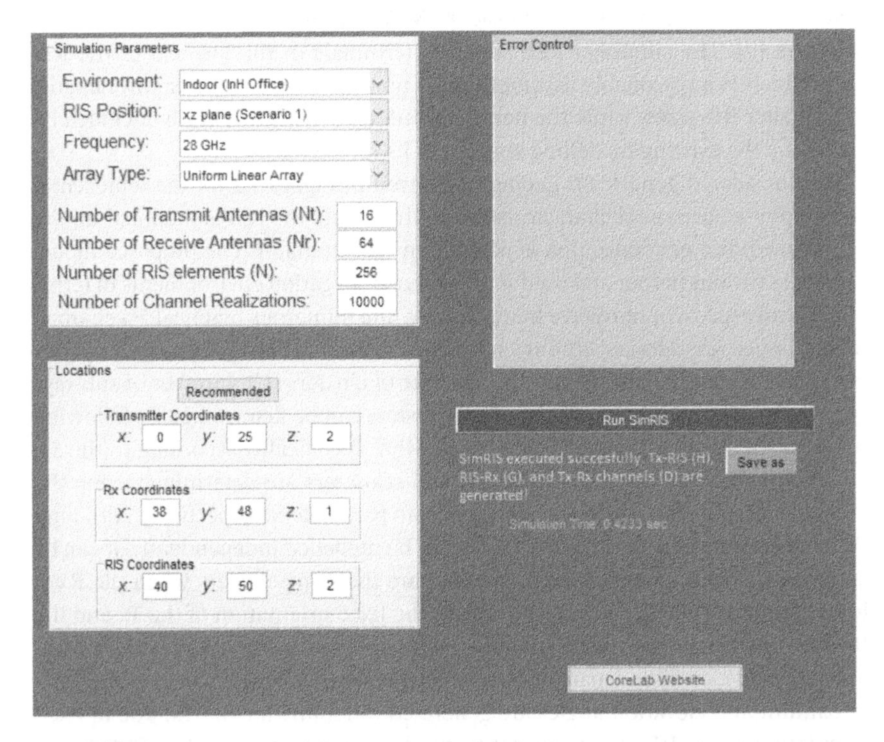

Figure 7.8 Graphical user interface (GUI) of the SimRIS Channel Simulator v2.0. Source: CoreLab/CC BY-NC-SA 4.0.

N_r (the number of Rx antennas), N (the number of RIS elements), and the number of channel realizations are also user-selectable input parameters and these options offer a flexible and versatile channel modelling opportunity to the users. Considering these input parameters, this simulator produces **H**, **G**, and **D** channel matrices by conducting Monte Carlo simulations for the specified number of realizations under 28 and 73 GHz mmWave frequencies. The general expressions of these channel matrices are given in Table 7.2 for the interested readers. Here, the double summation terms stem from the random number of clusters and scatterers and the LOS components might be equal to zero with a certain probability for increasing distances.

Consequently, the open-source SimRIS Channel Simulator package considers a narrowband channel model for RIS-empowered communication systems for both indoor and outdoor environments and it takes into account various physical characteristics of the wireless propagation environment. The open-source nature of our simulator, which is written in the MATLAB programming environment, encourages all researchers to use and contribute to the development of its future versions by exploring the interesting use-cases of the RIS in the transmission.

7.6 Performance Analysis Using SimRIS Channel Simulator

This section provides numerical results that are conducted via *SimRIS Channel Simulator* MATLAB package for the detailed evaluation of how RISs can be effectively used in future wireless networks to enrich and improve the existing communication systems. The system configurations and computer simulation parameters of the considered setups are given in Table 7.4, which will be used for the numerical results in the following. All computer simulations in this section are conducted in an InH Indoor Office environment at an operating frequency of 28 GHz, while the considered system parameters are also valid for 73 GHz and outdoor environments. The positions of the terminals are given in the 3D Cartesian coordinate system, and the noise and transmit powers are assumed to be −100 and 30 dBm, respectively.

In Figure 7.9, the achievable rate performance is investigated for seven different positions of the Rx in an InH Indoor Office environment under SISO case. Here, $(x^{\text{Rx}}, y^{\text{Rx}})$ coordinates of the test points are marked on Figure 7.9, while z^{Rx} is fixed to 1 m for all points. For this analysis, the first setup parameters in Table 7.4 are considered. In order to observe the effect of RIS in transmission, achievable rates of these seven reference positions of the Rx are calculated for three cases: RISs are not used, one RIS is used, and two RISs are used. In the case of only RIS1 being operated, an average increase of 26.91% is achieved in the achievable rate of

Table 7.4 System configurations and simulation parameters.

Setup	First	Second	Third
Frequency	28 GHz	28 GHz	28 GHz
Environments	InH Indoor Office	InH Indoor Office	InH Indoor Office
N	256	Varying	64
$N_t = N_r$	1	Varying	4
Array type	—	UPA	ULA
Tx position	[0, 25, 2]	[0, 25, 2]	[0, 25, 2]
Rx position	Varying	[45, 45, 1]	Varying
RIS positions	RIS1: [40, 50, 2] RIS2: [60, 40, 2.5]	[40, 50, 2]	RIS1: [40, 46, 2] RIS2: [62, 30, 2]

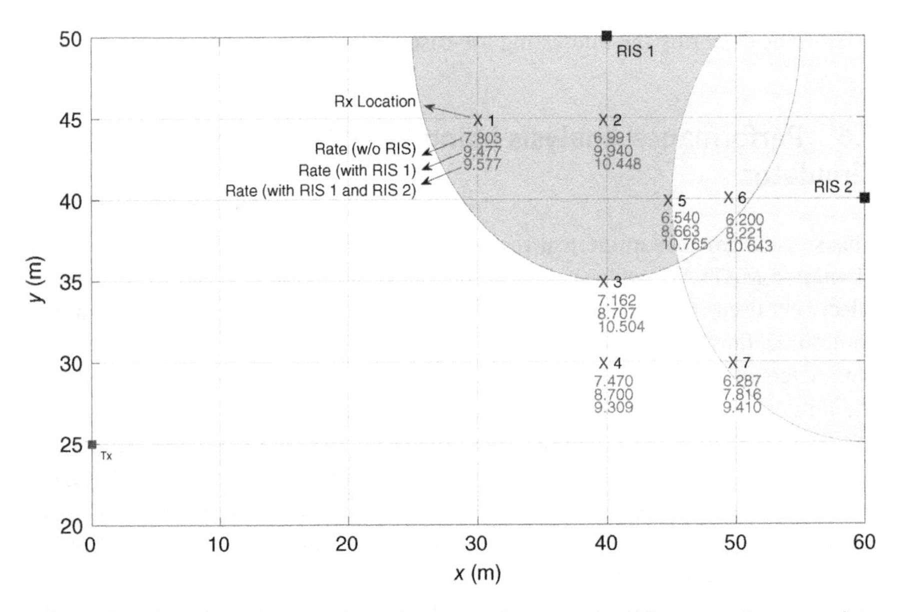

Figure 7.9 Top view of the considered transmission scenario with seven reference points along with the achievable rate values.

the reference points, while an increase of 45.72% is achieved when RIS1 and RIS2 are jointly activated. Furthermore, when RISs are deactivated, reference points 1, 3, and 4 will have a higher achievable rate since they are close to the Tx compared to other points, while all points will have approximately similar achievable rate performance when RIS1 and RIS2 are activated. As observed from Figure 7.9, we concluded that a significant improvement is obtained in achievable rate with RISs, particularly for the test points closer to the RIS.

In order to get a more precise understanding of the impacts of the reflecting elements and number of Tx/Rx antennas on the system performance, the combined effect of these two parameters on the achievable rate is analyzed in Figure 7.10. The second setup parameters in Table 7.4 are considered for this analysis, and pseudoinverse (pinv)-based algorithm [34] is used for adapting the phase shifts of an RIS-assisted MIMO transmission system. As observed from the Figure 7.10, doubling N_t and N_r values for $N = 25$ provides approximately 2.09 bit/s/Hz increase in achievable rate, while doubling N values for $N_t = N_r = 20$ provides a roughly 2.01 bit/s/Hz increase in achievable rate. In order to meet the increasing data demand in next-generation wireless networks, it is foreseen to use a large number of transmit and receive antennas. Although, using a large number of antennas enhance the achievable rate performances, the cost of signal processing and hardware in will be notably ascended. Using RISs in the transmission also provides enhanced achievable rate, while alleviating the cost of physical implementation and signal processing. From the obtained results in Figure 7.10, it can be said that the demand that will emerge in future wireless networks can be satisfied effectively by increasing the N, since doubling the number of reflectors is much less costly than doubling the number of Tx/Rx antennas.

By considering the third setup parameters in Table 7.4, the effect of the changing Rx positions on the xy-plane to its achievable rate performance is analyzed in Figure 7.11 in the presence of a single and two RISs under MIMO transmission

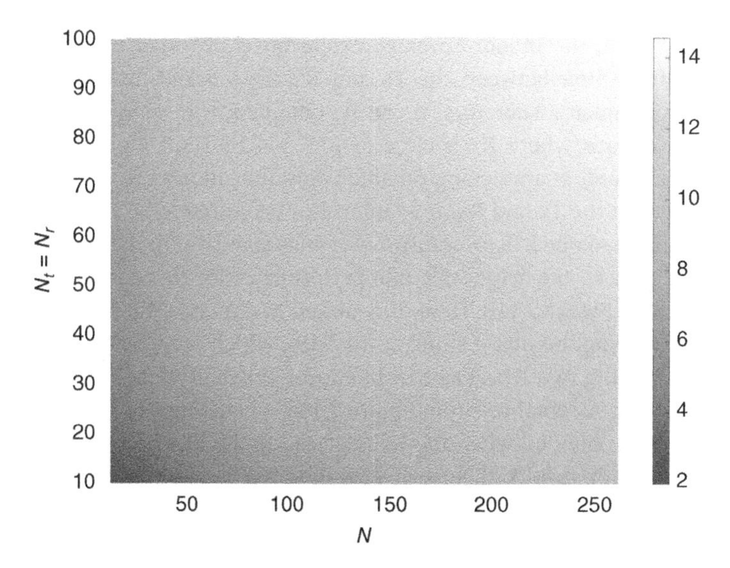

Figure 7.10 Achievable rate analysis in the presence of an RIS for changing N_t/N_r and N.

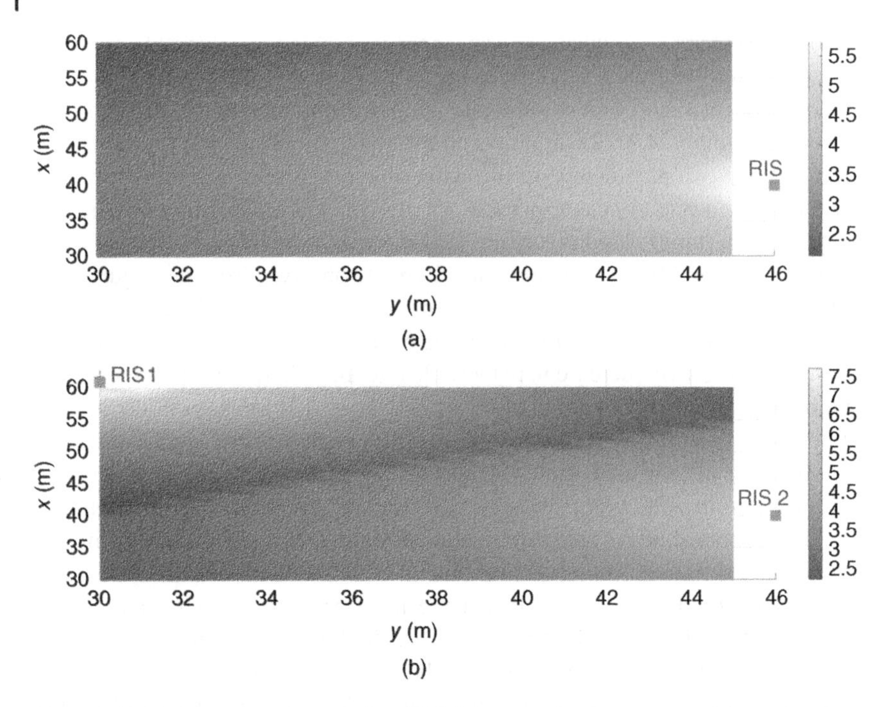

Figure 7.11 Achievable rate analysis of MIMO communication system with varying Rx positions in the presence of (b) A single RIS and (c) Two RISs. Note: The squares indicate the positions of the RISs on the *x* and *y*-axes.

scenario. In Figure 7.11a, the indoor coverage extension is provided by using an RIS when the direct link between the Tx and Rx are blocked due to the obstacles in the environment. Therefore, it can be obtained from Figure 7.11a that particularly in regions where Rx is close to RIS, a significant increase in achievable rate is achieved, guaranteeing reliable communication even if there is no direct link between the Tx and Rx. If a larger area is desired to be covered, the idea of using more than one RIS for transmission may stand out in the system design. Within this context, the achievable rate performance for the varying Rx locations is observed in Figure 7.11b. Here, it is aimed to enhance the received signal quality by modifying the phase shifts of the RISs, which are closer to the Rx. Particularly, employing two RISs create a handover capability for the Rx to increase its signal power. As obtained from Figure 7.11b, a considerable increase in the achievable rate is obtained when the Rx is placed to the proximity of any of the RISs. From the given results, the use of RISs in transmission boosts signal quality even in dead zones or cell edges and provides a coverage extension by providing low-cost and energy-efficient solutions by alleviating a large number of antenna requirements in future wireless networks.

7.7 Summary

This chapter presents a general framework on channel modelling strategies for the RIS-empowered communications systems considering the state-of-the-art channel and propagation modelling efforts in the literature. Another objective of this chapter is to draw attention to open-source and standard-compliant physical channel modelling efforts to provide comprehensive insights regarding the practical use-cases of RISs in future wireless networks. Within this context, a framework on the cluster-based statistical channel model is summarized for both sub-6 GHz and mmWave bands. Moreover, an open-source *SimRIS Channel Simulator* package that considers a narrowband channel model for RIS-empowered communication systems is presented. Finally, using *SimRIS Channel Simulator*, performances of the RIS-assisted communication system are evaluated, and efficient use-cases of RISs for next-generation wireless systems are investigated.

Funding Acknowledgment

This work was supported by the Scientific and Technological Research Council of Turkey (TUBITAK) under Grant 120E401.

References

1 SAMSUNG Research (2020). Samsung's 6G white paper lays out the company's vision for the next generation of communications technology, July 2020. https://research.samsung.com/next-generation-communications (accessed 21 June 2022).
2 Rajatheva, N., Atzeni, I., Bjornson, E. et al. (2020). White paper on broadband connectivity in 6G. April 2020. http://arxiv.org/abs/2004.14247.
3 Basar, E., Di Renzo, M., De Rosny, J. et al. (2019). Wireless communications through reconfigurable intelligent surfaces. *IEEE Access* 7 116753–116773.
4 Wu, Q., Zhang, S., Zheng, B. et al. (2021). Intelligent reflecting surface aided wireless communications: a tutorial. *IEEE Trans. Commun.* 69 (5): 3313–3351. https://doi.org/10.1109/TCOMM.2021.3051897.
5 Basar, E. (2019). Reconfigurable intelligent surface-based index modulation: a new beyond MIMO paradigm for 6G. April 2019. arXiv:1904.06704.
6 Yildirim, I., Kilinc, F., Basar, E., and Alexandropoulos, G.C. (2021). Hybrid RIS-empowered reflection and decode-and-forward relaying for coverage extension. *IEEE Commun. Lett.* 25 (5): 1692–1696. https://doi.org/10.1109/LCOMM.2021.3054819.

7 Yildirim, I., Uyrus, A., and Basar, E. (2021). Modeling and analysis of reconfigurable intelligent surfaces for indoor and outdoor applications in future wireless networks. *IEEE Trans. Commun.* 69 (2): 1290–1301. https://doi.org/10.1109/TCOMM.2020.3035391.

8 Arslan, E., Yildirim, I., Kilinc, F., and Basar, E. (2021). Over-the-air equalization with reconfigurable intelligent surfaces. June 2021. https://arxiv.org/abs/2106.07996.

9 Basar, E. and Yildirim, I. (2021). Reconfigurable intelligent surfaces for future wireless networks: a channel modeling perspective. *IEEE Wireless Commun.* 28 (3): 108–114. https://doi.org/10.1109/MWC.001.2000338.

10 Di Renzo, M., Danufane, F.H., Xi, X. et al. (2020). Analytical modeling of the path-loss for reconfigurable intelligent surfaces - anomalous mirror or scatterer? *IEEE International Workshop on Signal Processing Advances in Wireless Communications (SPAWC)*, pp. 1–5. https://doi.org/10.1109/SPAWC48557.2020.9154326.

11 Garcia, J.B., Sibille, A., and Kamoun, M. (2020). Reconfigurable intelligent surfaces: bridging the gap between scattering and reflection. *IEEE J. Sel. Areas Commun.* 38 (11): 2538–2547.

12 Ellingson, S.W. (2021). Path loss in reconfigurable intelligent surface-enabled channels. November 2021. http://arxiv.org/abs/1912.06759.

13 Ozdogan, O., Bjornson, E., and Larsson, E.G. (2020). Intelligent reflecting surfaces: physics, propagation, and pathloss modeling. *IEEE Wireless Commun. Lett.* 9 (5): 581–585.

14 Tang, W., Chen, M.Z., Chen, X. et al. (2021). Wireless communications with reconfigurable intelligent surface: path loss modeling and experimental measurement. *IEEE Trans. Wireless Commun.* 20 (1): 421–439. https://doi.org/10.1109/TWC.2020.3024887.

15 Danufane, F.H., Di Renzo, M., de Rosny, J., and Tretyakov, S. (2021). On the path-loss of reconfigurable intelligent surfaces: an approach based on green's theorem applied to vector fields. *IEEE Trans. Commun.* 69 (8): 5573–5592. https://doi.org/10.1109/TCOMM.2021.3081452.

16 Björnson, E. and Sanguinetti, L. (2020). Power scaling laws and near-field behaviors of massive MIMO and intelligent reflecting surfaces. *IEEE Open J. Commun. Soc.* 1: 1306–1324. https://doi.org/10.1109/OJCOMS.2020.3020925.

17 Najafi, M., Jamali, V., Schober, R., and Poor, H.V. (2021). Physics-based modeling and scalable optimization of large intelligent reflecting surfaces. *IEEE Trans. Commun.* 69 (4): 2673–2691. https://doi.org/10.1109/TCOMM.2020.3047098.

18 Gradoni, G. and Di Renzo, M. (2021). End-to-end mutual coupling aware communication model for reconfigurable intelligent surfaces: an electromagnetic-

compliant approach based on mutual impedances. *IEEE Wireless Commun. Lett.* 10 (5): 938–942. https://doi.org/10.1109/LWC.2021.3050826.

19 Qian, X. and Di Renzo, M. (2021). Mutual coupling and unit cell aware optimization for reconfigurable intelligent surfaces. *IEEE Wireless Commun. Lett.* 10 (6): 1183–1187. https://doi.org/10.1109/LWC.2021.3061449.

20 Abrardo, A., Dardari, D., Di Renzo, M., and Qian, X. (2021). MIMO interference channels assisted by reconfigurable intelligent surfaces: mutual coupling aware sum-rate optimization based on a mutual impedance channel model. February 2021. https://arxiv.org/abs/2102.07155.

21 Sun, Y., Wang, C.-X., Huang, J., and Wang, J. (2020). A 3D non-stationary channel model for 6G wireless systems employing intelligent reflecting surface. *International Conference on Wireless Communications and Signal Processing (WCSP)*, pages 19–25, Nanjing, China, October 2020. https://doi.org/10.1109/WCSP49889.2020.9299848.

22 Sun, Y., Wang, C.-X., Huang, J., and Wang, J. (2021). A 3D non-stationary channel model for 6G wireless systems employing intelligent reflecting surfaces with practical phase shifts. *IEEE Trans. Cognit. Commun.* 7 (2): 496–510. https://doi.org/10.1109/TCCN.2021.3075438.

23 Sun, G., He, R., Ma, Z. et al. (2021). A 3D geometry-based non-stationary MIMO channel model for RIS-assisted communications. *IEEE Vehicular Technology Conference (VTC2021-Fall)*, pages 1–5, Norman, OK, USA. https://doi.org/10.1109/VTC2021-Fall52928.2021.9625374.

24 Jiang, H., Ruan, C., Zhang, Z. et al. (2021). A general wideband non-stationary stochastic channel model for intelligent reflecting surface-assisted MIMO communications. *IEEE Trans. Wireless Commun.* 20 (8): 5314–5328. https://doi.org/10.1109/TWC.2021.3066806.

25 Xiong, B., Zhang, Z., Jiang, H. et al. (2021). A statistical MIMO channel model for reconfigurable intelligent surface assisted wireless communications. *IEEE Trans. Commun. (Earl Access)* 1. https://doi.org/10.1109/TCOMM.2021.3129926.

26 Basar, E. and Yildirim, I. (2020). SimRIS channel simulator for reconfigurable intelligent surface-empowered mmWave communication systems. *Proceedings of IEEE Latin-American Conference Communications (LATINCOM 2020).*

27 Basar, E., Yildirim, I., and Kilinc, F. (2021). Indoor and outdoor physical channel modeling and efficient positioning for reconfigurable intelligent surfaces in mmWave bands. *IEEE Trans. Commun.* 69 (12): 8600–8611. https://doi.org/10.1109/TCOMM.2021.3113954.

28 Kilinc, F., Yildirim, I., and Basar, E. (2021). Physical channel modeling for RIS-empowered wireless networks in sub-6 GHz bands. November 2021. https://arxiv.org/abs/2111.01537.

29 3GPP TR 38.901 V16.1.0 - Study on channel model for frequencies from 0.5 to 100 GHz, December 2019.

30 Nayeri, P., Yang, F., and Elsherbeni, A.Z. (2018). *Reflectarray Antennas: Theory, Designs, and Applications*. USA: Wiley.

31 5G channel model for bands up to 100 GHz, October 2016. http://www.5gworkshops.com/5GCMSIG_White%20Paper_r2dot3.pdf (accessed 21 June 2022).

32 3GPP TR 36.873 V12.7.0 - Study on 3D channel model for LTE, December 2017.

33 Kyösti, P. (2008). WINNER II channel models. *IST-4-027756 WINNER II D1.1.2 V1.2*, February 2008.

34 Hou, T., Liu, Y., Song, Z. et al. (2019). MIMO assisted networks relying on large intelligent surfaces: a stochastic geometry model. October 2019. https://arxiv.org/abs/1910.00959.

8

Intelligent Reflecting Surfaces (IRS)-Aided Cellular Networks and Deep Learning-Based Design

Taniya Shafique[1], Amal Feriani[1], Hina Tabassum[2], and Ekram Hossain[1]

[1]*Department of Electrical and Computer Engineering, University of Manitoba, Manitoba, Winnipeg, Canada*
[2]*Department of Electrical Engineering and Computer Science, York University, Ontario, Toronto, Canada*

8.1 Introduction

Intelligent reflecting surfaces (IRSs) are emerging as a cost-effective solution to enhance the energy efficiency, security, coverage, and data rates of the future-generation wireless communication systems [1]. A given IRS consists of many antenna elements (a.k.a. metasurfaces) [2]. The intelligent functionality of each element includes reflection, refraction, transmittance, and absorption [3]. These functionalities can be used all together or in separate based on the application requirement and objectives. Since meta-surfaces contain low-cost polymer diode/switch [2, 4], IRSs are energy-efficient than conventional relays. Unlike conventional relays that require active transmission and reception, the IRSs do not require any additional radio channel/frequency for signal transmission or reception which makes them cost-effective.

Along another note, IRS technology is one of the potential low-cost solutions that can extend the communication system performance by supporting more users with three possible transmission modes, i.e. (i) *Joint Transmission*: in which a user receives the IRS signals combined with the direct signal from the base-stations (BSs), (ii) *IRS-only Transmission*: in which a user receives only IRS transmissions and the direct transmissions get blocked, and (iii) *Direct Transmission*: in which a user gets served only through direct transmissions.

Selecting the mode of transmission is a fundamental question in a large-scale IRS-assisted network with multiple users, BSs, and IRSs [5]. For instance, given the fact that the direct transmissions from BSs may be impacted by the presence of IRSs, it is important to study the performance of sole direct transmissions in

Intelligent Reconfigurable Surfaces (IRS) for Prospective 6G Wireless Networks, First Edition.
Edited by Muhammad Ali Imran, Lina Mohjazi, Lina Bariah, Sami Muhaidat,
Tie Jun Cui, and Qammer H. Abbasi.

a large-scale IRS-assisted network. Since in most cases, the contribution of IRS is incremental compared to direct transmissions, IRS transmissions are relatively beneficial for users with no access to direct transmissions instead of users having access to both IRS and direct transmissions. Therefore, in this chapter, we advocate the IRS-only and direct transmission modes instead of joint mode of transmission. Also, joint transmission may suffer from incoherent multi-path delays rendering sophisticated synchronization, detection, and co-phasing techniques necessary.

8.2 Contributions

We maximize the achievable rate by optimizing the mode selection of users (i.e. considering direct transmission mode and indirect IRS-assisted transmission mode) and phase shifts of the nearest IRS of IRS-assisted users in a large-scale, multi-user, multi-BS, multi-IRS network. Given that the joint optimization problem is non-convex, we solve the problem using an alternating optimization algorithm. In particular, we decompose the original problem into two sub-problems, i.e. phase-shift optimization and mode selection. For mode selection, the user decides to associate to the mode that has highest achievable capacity. On the other hand, for phase shift-optimization, we consider semi-definite programming (SDP)-based optimization as well as proximal policy optimization (PPO) and Deep Deterministic policy gradient (DDPG) learning approaches. To date, most of the IRS-related optimization problems such as passive and active beamforming involve solving non-convex optimization problems through traditional alternating optimization resulting in sub-optimal solutions with high time complexity.

For both alternating optimization-based technique and deep reinforcement learning (DRL)-based technique, the computation of interference at the users with direct transmission is done in three ways: (i) *Random scenario*: assuming that all interfering IRSs have randomly distributed phase shifts, (ii) *Optimal scenario*: by optimizing the phase shifts of nearest IRS of the direct user such that the interference to direct transmission can be minimized, and (iii) *Worst-case scenario*: by optimizing the phase shifts to maximize the IRS-assisted transmission gain.

Numerical results show that the DDPG provides competitive results to traditional alternating optimization-based approach at significantly reduced inference time for a large-scale setup. Also, results demonstrate that the achievable capacity can be significantly enhanced when IRS channel and direct channel are used as separate communication modes. In addition, phase optimization to support direct mode also improves the system performance.

The chapter organization is as follows: Literature review and system model are provided in Sections 8.3 and 8.4, respectively. The problem formulation and solution approach are provided in Sections 8.5 and 8.6, respectively. Then, in Section 8.7, the numerical results are presented followed by conclusions in Section 8.8.

8.3 Literature Review

The authors in [6, 7] provided association with IRS-assisted communication. In [6], the authors derived the outage probability of a single BS and single-user scenario in the presence of Poisson distributed blockages. It is also assumed that a fraction of blockages are deployed with IRS surfaces. The communication between the BS and the user is possible in two ways. First, when the user has joint channel (i.e. direct channel and IRS-assisted channel) or when the user only has IRS-assisted channel due to blocked direct link. The authors provided different association schemes such as random association, closest IRS association, and all IRS association and derived the average outage probability for each association scheme. However, an IRS is used to extend the coverage to blind spot in multi-BS and multi-user downlink communication system in [7]. The authors solved the optimization problem that maximized the sum rate by joint optimization of IRS phase shifts, power allocation, and user association. The authors used iterative algorithm, e.g. alternating optimization, sequential fractional programming, and forward-reverse auction. Our proposed work is different from the [6, 7] because it has multiple BSs and multiple IRSs, that involves interference from all the IRSs and BSs.

8.3.1 Optimization

A plethora of research works has optimized IRS beamforming for a single-IRS network [2, 3, 8–12]. For instance, Wu and Zhang [3] maximized the total received signal power at the user by jointly optimizing the transmit beamforming at the BS and beamforming at the IRS considering a single IRS and single BS network. The authors provided two solution approaches for the joint optimization based on semi-definite relaxation (SDR) technique and alternating optimization. Abeywickrama et al. [8] maximized the rate by jointly optimizing the BS transmit beamforming and the IRS reflect beamforming for a single BS and single IRS setup. The authors provided an alternating optimization-based sub-optimal solution. Kammoun et al. [9] maximized the minimum user signal-interference-plus-noise ratio (SINR) under power constraint and beamforming using the algorithm based on projected gradient ascent. Huang et al. [2] maximized energy-efficiency and optimized transmit power allocations, IRS phase shifts constraint considering a single BS, single IRS, and multi-user setup. An iterative algorithm was developed on sequential fractional programming and conjugate gradient search. Rehman et al. [11] jointly optimized the active and passive beamforming in order to maximize spectral efficiency using vector approximate message passing for single IRS, single BS, and multi-user network.

The aforementioned research works considered IRS phase optimization for single IRS and single BS setup and the impact of the interference from IRSs was not considered. Recently, He et al. [10] maximized the total achievable rate by jointly optimizing the transmit power and the IRS phase shift matrix for a single source, single destination, and multiple IRS setup. The authors solved the optimization problem by transforming it to two multiple-ratio fraction programming sub-problems, which is further transformed into bi-convex problem. In addition, Omid et al. [13] optimized the active and passive beamforming considering a single IRS, multi-cell, and multi-user set-up. The optimization solution is obtained using low-overhead trellis-based solution. An SDR-based solution is used for benchmarking. The authors in [2, 9, 10] did not consider the direct link from the user to the BS, whereas [3, 8, 11–13] considered both direct and indirect channels. A summary is provided in Table 8.1.

8.3.2 Deep Learning

DRL methods have achieved tremendous success in solving complex and high-dimensional problems in different fields including wireless communication (e.g. [14, 15]). Notably, several DRL-based approaches are proposed in the literature for IRS-related beamforming problems. For instance, phase shift optimization is largely approached using DRL methods since the latter provides a faster and real-time alternative compared to classical optimization methods [16].

Learning the reflection coefficients of the IRS elements in single-input, single-output systems was studied in [17] to avoid training overhead. The proposed algorithm estimated the channel state using two orthogonal uplink pilot signals and a deep Q-network (DQN) [18] is trained to output the optimal reflectors such that the achievable rate is maximized. Feng et al. proposed an approach based on Deep Deterministic Policy Gradient (DDPG) [19] for passive beamforming in a multiple-input single-output (MISO) system and showed a better performance compared to the fixed-point iteration algorithm and closer results to an SDR upper bound [20]. The joint active and passive beamforming is also investigated in several works. As an example, a unified DRL algorithm that outputs the transmit beamforming and the phase shift simultaneously is proposed for a multi-user MISO system [21]. Alternatively, DRL can be applied to improve the physical layer security of IRS-assisted systems against eavesdroppers or active jammers [22]. For instance, an RL policy is learned to minimize the secrecy rate while maintaining the receivers' quality-of-service (QoS) by jointly

Table 8.1 Optimization literature review

References	Objective	Variables	System model	Solution technique
Wu and Zhang [3]	Received signal power maximization	Transmit beamforming and IRS phase shifts	Single BS, single IRS, Multi-user, both direct and IRS link	Joint optimization using SDR and alternating optimization
Abeywickrama et al. [8]	Rate Maximization	Transmit beamforming and phase shifts, both direct and IRS link	Single multi-antenna BS, single user	Alternating optimization
Kammoun et al. [9]	Minimum user SINR maximization	Transmit power constraint and IRS phase shifts	Single multi antenna BS, single IRS, Multi-user, no direct link	Algorithm based on projected gradient ascent
Huang et al. [2]	Energy efficiency maximization	Transmit power, IRS phase shifts and individual link budget	Single multi antenna BS, single IRS, Multi-user, no direct link	Algorithm based on sequential fractional programming and conjugate gradient
Rehman et al. [11]	Spectral efficiency maximization	transmit beamforming and IRS phase shifts	Single multi-antenna BS, Single IRS, multi-user, both direct and IRS link	Vector approximate message passing
He et al. [10]	Rate maximization	Transmit power and IRS phase shift	Single BS, single user, multiple IRS, no direct link	multiple-ratio fraction programming transformed into biconvex problem
Omid et al. [13]	Multiuser interference power minimization	Beamforming at the BS and IRS phase shifts	Multi-BS, single IRS, multi-user, both direct and IRS link	Low-overhead trellis-based solution

optimizing the beamformers at the transmitter and the IRS. Furthermore, the energy efficiency of IRS-assisted networks is investigated using DRL [23], where the transmit power allocation and IRS configuration are computed to maximize the energy efficiency.

DRL techniques are attractive since (i) they can be trained without a perfect knowledge of the system dynamics which can be detrimental for certain optimization techniques, (ii) they offer a better alternative in terms of computational cost and decision time, and (iii) they can more robust to changes in the environment compared to other optimization techniques [16]. However, the training of RL algorithms can suffer from slow convergence or/and require a large amount of interaction experience to achieve well-performing solutions.

8.4 System Model

In this section, we present the network, transmission signal, and interference models for users who are served by direct BS transmissions and those served by IRS-assisted transmissions.

8.4.1 Transmission Model

We consider a two-tier downlink cellular network consisting of IRS surfaces, single antenna BSs, and single antenna users. The locations of the BSs and IRSs follow a two-dimensional (2D) homogeneous Poisson Point Process (PPP) denoted as Φ_{B} with intensity λ_B and Φ_{R} with intensity λ_R, respectively. We assume that the IRSs are deployed at a fixed height H_{R} and are equipped with N elements each, whereas all the BSs have a fixed height H_B. We assume that there are two different types of users in the considered multi-BS and multi-IRS network, i.e.

- **Direct users:** who are served by direct BS transmissions, and
- **IRS-assisted users:** who are served by indirect IRS-assisted transmissions.

We also consider $\mathcal{A}$ IRS-assisted users and $1 - \mathcal{A}$ direct users in the system. For direct transmission from the BS, the user is associated with the nearest BS. In the indirect IRS-assisted transmission mode, the user associates to the nearest IRS, and then, that nearest IRS associates to the nearest BS (as illustrated in Figure 8.1). The main goal is to provide service to the maximum number of users both through IRS link and direct link.

We assume that an IRS can relay information from only one BS to only one user at a predefined time/frequency resource to maintain orthogonality. We consider that the *direct* communication (i.e. BS to the given user) and *indirect* IRS-assisted communication (i.e. BS to IRS and IRS to the given user) share different frequency spectrum such that a BS can serve both the direct and indirect IRS-assisted users.

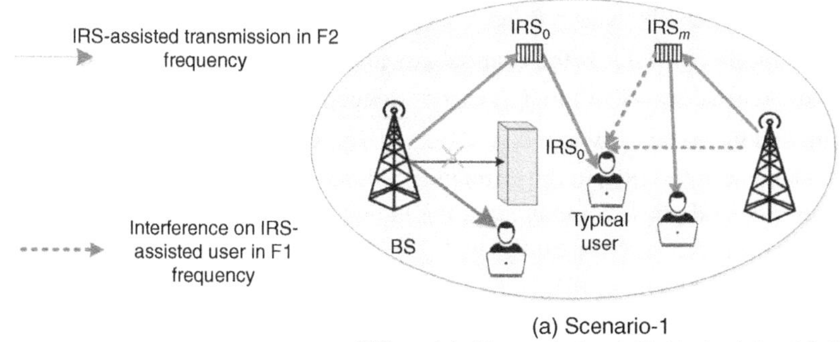

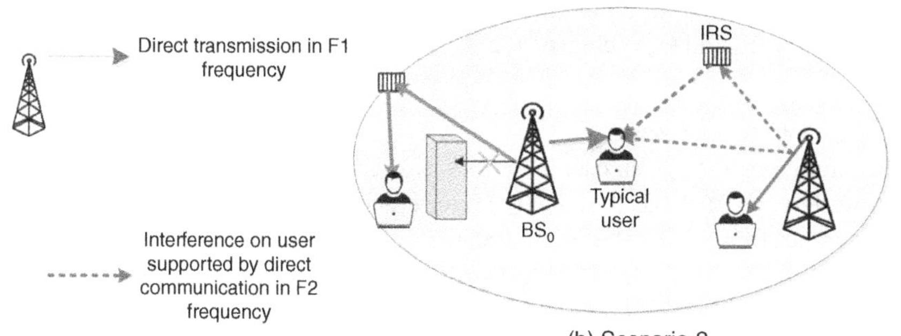

Figure 8.1 System model for direct and IRS-assisted communication in multi-IRS and multi-BS setup: (a) Scenario 1: when the user is connected with a BS through the nearest IRS and the direct link to the nearest BS is blocked, and (b) Scenario 2: when the user is connected with a nearest BS in the presence of a weak IRS link.

8.4.2 IRS-Assisted Transmission

8.4.2.1 Desired Signal Power

The signal power received at a given user from the nearest IRS (IRS_0) is given as [24, 25]:

$$S_R = p_t\ |\hat{\mathbf{g}}_{0,0}^H \Theta_0\ \hat{\mathbf{f}}_{0,j}|^2 = p_t \left| \sum_{n=1}^{N} r_{0,0_n}^{-\alpha/2} t_{0_n j}^{-\alpha/2} f_{0_n j}\, g_{0,0_n} e^{j\theta_{0_n}} \right|^2 \tag{8.1}$$

where p_t is the transmission power of the BSs in IRS-assisted mode, $g_{0,0_n} = |g_{0,0_n}| e^{-j\phi_{0,0_n}}$ is the Rayleigh fading channel[1] gain from the given user to the nth

1 The assumption of Rayleigh fading from BS to IRS and IRS to user link demonstrated the worst-case propagation scenario as in [26].

element of IRS$_0$, thus $\hat{g}_{0,0_n} = \beta(r_{0,0_n})^{-\alpha/2} g_{0,0_n}$, where $\alpha \geq 2$ represents the path-loss exponent, $\beta = \left(\frac{4\pi f_c}{c}\right)^{-2}$ is the channel power gain on free-space path-loss model at a reference distance of 1 m, f_c is carrier frequency, and c represents the speed of light, and $\hat{\mathbf{g}}_{0,0} \in \mathbb{C}^{1\times N}$, where $r_{0,0_n} = \sqrt{\ell_{0_n}^2 + H_R^2}$ represents the distance from the nth element of the IRS$_0$ to the given user. Note that $|g_{0,0_n}|$ and $\phi_{0,0_n}$ represent the magnitude and phase component of the fading channel from the nth element of IRS$_0$ to the given receiver. Similarly, $f_{0_n,j} = |f_{0_n,j}|e^{-j\psi_{0_n,j}}$ is the fading channel gain from the nth element of IRS$_0$ to jth BS, thus $\hat{f}_{0_n,j} = \beta(t_{0_n,j})^{-\alpha/2} f_{0_n,j}$ and $\hat{\mathbf{f}}_{0,j} \in \mathbb{C}^{N\times 1}$, where $t_{0_n,j}$ represents the distance from the nth element of IRS$_0$ to the given user. Note that $|f_{0_n,j}|$ and $\psi_{0_n,j}$ represent the magnitude and phase component of the fading channel from jth BS to nth element of IRS$_0$. Finally, Θ_0 denotes the phase shift of the IRS$_0$ and $\Theta_0 = \text{diag}\{e^{j\theta_{0_1}}, e^{j\theta_{0_2}}, \ldots, e^{j\theta_{0_N}}\}$.

8.4.2.2 Interference Power

The interference at a given user in the IRS-assisted mode is composed of two parts (i) interference from the BSs, and (ii) interference from the IRSs. The aggregate interference from all the BSs (excluding the nearest BS) is given as follows:

$$I_B = \sum_{j\in\Phi_B\backslash 0} p_t\beta^2|h_j|^2 d_j^{-\alpha} = \sum_{j\in\Phi_B\backslash 0} p_t\beta^2|h_j|^2(\ell_j^2 + H_B^2)^{-\alpha/2}. \tag{8.2}$$

On the other hand, the aggregate interference from the IRSs can be modelled as follows:

$$I_R = \sum_{j\in\Phi_B}\sum_{m=1}^{M\backslash 0} p_t\,|\hat{\mathbf{g}}_{0,m}^H\Theta_m\hat{\mathbf{f}}_{m,j}|^2 = \sum_{j\in\Phi_B}\sum_{m=1}^{M\backslash 0} p_t\left|\sum_{n=1}^{N} r_{0,m_n}^{-\alpha/2} t_{m_n,j}^{-\alpha/2} f_{m_n,j}\, g_{0,m_n}\, e^{j\theta_{m_n}}\right|^2 \tag{8.3}$$

where $g_{0,m_n} = |g_{0,m_n}|e^{-j\phi_{0,m_n}}$ is the fading channel gain from the given user to the nth element of IRS m, thus $\hat{g}_{0,m_n} = \beta(r_{0,m_n})^{-\alpha/2} g_{0,m_n}$, and $\hat{\mathbf{g}}_{0,m} \in \mathbb{C}^{1\times N}$, where $r_{0,m_n} = \sqrt{\ell_{m_n}^2 + H_R^2}$ represents the distance from nth element of mth IRS to the given user. Note that $|g_{0,m_n}|$ and ϕ_{0,m_n} represent the magnitude and phase component of the fading channel from nth element of mth IRS to the given receiver. Similarly, $f_{m_n,j} = |f_{m_n,j}|e^{-j\psi_{m_n,j}}$ is the fading channel gain from the nth element of IRS m to jth BS, thus $\hat{f}_{m_n,j} = \beta(t_{m_n,j})^{-\alpha/2} f_{m_n,j}$ and $\hat{\mathbf{f}}_{m,j} \in \mathbb{C}^{N\times 1}$, where $t_{m_n,j}$ represents the distance from nth element of mth IRS to the given user. Note that $|f_{m_n,j}|$ and $\psi_{m_n,j}$ represent the magnitude and phase component of the fading channel from jth BS to nth element of mth IRS. Finally, Θ_m denotes the phase shift of the IRS and $\Theta_m = \text{diag}\{e^{j\theta_{m_1}}, e^{j\theta_{m_2}}, \ldots, e^{j\theta_{m_N}}\}$.

8.4.3 Direct Transmission

8.4.3.1 Desired Signal Power

The signal power from the desired BS to the given user is

$$S_D = \hat{p}_t \beta^2 |h_0|^2 d_0^{-\alpha} = \hat{p}_t \beta^2 |h_0|^2 (\ell_0^2 + H_B^2)^{-\alpha/2} \tag{8.4}$$

where $\hat{p}_t$ is the transmission power of the BSs in direct mode, h_0 and d_0 are the small-scale fading channel and the distance between the given user to the nearest BS, respectively.

8.4.3.2 Interference Power

The interference at a given user in the direct mode is composed of two parts (i) interference from the BSs, and (ii) interference from the IRSs. The aggregate interference from the BSs (excluding the desired BS) is given as follows:

$$\hat{I}_B = \sum_{j \in \Phi_B \backslash 0} \hat{p}_t \beta^2 |h_j|^2 d_j^{-\alpha} = \sum_{j \in \Phi_B \backslash 0} \hat{p}_t \beta^2 |h_j|^2 (\ell_j^2 + H_B^2)^{-\alpha/2} \tag{8.5}$$

where h_j and d_j are the small-scale fading channel and the distance between the given user to the nearest BS, respectively. On the other hand, the aggregate interference from all IRSs can be modelled as follows:

$$\hat{I}_R = \sum_{j \in \Phi_B} \sum_{m=1}^{M} \hat{p}_t \, |\hat{\mathbf{g}}_{0,m}^H \Theta_m \, \hat{\mathbf{f}}_{m,j}|^2 = \sum_{j \in \Phi_B} \sum_{m=1}^{M} \hat{p}_t \left| \sum_{n=1}^{N} r_{0,m_n}^{-\alpha/2} t_{m_n j}^{-\alpha/2} f_{m_n j} g_{0,m_n} e^{j\theta_{m_n}} \right|^2. \tag{8.6}$$

In (8.6), we split the aggregate interference from all IRSs to the nearest IRS and the other interfering IRSs.

8.4.4 SINR and Achievable Rate

The SINR of the user associated with nearest IRS is given as follows:

$$\gamma_{\text{ID}} = \frac{S_R}{I_B + I_R + N_0}. \tag{8.7}$$

Similarly, the SINR of the user with the direct BS communication is given as follows:

$$\gamma_D = \frac{S_D}{\hat{I}_B + \hat{I}_R + N_0}. \tag{8.8}$$

The achievable data rate for the IRS-assisted user R_{ID} and the direct user R_{D} can be as follows:

$$R_{\text{ID}} = B \log_2(1 + \gamma_{\text{ID}}), \quad R_{\text{D}} = B \log_2(1 + \gamma_D). \tag{8.9}$$

The overall achievable data rate of a user is derived as follows:

$$R = (1 - \mathcal{A})R_{\mathrm{ID}} + \mathcal{A}R_{\mathrm{D}} \tag{8.10}$$

where $\mathcal{A}$ represents the user association with the IRS-assisted indirect communication mode defined as follows:

$$\mathcal{A} = \begin{cases} 1 \text{ if user}_0 \text{ is associated to IRS-assisted indirect link} \\ 0 \text{ otherwise} \end{cases}. \tag{8.11}$$

8.5 Problem Formulation

We consider maximizing the overall rate of a user by jointly optimizing its mode of transmission (or its association) $\mathcal{A}$ and phase shifts of its nearest IRS deployed, i.e.

$$\begin{aligned} \mathbf{P1}: \quad & \max_{\Theta_n \forall n \in \{1,N\},\ \mathcal{A}} \quad R = (1 - \mathcal{A})R_{\mathrm{ID}} + \mathcal{A}R_{\mathrm{D}} \\ \text{s.t.} \quad & \mathbf{C1} \quad \mathcal{A} \in \{0,1\} \\ & \mathbf{C2} \quad \theta_n \in [-\pi, \pi), \forall n = 1, \dots, N. \end{aligned} \tag{8.12}$$

Here the constraint (C1) represents the typical user association which is a binary integer. In (C2), we assume that the phase shifts varies continuously in $[-\pi, \pi)$ as in [3]. Finally, we consider that each user can associate with one IRS. The optimization problem in **P1** is a mixed integer (binary) non-linear programming (MINLP) problem, which is clearly non-convex. In this chapter, we propose a two-stage optimization algorithm comprised of

- **Phase optimization stage:** The phase shifts of the nearest IRS to the user are computed via classical optimization and DRL techniques.
- **Mode selection stage:** After obtaining the phase shifts, the users will decide on choosing the direct or IRS-assisted mode of transmission based on highest achievable rate. Based on the joint capacity in (8.10), users choose to associate with the IRS-assisted link when

$$\mathcal{A} = \begin{cases} 1 \text{ if } C_{\mathrm{ID}} > C_{\mathrm{D}} \\ 0 \text{ otherwise} \end{cases}. \tag{8.13}$$

In the next sections, we detail the proposed phase shift optimization algorithms using optimization and DRL theory.

8.6 Phase Shifts Optimization

In this section, we propose two different solutions for the phase shift optimization of the nearest IRS. First, we present the optimization-based approach and next we present the DRL-based.

8.6.1 Optimization-based Approach

In this section, we provide the IRS phase shift optimization solution approach to maximize the capacity of IRS-assisted user R_{ID} first and then to maximize the capacity for the direct user R_{D}.

(i) Capacity optimization of IRS-assisted user:

$$\begin{aligned} \textbf{P1:} \quad & \max_{\Theta_n \forall n \in \{1,N\}} \quad R_{\text{ID}} = B \log_2 \left(1 + \frac{S_R}{I_{\text{B}} + I_R + N_0}\right) \\ \text{s.t.} \quad & \textbf{C1} \quad \theta_n \in [-\pi, \pi), \forall n = 1, \dots, N. \end{aligned} \tag{8.14}$$

Here, we note that S_{R_0} in the numerator is only function of θ_0; therefore, we can ignore the logarithm and constants for simplification. Now, the problem can then be formulated to maximize the received signal power in Eq. (8.1) and obtain the optimal phase-shifts:

$$\begin{aligned} \textbf{P2:} \quad & \max_{\theta_n, \forall n} \text{S}_{\text{R}} = p_t \, |\hat{\mathbf{g}}_{0,0}^H \Theta_0 \, \hat{\mathbf{f}}_{0,j}|^2 \\ \text{s.t.} \quad & \textbf{C1} \quad \theta_{0_n} \in [-\pi, \pi), \forall n = 1, \dots, N. \end{aligned} \tag{8.15}$$

Since p_t is independent of the optimization variable, we can discard this term. Now, we transform the objective to equivalent matrix form as $|\tilde{\mathbf{g}}_{0,0}^H \mathbf{B}_0 \tilde{\mathbf{f}}_{0,j}|^2$, where $\tilde{\mathbf{g}}_{0,0} \in \mathbb{R}^{1 \times N}$, $\tilde{\mathbf{f}}_{0,j} \in \mathbb{R}^{N \times 1}$, $\mathbf{B}_0 = \text{diag}\{e^{j\beta_{0_1}}, e^{j\beta_{0_2}}, \dots, e^{j\beta_{0_N}}\}$, and $\beta_{0_n} = \theta_{0_n} - \phi_{0,0_n} - \psi_{0_n,j}$. Since the objective function is a scalar, we can convert absolute square to norm square as $\|\tilde{\mathbf{g}}_{0,0}^H \, \mathbf{\Delta} \, \tilde{\mathbf{f}}_{0,j}\|^2$. Finally, defining $\mathbf{v} = [v_1, \dots, v_n]^H$, where $v_n = e^{j\beta_{0_n}}, \forall n$, and $\Phi = \text{diag}(\tilde{\mathbf{g}}_{0,0}^H)\tilde{\mathbf{f}}_{0,j}$, we reformulate $\|\tilde{\mathbf{g}}_{0,0}^H B_0 \, \tilde{\mathbf{f}}_{0,j}\|^2 = \|\mathbf{v}^H \Phi\|^2$. The problem **P2** can thus be reformulated as follows:

$$\begin{aligned} \textbf{P3:} \quad & \max_{\mathbf{v}} \; \mathbf{v}^H \Phi \Phi^H \mathbf{v} \\ \text{s.t.} \quad & \textbf{C1} \quad |v_n|^2 = 1, \forall n = 1, \dots, N. \end{aligned} \tag{8.16}$$

P3 is non-convex quadratically constrained quadratic program (QCQP) in the homogeneous form with the rank one constraint. Now, defining $\mathbf{V} = \mathbf{v}\mathbf{v}^H$, we apply SDR to relax the constraint as follows:

$$\begin{aligned} \textbf{P4}: \quad & \max_{\mathbf{V}} \; \text{Tr}\,(\Phi \Phi^H \mathbf{V}) \\ \text{s.t.} \quad & \textbf{C1} \quad \mathbf{V}_{n,n} = 1, \forall n = 1, \dots, N, \quad \mathbf{V} \geq 0, \\ & \textbf{C2} \quad \mathbf{V} > 0 \end{aligned} \tag{8.17}$$

Since the problem is now transformed in to a convex semi-definite program (SDP), we solve it for the optimal value using CVX by following the similar approach provided in [3].

(ii) Capacity optimization of direct user:

$$\begin{aligned} \textbf{P1:} \quad & \max_{\Theta_{0_n} \forall n \in \{1,N\}} \quad R_{\text{D}} = B \log_2 \left(1 + \frac{S_{D_0}}{\hat{I}_B + \hat{I}_R + N_0}\right) \\ \text{s.t.} \quad & \textbf{C1} \quad \theta_{0_n} \in [-\pi, \pi), \forall n = 1, \dots, N, \\ & \textbf{C2} \quad \mathbf{V} > 0 \end{aligned} \tag{8.18}$$

We realize that the IRS the only function of theta is $\hat{I}_R$, which can be rewritten as $\hat{I}_R = S_R + \hat{I}_{R\backslash 0}$ using *assumption 1*. Similar to (8.14), we can ignore the logarithm and constants for simplification and now the problem can then be formulated to minimize the received signal power (given in (8.1)) and obtain the optimal phase-shifts as follows:

$$\begin{aligned} \textbf{P2:}\ & \min_{\theta_{0_n},\forall n}\ \mathrm{S_R} = p_t\ |\hat{\mathbf{g}}_{0,0}^H \Theta_0\ \hat{\mathbf{f}}_{0,j}|^2 \\ \text{s.t.}\ & \theta_{0_n} \in [-\pi,\pi), \forall n = 1,\dots,N. \end{aligned} \tag{8.19}$$

Using the transformation in (8.19), this can be equivalently written as a follows:

$$\begin{aligned} \textbf{P3:}\ & \max_{\mathbf{v}}\ -\mathbf{v}^H \Phi \Phi^H \mathbf{v} \\ \text{s.t.}\ & |v_n|^2 = 1, \forall n = 1,\dots,N \end{aligned} \tag{8.20}$$

which is similar to (8.16) and can be solved similarly. Now, we can substitute Θ_0 obtained from (8.17) and (8.20) to maximize capacity of direct and IRS-assisted modes.

8.6.2 DRL-based Approach

8.6.2.1 Backgound

The standard reinforcement learning (RL) consists of an agent learning to maximize its expected cumulative rewards through interacting with an environment. At each interaction or time step t, given an observation s_t, the agent chooses an action a_t and receives an instantaneous reward r_t, and the system transits to a new state s_{t+1}. The tuples $\{s_t, a_t, r_t, s_{t+1}\}$ constitute the agent experiences used for learning the optimal policy $\pi : S \mapsto \mathcal{P}(A)$. This learning problem is often modelled as a Markov Decision Process (MDP) [27], described by the tuple (S, A, P, R, η), when the state space is fully observable. S and A define the state and the action spaces, respectively; $P := S \times A \mapsto [0,1]$ denotes the transition probability function, $R := S \times A \mapsto R$ is the reward function, and $\eta \in [0,1]$ is a discount factor that trades-off the immediate and upcoming rewards. We define the agent's expected return as the sum of the discounted future rewards $R_t = \mathbb{E}\left[\sum_{i=t}^{T} \eta^{(i-t)} R(s_i, a_i) | a_i \sim \pi(.|s_i)\right]$. Another well-known function used to measure the agent's returns is the action-value function Q. It measures the expected accumulated rewards after executing an action a_t at a state s_t and following the policy π thereafter:

$$Q^{\pi}(s,a) = \mathbb{E}\left[\sum_{t=0}^{\infty} \eta^t R(s_t, a_t, s_{t+1}) | a_t \sim \pi(.|s_t), s_0 = s, a_0 = a\right]$$

In this part, we will detail the DRL agent aiming to maximize the capacity of the IRS-assisted user. On the contrary to the SDR method presented above, no simplifications are done for the DRL approach. We start by formulating the problem as an MDP and then we present the DRL algorithms to learn the policy.

8.6.2.2 MDP Formulation

- **States**: at each timestep t, the observation vector consists of the received SINR from the direct and indirect communication modes $\gamma_{\mathrm{ID}}(t-1)$ and $\gamma_D(t-1)$ and the phase shifts $\theta(t-1)$ of the previous timestep $t-1$. Thus, the state space is continuous and has $N+2$ elements:

$$s_t = \left[\gamma_{\mathrm{ID}}(t-1), \gamma_D(t), \theta_0(t)\right].$$

- **Actions**: after receiving a state s_t, the agent selects the vector of phase shifts based on the obtained information. The action vector $a_t \in \mathbb{R}^N$ is given by

$$a_t = \left[\theta_1, \ldots, \theta_N\right].$$

- **Rewards**: The DRL agent aims to maximize the user indirect/direct SINR. Hence, the reward function is given by $r_t = R_{\mathrm{ID}}$ for indirect communication mode and $r_t = R_D$ for direct communication path.

8.6.2.3 Training Procedure

Since our DRL problem involves a continuous action space, we consider (i) DDPG [19] where the agent learns a deterministic policy and (ii) PPO [28] which produces a stochastic policy. DDPG is an off-policy actor critic method and PPO is an on-policy algorithm. The main disadvantage of PPO compared to DDPG is its sample inefficiency. We refer the interested reader to [29] for an overview of DRL methods. In what follows, we provide details about these two algorithms.

8.6.2.4 Proximal Policy Optimization (PPO)

This is a policy optimization algorithm where the objective is to directly search for the optimal policy. The policy is often parametrized with a deep neural network with parameters ϕ. The optimal parameters are computed by performing a gradient ascent on the agent's expected long-term reward J as in

$$J^{\pi} = \mathbb{E}\left[\sum_{t=0}^{H} \eta^t R(s_t, a_t) | s_0 \sim \rho_0, a_t \sim \pi_{\phi}\right] \tag{8.21}$$

where H is the episode horizon and ρ_0 is the initial state distribution. The policy gradients can be expressed as follows:

$$\nabla_{\phi} J^{\pi}(\phi) = \mathbb{E}\left[\sum_{t=0}^{H} \nabla \log \pi_{\phi}(a_t|s_t) A^{\pi_{\phi}}(s_t, a_t) | s_0 \sim \rho_0, a_t \sim \pi_{\phi}\right] \tag{8.22}$$

where $A^{\pi_{\phi}}$ is the advantage function under the policy π_{ϕ} defined as $A^{\pi_{\phi}} = Q^{\pi_{\phi}} - V^{\pi_{\phi}}$. $V^{\pi_{\phi}}$ and $Q^{\pi_{\phi}}$ denote the state-value and the state-action value functions, respectively. PPO algorithm solves the problem defined in (8.21) under a constraint requiring the new updated policy to be close to the old one. To do so, a new surrogate loss is introduced in PPO as follows:

$$L_{\mathrm{PPO}}(\phi) = \min_{\phi} \left(\frac{\pi_{\phi}(a|s)}{\pi_{\phi_k}(a|s)} A^{\pi_{\phi_k}}(s,a), \right.$$
$$\left. \mathrm{clip}\left(\frac{\pi_{\phi}(a|s)}{\pi_{\phi_k}(a|s)}, 1-\delta, 1+\delta \right) A^{\pi_{\phi_k}}(s,a) \right), \tag{8.23}$$

where 'clip' is a function used to keep the value of the ratio $\frac{\pi_{\phi}(a|s)}{\pi_{\phi_k}(a|s)}$ between $1-\delta$ and $1+\delta$ to penalize the new policy if it gets far from the old one.

8.6.2.5 Deep Deterministic Policy Gradient (DDPG)

Value-based DRL methods, notably Q–learning [30], rely on the Temporal Difference (TD) formulation [27] to learn the Q function as follows:

$$Q(s_t, a_t) = (1-\alpha)Q(s_t, a_t) + \alpha \left[r(s_t, a_t) + \gamma \max_{a'} Q(s_{t+1}, a') \right] \tag{8.24}$$

where α is a learning rate. The main drawback of this formulation is the max operator, which restricts its application to discrete action spaces. To overcome this limitation, Deterministic Policy Gradient (DPG) algorithm [31] proposes to learn a separate function μ to estimate arg $\max_{a'} Q(.,a')$. Hence, μ can be viewed as a deterministic policy that predicts the action with the highest Q value. This is why, the DPG algorithm is an actor-critic method where a Q-function (i.e. critic) and a policy (i.e. actor) μ are learned concurrently. DDPG method follows the same technique as in DPG except the critic and the actor are both modelled as deep neural networks with parameters ϕ and ψ, respectively. The critic parameters are learned by minimizing the Bellman error defined as follows:

$$\phi^* = \arg\min_{\phi} \frac{1}{2} \sum_{(s,a,r,s')} ||Q_{\phi}(s,a) - y||^2$$
$$\text{where } y = R(s,a) + \eta \hat{Q}_{\phi'}(s', \hat{\mu}_{\psi'}(s'))$$

where $\hat{Q}$ and $\hat{\mu}$ are copies of Q and μ, called target networks and are updated using a Polyak update with a parameter τ. The actor is optimized using gradient ascent with respect to the critic parameters θ to maximize the following objective:

$$J(\psi) = \mathbb{E}_{s \sim D} \left[Q_{\phi}(s, \mu_{\psi}(s)) \right]$$

in which D is a replay buffer $\{(s,a,r,s')\}$.

8.7 Numerical Results

8.7.1 Experimental Setup

The simulation parameters are listed herein. The heights of IRSs and BSs are set to $H_R = 5$ m, and $H_B = 10$ m, respectively. The transmission power for IRS-assisted mode and direct mode is $p_t = 1$ W and $\hat{p}_t = 31$ W, respectively. The total number of IRS elements is $N = 50$, BS and IRS intensity within the coverage area is $\lambda_B = 10^{-5}$, and $\lambda_R = 10^{-4}$, respectively. Also, path-loss exponent is $\alpha = 3$, and noise power spectral density is $N_0 = 10^{-10}$ W/Hz. The results are obtained by averaging over 500 iterations for SDR method and 500 episodes with 100 time steps each for RL. The policy networks for both PPO and DDPG algorithms is a fully connected network with two hidden layers with 64 units in each layer. The training parameters for the PPO and DDPG algorithms are reported in Tables 8.2 and 8.3, respectively.

Table 8.2 PPO hyperparameters

Parameter	Value
Timesteps per update	100
Learning rate	$3e^{-4}$
Discount factor η	0.99
GAE λ	0.95
PPO epochs	10

Table 8.3 DDPG hyperparameters

Parameter	Value
Rollout steps	100
Actor learning rate	$1e^{-4}$
Critic learning rate	$1e^{-3}$
Discount factor η	0.99
Buffer size	50 000
Action noise std	0.1
Action noise distribution	Normal

8.7.2 Baselines

We consider the following baselines to evaluate the performance of DRL agent:

- **Theory baseline**: where the optimal IRS phase shifts for IRS-assisted mode is obtained by substituting the channel phases $\beta_{0_n,j}$ by zero $\forall n \in \{1, \dots, N\}$. It can be clearly seen from (8.17) that the maximum IRS-assisted signal power is obtained by taking $\beta_{0_n,j} = 0$ which maximizes the exponential term to unity in S_R of (8.17) [32]. The optimal scenario is not provided for the theory baseline since there exist no upper bound for nearest IRS that can maximize the direct mode capacity.
- **SDR baseline**: The alternating optimization-based techniques are performed using SDR optimization. The SDR-based phase shift optimization of IRS assisted mode is solved by following Section 8.6.1(i) and the SDR-based phase shift optimization for the nearest IRS of direct mode is solved using the steps explained in Section 8.6.1.

8.7.3 Results

To evaluate the considered methods, we consider two network configurations depending on the signal strength of the indirect communication mode. We denote by **weak (strong)** the network configuration with a dominant direct (indirect) path. In the strong configuration, we expect that the association probability to the indirect mode to be considerably high and vice versa for the weak configuration. Furthermore, we consider three interference scenarios to evaluate the performance of the different approaches:

- **Worst-case scenario**: considers that the phase shifts are optimized to maximize the IRS-assisted communication mode.
- **Random scenario**: assumes that the direct communication mode uses random phase shifts to compute the interference from the indirect path through the IRS. This scenario is generally applicable since the direct user does not have the flexibility to optimize the IRS phases to minimize interference and, hence, the direct user sees the phase shifts of all interfering IRSs including the nearest IRS as randomly distributed.
- **Optimal scenario**: suggests that the direct communication mode uses the optimal phase shifts obtained by minimizing the interference from the indirect path through the IRS. This scenario is applicable when the closest IRS to the direct user is not associated with any other user and can be configured by the direct user to minimize the direct user interference.

In the rest of this section, we compare the performance of DRL methods to the considered baselines. Figures 8.2, 8.3, and 8.4 illustrate the achieved capacity and the

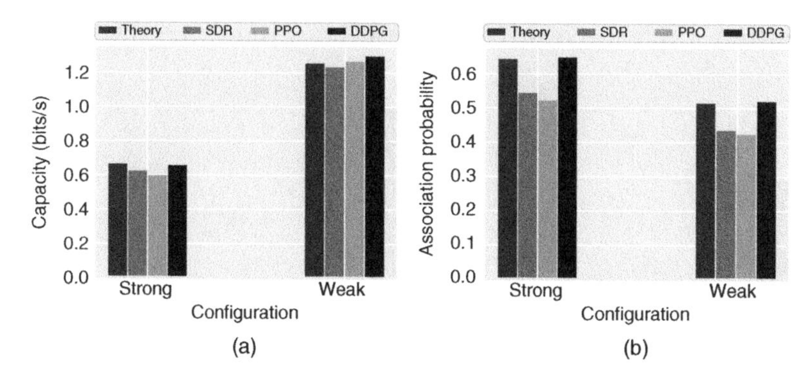

Figure 8.2 **Worst-case scenario**: Comparison of the user achievable rate and the association probability by optimizing IRS phases for the indirect communication mode and using it in the direct communication mode. (a) Achievable rate and (b) association probability.

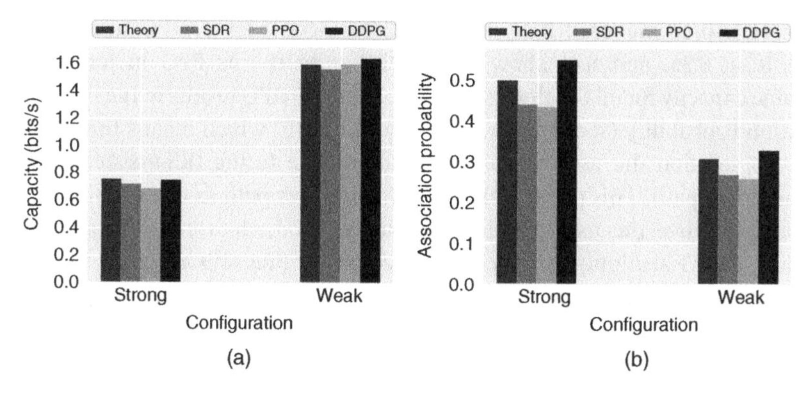

Figure 8.3 **Random scenario**: Comparison of the user achievable rate and the association probability by optimizing IRS phases for the indirect communication mode and using random phases in the direct communication mode. (a) Capacity and (b) association probability.

association probability for the different considered scenarios (worst-case, random, and optimal).

We start by investigating the impact of selecting the phase shifts of the direct communication mode. The results show that the worst-case scenario yields the lowest achievable capacity compared with other scenarios. The association probability is also higher for the worst-case scenario. This highlights the importance of interference computation for the direct mode when both indirect and direct modes are considered as separate propagation paths. Also, the association probability in the strong configuration is higher than when the indirect path is weak which is as expected.

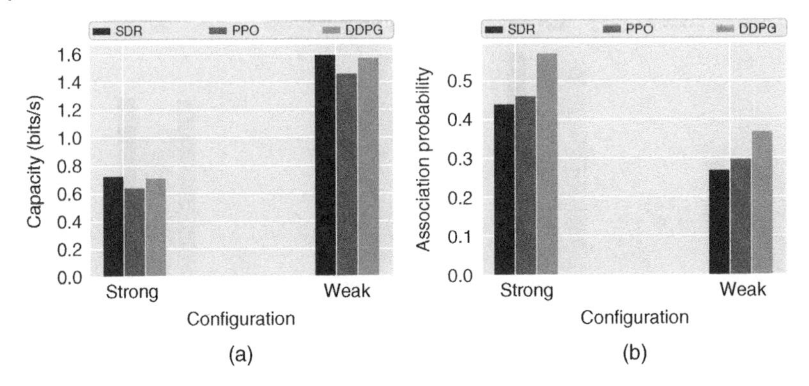

Figure 8.4 **Optimal scenario**: Comparison of the user-achievable rate and the association probability by optimizing IRS phases for both communication modes: (a) Achievable rate and (b) association probability.

In this chapter, we used two different DRL algorithms: DDPG and PPO. Figures 8.2a, 8.3a, and 8.4a show that DDPG outperforms PPO in terms of achievable capacity for all the three scenarios. In fact, DDPG results in the highest association probability (see Figures 8.2b, 8.3b, and 8.4b) which means that with the DDPG solution the user is most likely to connect to the IRS-assisted path than the direct path. This proves that using the indirect path as communication mode can improve the user's achieved capacity. For both worst and random scenarios, DDPG also outperforms the SDR baseline that fails to reach results comparable to the theoretical baseline in terms of achieved capacity. Also, for these same scenarios, DDPG meets the theoretical baseline closely, which shows the accuracy of DDPG for both capacity and association probability; however, other methods (SDR) fail to reach the theoretical result. Recall that the theory baseline provides an upper bound on the association probability for the worst-case scenario.

For the optimal scenario, two DRL agents are trained separately for the indirect and direct modes. The indirect phase optimization follows the same MDP formulation as in Section 8.6.2 and the direct phase optimization uses the same state and action definitions but the reward function is changed to minimized the indirect interference. We observe from Figure 8.4a that performance of DDPG is slightly lower than SDR. One plausible explanation for this slight decrease of the DDPG performance is because the DDPG agents were trained independently which motivate the joint optimization of the phase shifts and the association probability that we will study in a future work. Consequently, our results prove that DDPG algorithm is a competitive solution for phase shift optimization in large-scale systems.

Next, we further motivate the DRL-based approaches by comparing the inference time. The results in Table 8.4 demonstrate that the DRL methods

Table 8.4 Inference time in ms

Configuration	SDR	PPO/DDPG
Strong	950	0.4
Weak	940	0.5

(PPO/DDPG) significantly reduce the computational costs and allow real-time computation of phase shifts. This is one of the main motivation to adapt ML algorithm for wireless communication problems.

8.8 Conclusion

In this chapter, we formulated the problem to maximize the achievable rate by jointly optimizing the mode selection of users (i.e. considering direct transmission mode and indirect IRS-assisted transmission mode) and phase shifts of the nearest IRS of IRS-assisted users in a large-scale multi-user, multi-BS, multi-IRS network. Given that the joint optimization problem is non-convex, we solve the problem using an alternating optimization and DRL-based algorithms (DDPG and PPO). The numerical results show that more users can get service when IRS-assisted and direct modes are considered separately rather then jointly combining both links. Also, phase optimization for the direct user improves the achievable capacity by minimizing the interference coming from other IRSs. Moreover, the DDPG algorithm provides results close to the theoretical ones in comparison to other traditional iterative optimization techniques at a low inference time. An interesting extension of this work is the consideration of energy efficient user scheduling along the lines of [33].

References

1 Wu, Q. and Zhang, R. (2020). Towards smart and reconfigurable environment: intelligent reflecting surface aided wireless network. *IEEE Commun. Mag.* 58 (1): 106–112.

2 Huang, C., Zappone, A., Alexandropoulos, G.C. et al. (2019). Reconfigurable intelligent surfaces for energy efficiency in wireless communication. *IEEE Trans. Commun.* 18 (8): 4157–4170.

3 Wu, Q. and Zhang, R. (2019). Intelligent reflecting surface enhanced wireless network via joint active and passive beamforming. *IEEE Trans. Commun.* 18 (11): 5394–5409.

4 Liaskos, C., Nie, S., Tsioliaridou, A. et al. (2018). A new wireless communication paradigm through software-controlled metasurfaces. *IEEE Commun. Mag.* 56 (9): 162–169.

5 Shafique, T., Tabassum, H., and Hossain, E. (2022). Stochastic geometry analysis of IRS-assisted downlink cellular networks. *IEEE Trans. Commun.* 23 (2): 1226–1252.

6 Psomas, C., Suraweera, H.A., and Krikidis, I. (2021). On the association with intelligent reflecting surfaces in spatially random networks. *ICC 2021-IEEE International Conference on Communications*, pp. 1–6. IEEE.

7 Zhao, D., Lu, H., Wang, Y. et al. (2021). Joint power allocation and user association optimization for IRS-assisted mmWave systems. *IEEE Trans. Wireless Commun.*

8 Abeywickrama, S., Zhang, R., Wu, Q., and Yuen, C. (2020). Intelligent reflecting surface: practical phase shift model and beamforming optimization. *IEEE Trans. Commun.* 68 (9): 5849–5863.

9 Kammoun, A., Chaaban, A., Debbah, M., Alouini, M.-S. et al. (2020). Asymptotic max-min SINR analysis of reconfigurable intelligent surface assisted MISO systems. *IEEE Trans. Wireless Commun.* 19 (12): 7748–7764.

10 He, J., Yu, K., and Shi, Y. (2020). Coordinated passive beamforming for distributed intelligent reflecting surfaces network. *2020 IEEE 91st Vehicular Technology Conference (VTC2020-Spring)*, pp. 1–5. IEEE.

11 Rehman, H.U., Bellili, F., Mezghani, A., and Hossain, E. (2021). Joint active and passive beamforming design for IRS-assisted multi-user MIMO systems: a VAMP-based approach. *IEEE Trans. Commun.* 19 (12): 7748–7764.

12 Ibrahim, H., Tabassum, H., and Nguyen, U.T. (2021). Exact coverage analysis of intelligent reflecting surfaces with Nakagami-*M* channels. *IEEE Trans. Veh. Technol.* 70 (1): 1072–1076.

13 Omid, Y., Shahabi, S.M.M., Pan, C. et al. (2020). IRS-aided large-scale MIMO systems with passive constant envelope precoding. *arXiv preprint arXiv:2002.10965.*

14 Luong, N.C., Hoang, D.T., Gong, S. et al. (2018). Applications of deep reinforcement learning in communications and networking: a survey. *CoRR*, abs/1810.07862. http://arxiv.org/abs/1810.07862.

15 Li, M. and Li, H. (2020). Application of deep neural network and deep reinforcement learning in wireless communication. *PLoS One* 15 (7): e0235447.

16 Feriani, A., Mezghani, A., and Hossain, E. (2021). On the robustness of deep reinforcement learning in IRS-aided wireless communications systems. *arXiv preprint arXiv:2107.08293.*

17 Taha, A., Zhang, Y., Mismar, F.B., and Alkhateeb, A. (2020). Deep reinforcement learning for intelligent reflecting surfaces: towards standalone operation, pp. 1–5.

18 Mnih, V., Kavukcuoglu, K., Silver, D. et al. (2013). Playing Atari with deep reinforcement learning. *CoRR*, abs/1312.5602. http://arxiv.org/abs/1312.5602.

19 Lillicrap, T.P., Hunt, J.J., Pritzel, A. et al. (2016). Continuous control with deep reinforcement learning. In: *4th International Conference on Learning Representations, ICLR 2016, San Juan, Puerto Rico, May 2–4, 2016, Conference Track Proceedings* (ed. Y. Bengio and Y. Le Cun). http://arxiv.org/abs/1509.02971.

20 Feng, K., Wang, Q., Li, X., and Wen, C.-K. (2020). Deep reinforcement learning based intelligent reflecting surface optimization for MISO communication systems. *IEEE Wireless Commun. Lett.* 9 (5): 745–749.

21 Huang, C., Mo, R., and Yuen, C. (2020). Reconfigurable intelligent surface assisted multiuser MISO systems exploiting deep reinforcement learning. *IEEE J. Sel. Areas Commun.* 38 (8): 1839–1850.

22 Yang, H., Xiong, Z., Zhao, J. et al. (2021). Deep reinforcement learning-based intelligent reflecting surface for secure wireless communications. *EEE Trans. Commun.* 69 (10): 6734–6749. https://doi.org/10.1109/TWC.2020.3024860.

23 Lee, G., Jung, M., Kasgari, A.T.Z. et al. (2020). Deep reinforcement learning for energy-efficient networking with reconfigurable intelligent surfaces. *ICC 2020 - 2020 IEEE International Conference on Communications (ICC)*, pp. 1–6. https://doi.org/10.1109/ICC40277.2020.9149380.

24 Björnson, E. and Sanguinetti, L. (2020). Power scaling laws and near-field behaviors of massive MIMO and intelligent reflecting surfaces. *arXiv preprint arXiv:2002.04960*.

25 Peng, Z., Li, T., Pan, C. et al. (2021). Analysis and optimization for RIS-aided multi-pair communications relying on statistical CSI. *IEEE Trans. Veh. Technol.*

26 Lyu, J. and Zhang, R. (2020). Hybrid active/passive wireless network aided by intelligent reflecting surface: system modeling and performance analysis. *arXiv preprint arXiv:2004.13318*.

27 Sutton, R.S. and Barto, A.G. (1998). *Reinforcement Learning - An Introduction, Adaptive Computation and Machine Learning*. MIT Press.

28 Schulman, J., Wolski, F., Dhariwal, P. et al. (2017). Proximal policy optimization algorithms. *CoRR*, abs/1707.06347. http://arxiv.org/abs/1707.06347.

29 Feriani, A. and Hossain, E. (2021). Single and multi-agent deep reinforcement learning for AI-enabled wireless networks: a tutorial. *IEEE Commun. Surv. Tutorials* 23 (2): 1226–1252. https://doi.org/10.1109/COMST.2021.3063822.

30 Watkins, C.J.C.H. and Dayan, P. (1992). Q-learning. *Mach. Learn.* 8 (3–4): 279–292.

31 Silver, D., Lever, G., Heess, N. et al. (2014). Deterministic policy gradient algorithms. *Proceedings of the 31st International Conference on International Conference on Machine Learning*, pp. 387–395.

32 Basar, E., Di Renzo, M., De Rosny, J. et al. (2019). Wireless communications through reconfigurable intelligent surfaces. *IEEE Access* 7: 116753–116773.

33 Tabassum, H., Hossain, E., Hossain, M.J., and Kim, D.I. (2015). On the spectral efficiency of multiuser scheduling in RF-powered uplink cellular networks. *IEEE Trans. Wireless Commun.* 14 (7): 3586–3600.

9

Application and Future Direction of RIS

Jalil R. Kazim, James Rains, Muhammad Ali Imran, and Qammer H. Abbasi

Electronics and Nanoscale Division, James Watt School of Engineering, University of Glasgow, Glasgow, United Kingdom

9.1 Background

During the last four decades, mobile communication networks have gone through five generations. A paradigm shift towards a new generation of mobile networks has been witnessed every ten years. Each generation consists of upgraded technologies and capabilities to enable humans in improving their work and lifestyle. For instance, the Zeroth-Generation (0G) of mobile communication networks, which provided simple radio communication functionality with devices such as walkie-talkies before the 1980s, is known as the pre-cell phone era [1, 2]. In the 1980s, the First-Generation (1G) cellular networks became publicly and commercially available. These networks used analogue mobile technology to provide voice communication. The Second Generation (2G) of mobile communication networks marked the shift from analogue to digital in mobile networks. In addition to voice communication, it supported data services such as Short Message Services (SMS). With the massive demand for data services, the Third generation (3G) mobile broadband services were introduced, allowing for new applications such as Multimedia Message Services (MMS), mobile TV, and video calls. The Fourth-Generation (4G) [3] known as Long-Term Evolution (LTE) introduced improved mobile broadband services, Voice Over IP (VoIP), online gaming, and Ultra-High-Definition video (UHD).

Lately, Fifth-Generation (5G) mobile communication networks are already being deployed around the world [4]. A prominent technology that enables the programmability and dynamicity of the 5G network is network softwarization [5]. The capabilities of 5G have enabled novel applications such as Mixed Reality (MR), Virtual and Augmented Reality [6], Internet of Things (IoT) [7],

Intelligent Reconfigurable Surfaces (IRS) for Prospective 6G Wireless Networks, First Edition.
Edited by Muhammad Ali Imran, Lina Mohjazi, Lina Bariah, Sami Muhaidat,
Tie Jun Cui, and Qammer H. Abbasi.

autonomous vehicles [8], and Industry 4.0 [9]. Research is been carried out on the beyond 5G (B5G) and future Sixth-Generation (6G) wireless mobile communication systems that will offer seamless access, enhanced Mobile Broadband (eMBB) with 1000x higher data rates Ultra-Reliable and Low-Latency Communications (URLLC), i.e. 5x fewer delay optimized data [10–13]. Novel technologies such as Extended Reality (XR) [14], Holographic Telepresence [14], Unmanned Aerial Vehicles (UAVs) [15], smart grid 2.0 [16], and Industry 5.0 [17] are slowly emerging, but these applications require ultra-high data rates, extremely high reliability, extremely low latency, powerful computing resources, and precision localization and sensing.

9.2 Introduction

Unprecedented performance, for example in terms of available data rate and latency, is typically associated with the introduction of a new generation of mobile networks. Massive multiple-input multiple-output (mMIMO) and m (mmWave) communications are both critical enablers of 5G networks. To meet the requirements described in Section 9.1, the 6G technology is expected to rely not only on the conventional spectrum, i.e. sub-6 GHz and mmWaves but also on frequency bands in the THz spectrum. The adoption of THz frequency bands will address both the spectrum scarcity as well as the capacity challenges of current wireless systems. In addition to macro- and micro-scale applications, the THz frequencies will also enable wireless communication among nanomachines [18].

Nevertheless, a major challenge at mmWave and THz frequencies is the very high-propagation loss that drastically limits the propagation distance. Specifically, the free space path loss easily exceeds 100 dB over a communication distance of 10 m at THz frequencies. Furthermore, the molecular absorption loss due to the wave energy converted to the kinetic energy of the molecules in the medium also contributes to the path loss in the mmWave and THz bands.

A simple solution to enhance the propagation distance is to deploy multiple base stations which is conventionally adopted at sub-6 GHz. But this would be extremely costly and energy-consuming as this would require placing base stations every 50 m. Hence, a low-cost and passive node that could extend the propagation range and simultaneously track and serve the users should be developed. In this regard, a novel technology being proposed is addressed by various names, i.e. smart reflect-arrays [19], large intelligent surface (LIS) [20], large intelligent meta-surface (LIM) [21] and reconfigurable meta-surface [22], reconfigurable intelligent surface (RIS) [23], software-defined surface (SDS) [24], software-defined meta-surfaces (SDMs) [25], passive intelligent surface (PIS) [26], and passive intelligent mirrors [27].

9.2.1 Intelligent Reflective Surface

The intelligent reflective surface (IRS) or the RIS is considered for B5G and 6G. As discussed, the signal propagation is heavily attenuated and can be blocked completely by the impediments in the propagation environment if operated at mmWave or THz. Hence, the communication channel between the transmitter and receiver is modelled as sparse channel. The link between the transmitter and receiver is assumed to be completely blocked. In these scenarios, the ability to steer the beams using the RIS can enable the communication link between the transmitter and the receiver.

At frequencies below 10 GHz, the channel is not modelled as a sparse channel and the transmitter, and the receiver is not completely blocked due to obstacles. The communication link is usually established via multipath signals. Consequently, the role of the RIS is not to enable the communication between the transmitter and the receiver as at higher frequencies, but in some cases, the coverage enhancement by directing the transmitted signal behind an obstacle such as a building in an urban environment or hills in a sub-urban and rural environment.

Compared to reflectarrays [28] which are planar reflective surfaces typically used in satellite applications, RIS can exhibit unconventional electromagnetic (EM) properties by interacting with EM waves. This typically include beam steering, wave absorption, and beam focusing [29]. The evolution of the RIS is shown in Figure 9.1 [30]. At first, only a simple reflective surface is realized which consists of microstrip patches with some pre-defined periodic spacing. These are designed to operate on a single frequency and single function. Using

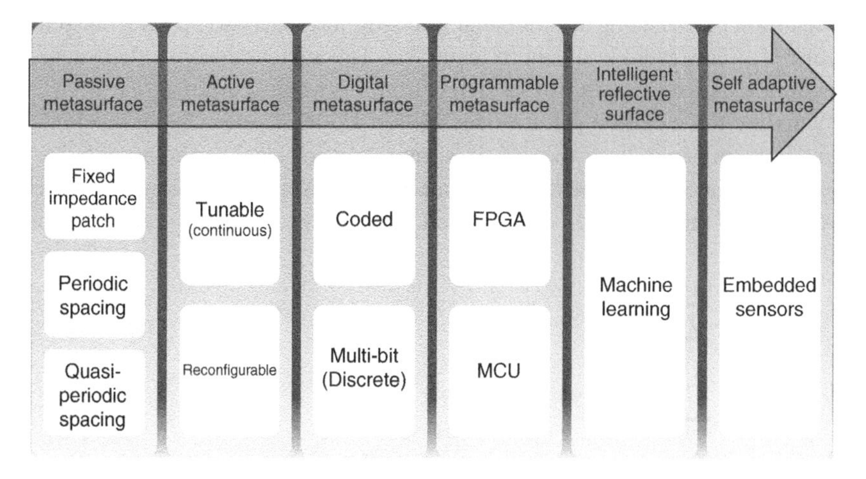

Figure 9.1 Evolution of intelligent reflective surface.

active elements, the reflective surface can be reconfigured and tuned at a different frequency, polarization, and perform a dual operation. With the introduction of programmable surfaces [31, 32], the functions of the EM surface could be mapped in the digital domain. This feature lessens the computational complexity which is typically associated with non-programmable surfaces. With the incorporation of machine learning and the programmable reflective surface, a more versatile and robust operation can be achieved. Using machine learning, the computational complexity could further be reduced paving the way for nearly passive intelligent reflective surfaces [33].

9.2.2 Analysis of RIS

A perfect electric conductor bounces off the EM waves according to Snell's law. It relates the angle of incidence with the angle of refraction, i.e. the change in the direction of propagation of the EM wave at the boundary of two dielectrics. The law states that the angle of the reflection and refraction depend on the properties of the material at both sides of the boundary between the dielectrics and the incident angle of the EM wave. The boundary between the two materials does not allow to steer the EM waves to any other directions. However, when a suitably designed metal sheet is placed between the two dielectrics, it is possible to manoeuvre the behaviour of the EM waves above and/or below the sheet more easily. These engineered thin sheets with pre-designed refractive, reflective, and absorption properties and are called meta-surfaces. The intelligent reflective surface is such a type of meta-surface that could be intelligently programmed to perform different functions, e. g., steering, focusing, and absorption.

The analysis of the RIS follows the conventional planar array theory. The RIS elements are placed along a rectangular grid with inter-element spacing taken as 'd_x' and 'd_y'. The layout is shown in Figure 9.2. The angular position of elements from the x-axis is represented by 'φ', while from the z-axis is taken as 'θ'.

To analyse the radiation pattern which is formed in 3D space by an incoming signal bouncing back from the RIS, it is important to calculate the array factor. The array factor of 'M' elements along the x-axis and 'N' elements along the y-axis shows a visual representation of the radiation property of any source and is given by [34].

$$\mathrm{AF}_{xm} = \sum_{m=1}^{M} A_m e^{j(m-1)(kd_x \sin\theta \cos\phi + \beta_x)} \tag{9.1}$$

$$\mathrm{AF}_{yn} = \sum_{n=1}^{N} A_n e^{j(n-1)(kd_y \sin\theta \sin\phi + \beta_y)} \tag{9.2}$$

Equations (9.1) and (9.2) represent the array factor of element's position along x- and y-axis. Distance 'd_x' and 'd_y' are inter-element spacing along the x- and

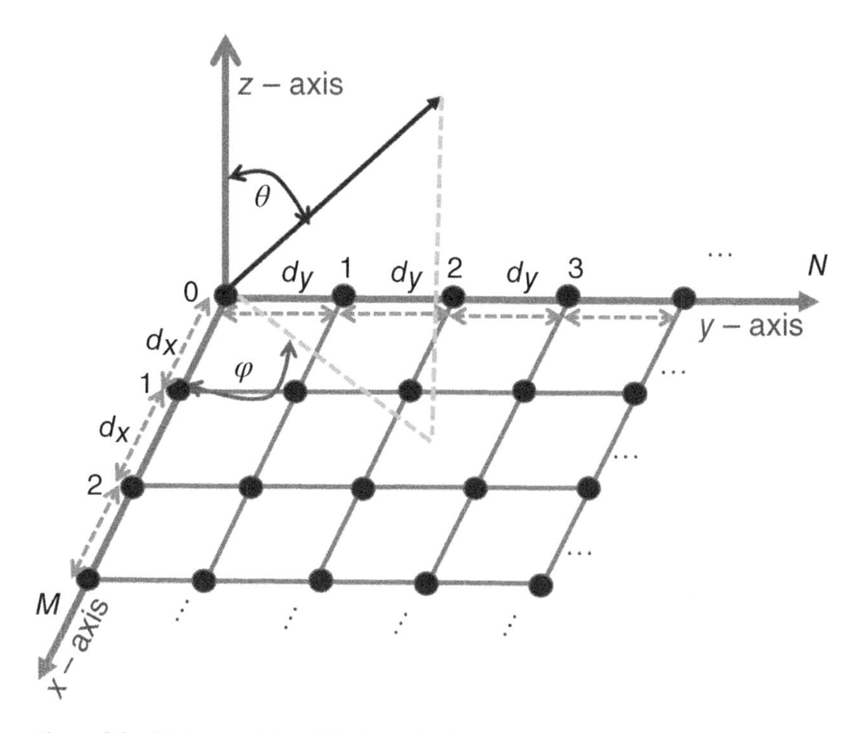

Figure 9.2 2D layout of the RIS elements along x- and y-axis.

y-axis. The amplitude coefficients are represented by 'A_m' and 'A_n', respectively. Additionally, 'k' represents the wavenumber which is cycles per unit length.

To steer the beam in a specific direction, it is important to apply a progress phase shift which can be calculated using Eqs. (9.3) and (9.4). It is important to note that the phase shift 'β_x' and 'β_y' are independent of each other, but for beam alignment, their values are kept the same.

$$\beta_x = -kd_x \sin\theta_0 \cos\phi_0 \tag{9.3}$$

$$\beta_y = -kd_y \sin\theta_0 \sin\phi_0 \tag{9.4}$$

Finally, the array factor of the 2D array is obtained by the product of array factors of the linear arrays along the x- and y-direction.

$$\mathrm{AF} = \mathrm{AF}_{xm} \cdot \mathrm{AF}_{yn} \tag{9.5}$$

To calculate maximum directivity achieved by the RIS surface, a general expression from [34] is given as follows:

$$D_0 = \frac{4\pi \cdot |\mathrm{AF}(\theta_0, \phi_0)|^2}{\int_0^{2\pi\pi} \int_0^{\pi} |AF(\theta, \phi)|^2 \sin\theta d\theta d\phi} \tag{9.6}$$

The expression in the numerator represents the radiation intensity in the direction of 'θ_0' and 'φ_0', while the denominator gives the radiation intensity averaged over all directions.

9.2.3 Basic Functions of RIS

The RIS can perform various functions such as beam focusing and/or beam steering based on the position of the transmitter and receiver location. If the transmitter and/or receiver are in the near-field region, Eqs. (9.1) and (9.2) need to be amended to include the spatial phase delay on the surface of the IRS. The total phase compensation 'ϕ_{rad}' consists of two parts; 'ϕ_{CT}' compensates for the spatial phase delay due to the transmitter located in the near field of the RIS and 'ϕ_{CR}' compensates the receiver distance if it is in the near field. Hence, Eqs. (9.1) and (9.2) can be rewritten as follows:

$$\text{AF}_{xm} = \sum_{m=1}^{M} A_m e^{j(m-1)(kd_x \sin\theta \cos\phi + \beta_x + \phi_{\text{rad}})} \tag{9.7}$$

$$\text{AF}_{yn} = \sum_{n=1}^{N} A_n e^{j(n-1)(kd_y \sin\theta \sin\phi + \beta_y + \phi_{\text{rad}})} \tag{9.8}$$

To determine 'ϕ_{rad}', let us consider three different cases such that (i) only the transmitter is in the near field, (ii) only receiver is in the near field, and (iii) both transmitter and receiver is in the near field.

We consider the position x_T, y_T, z_T of the transmitter in a rectangular coordinate system as follows:

$$\begin{aligned} x_T &= T\,\text{Cos}\Phi_T\text{Sin}\theta_T \\ y_T &= T\,\text{Cos}\Phi_T\text{Sin}\theta_T \\ z_T &= T\,\text{Cos}\theta_T \end{aligned} \tag{9.9}$$

where T is the radial distance of the transmitter from the centre element of the RIS at origin (0,0,0). 'Φ_T' and 'θ_T' are the azimuthal and elevation position of the transmitter relative to the RIS.

Similarly, the position of the receiver in space is taken as x_R, y_R, z_R such that

$$\begin{aligned} x_R &= R\,\text{Cos}\Phi_R\text{Sin}\theta_R \\ y_R &= R\,\text{Cos}\Phi_R\text{Sin}\theta_R \\ z_R &= R\,\text{Cos}\theta_R \end{aligned} \tag{9.10}$$

where R is the radial distance from the RIS to the receiver position. Φ_R and θ_R are the azimuthal and elevation position of the receiver relative to the RIS.

Case 1: Transmitter in the near-field while receiver in the far-field
The transmitter location in the near-field is a conventional example of RIS operating as a reflectarray. Assuming the position of the RIS elements as 'x_m' and 'y_n', the phase compensation with respect to transmitter and IRS element location is given as follows:

$$\Delta\phi_{CT} = k\left|\sqrt{(x_m - x_T)^2 + (y_n - y_T)^2 + {z_T}^2}\right| \tag{9.11}$$

where m and n is equal to 0, 1, 2, 3,…,(N−1) such N is a total number of elements on the RIS. As the receiver is in the far-field, 'ϕ_{CR}' is taken as 0.

Case 2: Transmitter in the far-field and receiver in the near-field
In the communication system, the base station will be located in the far-field, and due to the inherently large aperture of the RIS, the receiver communication might take place in the near-field of the RIS; hence, the RIS should be able to focus the beam towards the receiver. The incident wave will be a plane wave such that '$\phi_{\text{CT}} = 0$', and the focal spot towards the receiver could be calculated as follows:

$$\Delta\phi_{\text{CR}} = k\left|\sqrt{(x_m - x_R)^2 + (y_n - y_R)^2 + {z_R}^2}\right| \tag{9.12}$$

where m and n is equal to 0, 1, 2, 3,…,(N–1).

Case 3: Transmitter and receiver are both in the near field
In the event, where both transmitter and receiver are in the near-field of the RIS, the total phase correction will consist of the following, i.e. adjustment of the spatial phase delay from transmitter and beam focusing on the receiver. The RIS element's desired phase distribution would not be a linear progressive phase distribution to focus the beam in a desired position in the RIS near-field:

$$\phi_{\text{rad}} = \Delta\phi_{\text{TR}} + \Delta\phi_{\text{CR}} \tag{9.13}$$

Case 4: Transmitter and receiver are both in far field
With the transmitter and receiver located in the far-field of the RIS, 'ϕ_{rad}' will be zero. In this case, only the progressive phase shift is given in Eqs. (9.3) and (9.4) can be used to steer the beam in both azimuth and/or elevation planes.

9.3 RIS-assisted High-Frequency Communication

Existing network operators are always faced with a challenge to ensure seamless connectivity to end-users especially in harsh propagation conditions, i.e. involving tall buildings and trees. This will become more difficult with the adaptation of mmWave. It is anticipated that mmWave 5G communication, as well as future

THz 6G communication, will be affected by blind spots that will not bc adequately covered due to the severe blocking loss of such short-length waveforms. Furthermore, users tend to be distributed unevenly in the desired coverage area and users are increasing day by day putting an enormous load on the existing networks. Technologies such as massive MIMO [35], ultra-dense networks (UDN)[36] are being proposed to combat the aforementioned challenges, but they come at a significant cost. The massive MIMO requires a huge number of radio frequency (RF) chains with costly phase shifters and UDN is realized to have a very high cost of deployment.

This is where the RIS plays a substantial role in coverage enhancements of blind spots and extending the base station range. The RIS extended coverage for 5G and beyond 5G communications is one of the most promising applications. Additionally, RIS has no RF chain and operates in full-duplex mode. The cost of deployment is negligible compared to the massive MIMO system and the UDN.

A very basic application of RIS involves coverage enhancement. Figure 9.3 shows various application scenarios of RIS assisted wireless communication. In scenario A, the function of RIS1 simply involves steering beams towards users at cell edges that might be away from the base station. In the same scenario, RIS4 operates to perform signal cancellation towards an eavesdropper. Similarly, in scenario B, RIS2 is programmed to perform signal cancellation at node M2 such that M2 is not affected by interference signal coming from the base station 2. Finally, in scenario C, the RIS3 is designed to scatter the beam in a different direction, i.e. this involves creating a rich scattering channel to create multi-path. This is helpful where the channel is noisy and affects the link quality at the receivers.

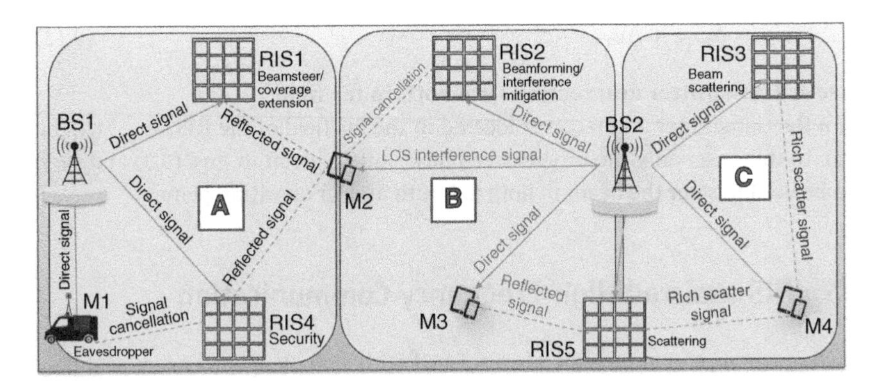

Figure 9.3 Different RIS deployment scenarios. Source: Adapted from Kazim et al. [37].

9.3.1 RIS-assisted Multi-User Communication

A basic setup of a communication link involves a transmitter and a single receiver. From a practical perspective, the transmitter is always connected with multiple receivers. Consequently, the transmitter antennas are implemented to perform beam switching or generate multiple beams. Existing technology employ phased arrays, that can generate multiple beams which are considered a costlier solution by network operators.

In this regard, the phase distribution profile on RIS could be programmed to generate multiple beams. This makes it suitable for multi-user communication[38]. As discussed in the previous section, the RIS is able to focus in the near-field region. With multi-focal spots, the beam could be reflected to the desired number of users in the near-field. In the far-field region, the aperture of the RIS can be divided such that a multi-aperture RIS could be utilized. With '*N*' number of sub-RIS apertures, we could generate '*N*' beams which is a conventional technique adapted in phased arrays.

9.4 RIS-assisted RF Sensing and Imaging

Over the years, EM sensing is becoming popular in the research community as it enables unobtrusive and non-contactless sensing of targets. The growing interest lies in the ability of the EM sensing systems to be ubiquitous, contactless, and privacy conservation. In this regard, human sensing based on wireless signals has gained significant attention from the research community [39]. Many applications including human activity monitoring [40] gesture recognition [41], vital signs monitoring, [42] and many more have been achieved. The basis of these applications lies in the fact that the movement in the environment would alter the propagation of signals, which consequently makes it feasible to extract information from the signal variation.

Apart from RF-sensing, visual sensing, e.g. using optical cameras, is limited to lighting conditions and issues concerning privacy do not make it a choice. Furthermore, RF sensing using radio detection and ranging (RADAR) is expensive for deployment on large scale. The RF sensing, in this case, is restricted only in the line-of-sight.

Existing research in the field of RIS is concentrated on the ability of the RIS to enhance the signal-to-noise ratio of the wireless communication links. A unique aspect of the RIS that is currently being explored is the ability of the RIS for healthcare applications [37]. Some of the various possible application of RIS in healthcare is shown in Figure 9.4. Mobility of users leads to variation in the propagation

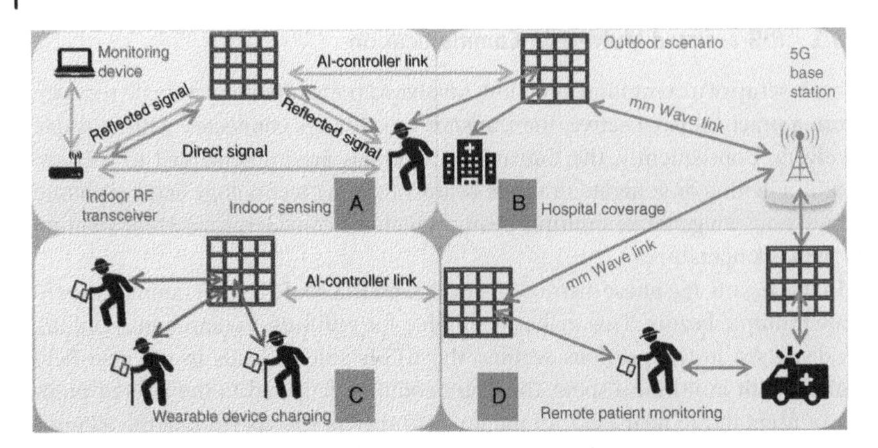

Figure 9.4 Various scenarios of RF-assisted health care application. Source: Kazim et al. [37]/with permission of IEEE.

environments. This in turn could change the Channel State Information (CSI) at the receiver. In normal scenarios, the variation in the signal is negligible at the receiver due to extremely subtle changes in the propagation environment.

The RIS can enhance the variation between the target and the receiver such that the movement becomes significantly distinguishable. A promising application that is demonstrated using the RIS is human posture recognition [43, 44]. The technology involves Wi-Fi frequency and RIS for instantaneous in situ imaging and recognition of hand signs.

9.5 RIS-assisted-UAV Communication

In the last decade, the global market for drones, i.e. UAVs have risen from 19.3 billion dollars in 2020 and will rise to 45.8 billion dollars in 2025 [45]. The gradually decreasing manufacturing expenditures and the increasing number of applications, e.g. monitoring and surveillance, law enforcement, agriculture, logistics, and emergency health care have driven the need for the adoption of UAVs.

Presently, UAV applications such as real-time video streaming and filming require very high data rates. At present, cellular technologies such as the B5G will play a critical role in UAV communication. The integration of UAV with B5G will enable UAV-assisted wireless communication [46]. While the benefits of UAV communication cannot be overlooked, the complex terrain and surroundings may cause blockage of air-to-ground wireless links. Furthermore, in the presence of eavesdropping, the information security of legitimate users may not be guaranteed.

The IRS can play a major role in UAV-assisted air-to-ground communication networks to obtain a favourable propagation environment and enhance the transmission quality of desired users to address these issues [47]. Meanwhile, by properly designing the beamforming mechanism, the IRS can cancel unwanted signals to curb the interference and prevent eavesdropping. Recently, investigations combining UAV and IRS to improve the performance of air-to-ground communication links have become apparent [47–49]. With the help of the IRS, the coverage of UAVs can be expanded, and thus various quality of service (QoS) requirements of users can be met. When the RIS is mounted on a UAV, it can have more deployment flexibility and a broader range of signal reflection compared to deployment on a fixed building.

9.6 RIS-assisted Wireless Power Transfer

With the shift from human-to-human communication, towards human–machine and machine-to-machine communication in the previous decade, the idea of IoT has evolved [50]. These devices a low power and are typically used for sensing and data collection application. Hence, maintaining millions of IoT devices operational is difficult. Due to rising maintenance costs, conventional methods using removable batteries and power cords are no longer suitable [2]. Wireless Power Transfer (WPT) using RF has recently emerged as a promising technology for resolving this problem [51, 52]. The RF WPT has a longer charging distance as compared to technologies that use inductive or magnetic resonance coupling. The efficiency of the transfer, on the other hand, rapidly decreases as the distance between the two points increases.

With RF WPT, much higher efficiency has recently been achieved by using beamforming technology. Beamforming can be used to focus an EM wave at the receiver using phase arrays. Regardless, RF components such as attenuators, phase shifters, and amplifiers should be installed in each radiating element. This leads to a system with a high level of complexity, implementation costs, and power consumption, especially in large-scale systems.

However, information transmission enabled simultaneous wireless information and power transfer (SWIPT) is an appealing technique for IoT networks [53], which will require a lot of power due to their energy consumption. More specifically, wireless signals will be transmitted to a group of devices by a base station with a constant power supply. Some devices, referred to as information nodes are designed to decode the information contained in a received signal, whereas others, referred to as energy nodes, are designed to harvest the signal energy contained in a received signal.

Hence, the RIS can simultaneously allow beam focusing for energy as well as information transfer without the use of power-hungry active components [54, 55]. It can combine the power from different RF sources and focus it towards the energy node [56]. In other words, when compared to existing technologies that use phased arrays, the RIS ensures lower loss in RF wireless energy transfer. Furthermore, RISs can be mass-produced at a very low cost. Then, to improve power transfer efficiency, many RISs can be easily deployed in the walls of a building or a room. The RIS can be installed in the ceiling of the building to assist the WPT system. The RIS reflects an EM beam from a power beacon before focusing the reflected wave on ground devices. As a result, the RIS aids the WPT system in improving power transfer efficiency and extending the range of power transfer. Because of these characteristics, it is ideal for RF wireless power transfer applications.

9.7 RIS-assisted Indoor Localization

Indoor localization has been an active research area for the past decade as compared to outdoor scenarios which employs global navigation position (GPS). The indoor localization falls in two categories, i.e. active and passive localization. The active localization involves a tag to locate position of the target while passive localization can detect location of target in a contactless and unobtrusive way. Typical use case scenarios of localization include intrusion detection [57], monitoring of elderly patients [39], and real-time positioning of criminals [58]. Techniques using vision-based technologies are limited by lighting conditions and privacy concerns for real-life deployment, while RADAR is not encouraged for indoor localization.

A Wi-Fi-based system is a cost-effective solution because of the widespread availability of Wi-Fi devices. Existing Wi-Fi-based methods primarily rely on the estimation of time-of-flight (ToF) and/or angle-of-arrival (AoA) to determine the target position. However, due to the limited bandwidth and the number of antennas available on Wi-Fi devices, the resolution of the estimated AoA and ToF is generally poor in comparison to higher-end devices [59]. In such a case, the accuracy of passive localization is poor, and as a result, the range of applications is restricted.

To estimate the AoA, ToA and Time-Difference-of-Arrival (TDoA), the conventional localization technology relies on GPS signals or base stations in existing cellular networks. But GPS and base station signals are frequently affected by blind spots, such as underground and various obstacles between transmitter and receiver antennas. Furthermore, the requirement for indoor localization are more stringent that outdoors.

To address the challenges, the RIS can play a significant role to improve the accuracy of localization [60]. With a large number of RIS elements, the resolution

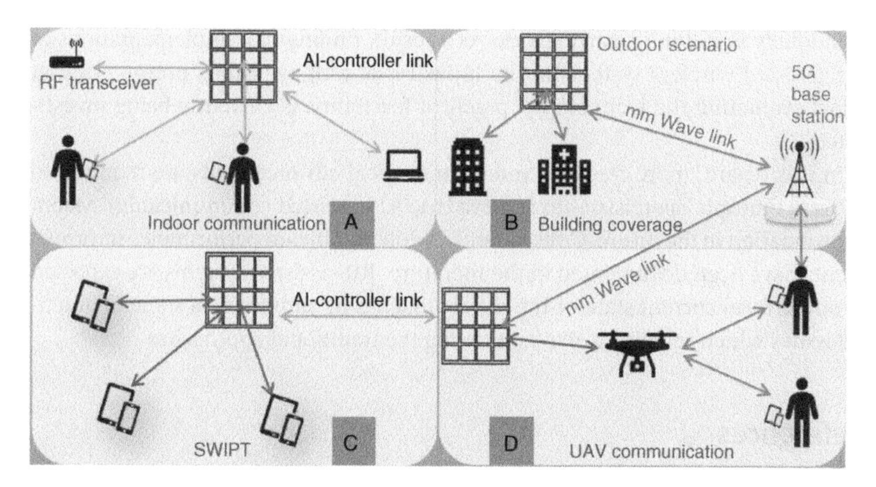

Figure 9.5 RIS multi-communication scenarios.

of the estimated AoA and ToF can be significantly improved resulting in a very high accuracy. With the ability of beam steerability using nearly passive low-cost elements, the RIS would play significant role in future indoor localization system, e.g. in a typical localization scenario involving the RIS, the access point (AP) will transmit signals to the user and the RIS. The RIS will reflect the signal towards the user coming from the AP. The signal could be transmitted to multiple users to estimate multiple locations [61]. Multiple scenarios of RIS-assisted communication is summarized in Figure 9.5.

9.8 Conclusion

With the amount of literature being published related to the RIS, it has become evident that the technology will play a significant role in future communication systems. RIS has emerged as a promising solution for future technologies such as B5G and 6G. The absence of active RF chains and the idea of controlling the propagation environment with arrays of passive elements that consume very little power is extremely appealing in terms of lowering the cost of the network. The RIS relies on its multi-functional role, i.e. beam focusing and beam steering both in the near-field and the far-field. RIS provides an intelligent paradigm to alter the propagation environment. This includes coverage extensions, covering blind spots and provides reliable signals to users at cell edges.

However, due to several challenges, such as hardware limitations, the difficulty of obtaining accurate channel information and the need to optimize IRS positioning depending on the communications scenario, practical deployment of this technology is still in its early stages. As a result, finding real implementations of IRS-assisted wireless systems in the industry, as well as testbed prototypes that allow evaluating the technology's practical feasibility, is currently being investigated.

In this regard, more research and technological advancements are required to address multiple open issues to achieve true RIS-assisted communication system consolidation in the future. Consequently, while significant performance improvements have been documented in the literature, RIS-assisted systems are expected to outperform current state-of-the-art approaches by addressing a some of practical issues which remains a challenge using the traditional approaches.

References

1 Pereira, V. and Sousa, T. (2004). Evolution of mobile communications: from 1G to 4G. *Academia* 4: 20.

2 Bhalla, M.R. and Bhalla, A.V. (2010). Generations of mobile wireless technology: a survey. *Int. J. Comput. Appl.* 5 (4): 26–32.

3 Datta, P. and Kaushal, S. (2014). Exploration and comparison of different 4G technologies implementations: a survey. In: *2014 Recent Advances in Engineering and Computational Sciences, RAECS*, 2014. Chandigarh, India: IEEE.

4 Popovski, P., Trillingsgaard, K.F., Simeone, O., and Durisi, G. (2018). 5G wireless network slicing for eMBB, URLLC, and mMTC: a communication-theoretic view. *IEEE Access* 6: 55765–55779.

5 Lake, D., Wang, N., Tafazolli, R., and Samuel, L. (2021). Softwarization of 5G networks-implications to open platforms and standardizations. *IEEE Access* 9: 88902–88930.

6 Siriwardhana, Y., Porambage, P., Liyanage, M., and Ylianttila, M. (2021). A survey on mobile augmented reality with 5G mobile edge computing: architectures, applications, and technical aspects. *IEEE Commun. Surv. Tutorials* 23 (2): 1160–1192.

7 Ejaz, W. et al. (2016). Internet of things (IoT) in 5G wireless communications. *IEEE Access* 4: 10310–10314.

8 Raissi, F., Yangui, S., and Camps, F. (2019). *Autonomous Cars, 5G Mobile Networks and Smart Cities: Beyond the Hype*, 180–185. Institute of Electrical and Electronics Engineers Inc.

9 Nakimuli, W., Garcia-Reinoso, J., Enrique Sierra-Garcia, J. et al. (2021). Deployment and evaluation of an Industry 4.0 use case over 5G. *IEEE Commun. Mag.* 59 (7): 14–20.

10 Giordani, M., Polese, M., Mezzavilla, M., et al. (2019). Towards 6G networks: use cases and technologies, arXiv preprint arXiv:1903.12216.

11 Huq, K.M.S., Busari, S.A., Rodriguez, J. et al. (2019). Terahertz-enabled wireless system for beyond-5G ultra-fast networks: a brief survey. *IEEE Network* 33 (4): 89–95.

12 Saad, W., Bennis, M., and Chen, M. (2019). A vision of 6G wireless systems: applications, trends, technologies, and open research problems. *IEEE Network* 34 (3): 134–142.

13 Tariq, F., Khandaker, M., Wong, K.-K., et al. (2019). A speculative study on 6G, arXiv preprint arXiv:1902.06700.

14 Martin, M. and Amin, E. (2021). XR in the 6G post-smartphone era. In: *Toward 6G: A New Era of Convergence*, 167–182. IEEE.

15 Mozaffari, M., Lin, X., and Hayes, S. (2021). Toward 6G with connected sky: UAVs and beyond. *IEEE Commun. Mag.* 59 (12): 74–80.

16 Cao, J. and Yang, M. (2014). *Energy Internet - Towards Smart Grid 2.0*, 105–110. IEEE Computer Society.

17 Nahavandi, S. (2019). Industry 5.0-a human-centric solution. *Sustainability (Switzerland)* 11 (16): 4371.

18 Akyildiz, I.F., Brunetti, F., and Blázquez, C. (2008). Nanonetworks: a new communication paradigm. *Comput. Networks* 52 (12): 2260–2279.

19 Tan, X., Sun, Z., Jornet, J.M., and Pados, D. (2016). Increasing indoor spectrum sharing capacity using smart reflect-array. In: *2016 IEEE International Conference on Communications (ICC)*, 1–6. IEEE.

20 Huang, C., Zappone, A., Alexandropoulos, G.C., et al. (2018). Large intelligent surfaces for energy efficiency in wireless communication, arXiv preprint arXiv:1810.06934 v1.

21 He, Z. and Yuan, X. (2020). Cascaded channel estimation for large intelligent metasurface assisted massive MIMO. *IEEE Wireless Commun. Lett.* 9 (2): 210–214.

22 Di Renzo, M. and Song, J. (2019). Reflection probability in wireless networks with metasurface-coated environmental objects: an approach based on random spatial processes. *EURASIP J. Wireless Commun. Networking* 2019 (1): 99.

23 Huang, C., Zappone, A., Alexandropoulos, G.C. et al. (2019). Reconfigurable intelligent surfaces for energy efficiency in wireless communication. *IEEE Trans. Wireless Commun.* 18 (8): 4157–4170.

24 Basar, E. (2019). Large intelligent surface-based index modulation: a new beyond MIMO paradigm for 6G, arXiv preprint arXiv:1904.06704.

25 Liaskos, C., Tsioliaridou, A., Nie, S. et al. (2019). An interpretable neural network for configuring programmable wireless environments. In: *2019 IEEE 20th International Workshop on Signal Processing Advances in Wireless Communications (SPAWC)*, 1–5. IEEE.

26 Mishra, D. and Johansson, H. (2019). Channel estimation and low-complexity beamforming design for passive intelligent surface assisted MISO wireless

energy transfer. In: *ICASSP 2019-2019 IEEE International Conference on Acoustics, Speech and Signal Processing (ICASSP)*, 4659–4663. IEEE.

27 Huang, C., Zappone, A., Debbah, M., and Yuen, C. (2018). Achievable rate maximization by passive intelligent mirrors. In: *2018 IEEE International Conference on Acoustics, Speech and Signal Processing (ICASSP)*, 3714–3718. IEEE.

28 Payam, N., Fan, Y., and Atef, Z.E. (2018). Introduction to reflectarray antennas. In: *Reflectarray Antennas: Theory, Designs, and Applications*, 1–8. IEEE.

29 Yang, H. et al. (2016). A programmable metasurface with dynamic polarization, scattering and focusing control. *Sci. Rep.* 6 (1): 35692.

30 Barbuto, M. et al. (2021). Metasurfaces 3.0: a new paradigm for enabling smart electromagnetic environments. *IEEE Trans. Antennas Propag.*

31 Cui, T.J., Qi, M.Q., Wan, X. et al. (2014). Coding metamaterials, digital metamaterials and programmable metamaterials. *Light Sci. Appl.* 3 (10): e218–e218.

32 Della Giovampaola, C. and Engheta, N. (2014). Digital metamaterials. *Nat. Mater.* 13 (12): 1115–1121.

33 Wang, J. et al. (2021). Interplay between RIS and AI in wireless communications: fundamentals, architectures, applications, and open research problems. *IEEE J. Sel. Areas Commun.* 39 (8): 2271–2288.

34 Balanis, C.A. (2015). *Antenna Theory: Analysis and Design*. Wiley.

35 Larsson, E.G. (2017). Massive MIMO for 5G: overview and the road ahead. In: *2017 51st Annual Conference on Information Sciences and Systems (CISS)*, 1–1. IEEE.

36 Andreev, S., Petrov, V., Dohler, M., and Yanikomeroglu, H. (2019). Future of ultra-dense networks beyond 5G: harnessing heterogeneous moving cells. *IEEE Commun. Mag.* 57 (6): 86–92.

37 Kazim, J.U.R. et al. (2021). Wireless on walls: revolutionizing the future of health care. *IEEE Antennas Propag. Mag.* 63 (6): 87–93.

38 Cao, X., Yang, B., Zhang, H. et al. (2021). Reconfigurable-intelligent-surface-assisted mac for wireless networks: protocol design, analysis, and optimization. *IEEE Internet Things J.* 8 (18): 14171–14186.

39 Shah, S.A. and Fioranelli, F. (2019). RF sensing technologies for assisted daily living in healthcare: a comprehensive review. *IEEE Aerosp. Electron. Syst. Mag.* 34 (11): 26–44.

40 Taylor, W., Shah, S.A., Dashtipour, K. et al. (2020). An intelligent non-invasive real-time human activity recognition system for next-generation healthcare. *Sensors* 20 (9): 2653.

41 Thariq Ahmed, H.F., Ahmad, H., and Aravind, C.V. (2020). Device free human gesture recognition using Wi-Fi CSI: a survey. *Eng. Appl. Artif. Intell.* 87: 103281.

42 Cardillo, E. and Caddemi, A. (2020). A review on biomedical mimo radars for vital sign detection and human localization. *Electronics (Switzerland)* 9 (9): 1–15.

43 Hu, J. et al. (2020). Reconfigurable intelligent surface based RF sensing: design, optimization, and implementation. *IEEE J. Sel. Areas Commun.* 38 (11): 2700–2716.

44 Li, L. et al. (2019). Intelligent metasurface imager and recognizer. *Light Sci. Appl.* 8 (1).

45 Available: https://www.marketsandmarkets.com/pdfdownloadNew.asp?id=662

46 Zeng, Y., Wu, Q., and Zhang, R. (2019). Accessing from the sky: a tutorial on UAV communications for 5G and beyond. *Proc. IEEE* 107 (12): 2327–2375.

47 Pang, X., Sheng, M., Zhao, N. et al. (2021). When UAV meets IRS: expanding air-ground networks via passive reflection. *IEEE Wireless Commun.* 28 (5): 164–170.

48 Li, J., Xu, S., Liu, J. et al. (2021). Reconfigurable intelligent surface enhanced secure aerial-ground communication. *IEEE Trans. Commun.* 69 (9): 6185–6197.

49 Sun, G., Tao, X., Li, N., and Xu, J. (2021). Intelligent reflecting surface and uav assisted secrecy communication in millimeter-wave networks. *IEEE Trans. Veh. Technol.* 70 (11): 11949–11961.

50 Ploennigs, J., Cohn, J., and Stanford-Clark, A. (2018). The future of IoT. *IEEE Internet Things Mag.* 1 (1): 28–33.

51 Zhou, J., Zhang, P., Han, J. et al. (2022). Metamaterials and metasurfaces for wireless power transfer and energy harvesting. *Proc. IEEE* 110 (1): 31–55.

52 Tian, S., Zhang, X., Wang, X. et al. (2022). Recent advances in metamaterials for simultaneous wireless information and power transmission. *Nanophotonics* 1697–1723.

53 Krikidis, I., Timotheou, S., Nikolaou, S. et al. (2014). Simultaneous wireless information and power transfer in modern communication systems. *IEEE Commun. Mag.* 52 (11): 104–110.

54 Wu, Q., Guan, X., and Zhang, R. (2022). Intelligent reflecting surface-aided wireless energy and information transmission: an overview. *Proc. IEEE* 110 (1): 150–170.

55 Tran, N.M., Amri, M.M., Park, J.H. et al. (2019). A novel coding metasurface for wireless power transfer applications. *Energies* 12 (23): 4488.

56 Yu, S., Liu, H., and Li, L. (2019). Design of near-field focused metasurface for high-efficient wireless power transfer with multifocus characteristics. *IEEE Trans. Ind. Electron.* 66 (5): 3993–4002.

57 Bassey, J., Adesina, D., Li, X. et al. (2019). Intrusion detection for IoT devices based on RF fingerprinting using deep learning. In: *2019 Fourth International Conference on Fog and Mobile Edge Computing (FMEC)*, 98–104. IEEE.

58 Tundis, A., Kaleem, H., and Mühlhäuser, M. (2020). Detecting and tracking criminals in the real world through an IoT-based system. *Sensors (Basel, Switzerland)* 20 (13): 3795.

59 Kandel, L.N. and Yu, S. (2019). Indoor localization using commodity Wi-Fi APs: techniques and challenges. In: *2019 International Conference on Computing, Networking and Communications (ICNC)*, 526–530. IEEE.

60 Zhang, H., Zhang, H., Di, B. et al. (2021). MetaLocalization: reconfigurable intelligent surface aided multi-user wireless indoor localization. *IEEE Trans. Wireless Commun.* 20 (12): 7743–7757.

61 Zhang, G., Zhang, D., He, Y., et al. (2022). Multi-Person Passive WiFi Indoor Localization with Intelligent Reflecting Surface, arXiv preprint arXiv:2201.01463.

10

Distributed Multi-IRS-assisted 6G Wireless Networks: Channel Characterization and Performance Analysis

Tri N. Do, Georges Kaddoum, and Thanh L. Nguyen

The Department of Electrical Engineering, the École de Technologie Supérieure (ÉTS), Université du Québec, Montréal, QC, Canada

10.1 Introduction

Intelligent reconfigurable surface (IRS) has been demonstrated to be one of the key technologies for achieving a smart radio environment (SRE) for the sixth-generation (6G) wireless communication systems [1–3]. In the literature, the term 'IRS' has also been referred to as reconfigurable intelligent surfaces [1], intelligent reflecting surfaces [2], or software-defined meta-surfaces [3], just to name a few. An IRS is composed of a large number of reflecting elements that are controlled and programmable to reflect and steer incident signals in the desired propagation direction. To accomplish this, the phase shift of each reflecting element is reconfigured, resulting in the direction of the scattered signal beam being adjusted. On the receiver side, coherent combining is implemented to constructively add all the multi-path received signals reflected from different elements of all distributed IRSs [3].

In the literature, channel modelling and performance analysis of single-IRS-assisted systems have received extensive attention, as depicted in [4–7]. In [4], Basar et al. derived the analytical expression of the optimal phase shift configuration, resulting in a maximum received signal-to-noise ratio (SNR). Considering a large IRS-assisted, point-to-point wireless system without a direct link, de Figueiredo et al. [5] demonstrated that the end-to-end (e2e) channel magnitude subjected to Rayleigh fading can be approximated by a Gamma distribution. In [6], when studying a single-IRS-assisted transmission (without the direct channel) and considering that the cascaded channel is subjected to Rician fading,

Intelligent Reconfigurable Surfaces (IRS) for Prospective 6G Wireless Networks, First Edition.
Edited by Muhammad Ali Imran, Lina Mohjazi, Lina Bariah, Sami Muhaidat,
Tie Jun Cui, and Qammer H. Abbasi.

Gan et al. relied on the central limit theorem (CLT) to show that the true distribution of the e2e channel amplitude can be approximated by a normal distribution. In [7], considering single-IRS-assisted systems, Ibrahim et al. showed that the true distribution of the e2e channel magnitude can be characterized by a Gamma distribution through the moment-matching method. The aforementioned-related works were based on the assumption that channels associated with different reflecting elements on the same IRS are independently and identically distributed (i.i.d.).

Next, we discuss recent related works that consider a more general system setting, i.e. multi-IRS-assisted wireless systems. In [8], considering a multi-IRS-assisted SISO system under i.i.d. Nakagami-m fading, where the direct link is available, Galappaththige et al. used the CLT to show that the e2e channel magnitude can approximately follow a normal distribution. In [9], considering a large IRS-assisted system subjected to Rician fading, where the IRS consists of multiple tiles, each consisting of a huge number of reflecting elements, Jung et al. showed that the true distribution of the squared channel magnitude can be approximated by a Gaussian distribution by resorting to the CLT and the law of large numbers. In [10], Yang et al. studied a point-to-point transmission aided by multiple IRSs where the direct link does not exist, and only the best IRS is chosen to assist the transmission. Moreover, the authors in [10] demonstrated that the e2e SNR follows a non-central, chi-square distribution when all Rayleigh fading channels between different IRSs are i.i.d. channels. In [11], Mei and Zhang presented an IRS selection technique that maximizes the e2e SNR when considering multi-hop, multi-IRS-assisted systems. In addition, the authors in [11], on the other hand, only considered the influence of path-loss and neglected the impact of fading. In [12], when a direct link between a source and a destination is absent, Yildirim et al. examined multi-IRS-assisted solutions for both interior and outdoor communications. Additionally, the authors in [12] developed an IRS selection technique that selects the IRS with the highest SNR to aid the communication in order to achieve low-complexity transmission. Small-scale fading, on the other hand, was disregarded, and no performance analysis of the IRS selection technique was conducted. In [13], Lyu and Zhang evaluated the throughput of multi-cell, IRS-aided systems, in which multiple IRSs are deployed to assist multiple users, and the communications are subjected to co-channel interference.

Despite the fact that previous works have made significant contributions to the study of multi-IRS-assisted systems, accurate fading model characterization remains a challenge. Indeed, on the one hand, to make performance analysis tractable, some previous studies took into account large-scale fading while ignoring small-scale fading, as in [11, 12]. In existing works that considered small-scale fading, the channels were assumed to be i.i.d. [8, 10] or deterministic [9]. It is noted that because the reflecting elements on an IRS are sub-wavelength in size and are implanted on a compact panel [1], channels associated with different elements of the same IRS can be reasonably assumed to be i.i.d. channels. However, channels between different distributed IRSs cannot be assumed to be i.i.d. because the distance between different IRSs is significant.

Based on the foregoing observations, in this work, we consider a distributed multi-IRS (DMI)-assisted system, in which all the IRSs with different geometric sizes and distributed locations participate in assisting the transmission of a pair of transceivers. In light of such a multi-IRS-assisted system, we raise a research question that has received little attention in the literature: *how to determine the statistical characterization of the DMI system's e2e channel magnitude?* Our work's distinct technological contribution is our unified analysis approach, which provides a thorough answer to the presented research topic.

Specifically, the unique contributions of our paper are as follows:

- We consider an independent but not identically distributed (i.n.i.d.) fading environment, which is more comprehensive and practical than an i.i.d. fading environment, for DMI-assisted wireless systems. Existing works such as [7] considered i.n.i.d. fading environments for single-IRS-assisted wireless systems. In our work, we consider i.n.i.d. fading environments for multi-IRS-assisted wireless communications. The consideration of the i.n.i.d. fading makes the analysis more involved.
- We introduce a framework for implementing the method of moments to statistically characterize the e2e channel of any wireless system, including multi-IRS-assisted wireless systems. From the proposed framework, we highlight a unique finding that both Gamma and log-normal distributions can be used to characterize the e2e fading channel of the considered multi-IRS-assisted wireless systems. This finding has not been reported in the literature.
- In our analysis, we derive general closed-form expressions of the kth moment of the considered random variables (RVs), which are the key components to carry out the method of moments. Thus, based on the derived general expressions and the analytical framework, one can evaluate different performance metrics, including the ergodic capacity (EC) and outage probability (OP).
- In our simulations, we consider realistic simulation settings, as shown in Table 10.1 of the revised manuscript, and use the symbolic Matlab toolbox to overcome the problem of Not a Number (NaN) values caused by Matlab itself. As a result, our simulation results are still corroborated with the analytical results when some simulation parameters, such as the locations of the nodes and the fading parameters, are set randomly.

The structure of the chapter is described as follows: In Section 10.2, we present in detail the signal modelling and mathematical description of the DMI system model. Next, in Section 10.3, we statistically characterize the e2e channel of the considered system and carry out the system performance analysis in terms of EC and OP. In Section 10.4, we present numerical results to validate the developed analysis and to give insights into the DMI system's performance. Finally, we summarize and give conclusions to the research presented in this chapter.

Notations: $\Gamma(\cdot)$ denotes the Gamma function [14, Eq. (8.310.1)]; $\gamma(\cdot,\cdot)$ denotes the lower incomplete Gamma function [14, Eq. (8.350.1)]; $\Gamma(\cdot,\cdot)$ denotes the upper incomplete Gamma function [14, Eq. (8.350.2)]; $K_v(z)$ denotes the modified Bessel function of the second kind [14, Eq. (8.407.1)]; erf$(\cdot)$ denotes the error function [15, Eq. (7.1.1)]; erfc$(\cdot)$ denotes the complementary error function [14, Eq. (8.250.4)]; $G_{p,q}^{m,n}[\cdot]$ denotes the Meijer-G function [16, Eq. (8.2.1.1)]; $\mathbb{E}[X]$ denotes the expectation of X, while $\overline{\mu}_X$ denotes the value of that expectation; $\overline{\mu}_X(k)$ denotes the kth moment of X; and VAR$[X]$ denotes the variance of X.

10.2 System Model

Let us consider a DMI system composed of one source, S, one destination, D, and N distributed IRSs, $R_n, n = 1, 2, \dots, N$, as illustrated in Figure 10.1. The direct link between S and D is available but may experience severe fading and heavy shadowing conditions and would need the additional assistance of the IRSs. S and D are equipped with a single antenna, whereas the nth IRS consists of L_n reflecting elements. In this work, we consider that the IRSs may have different numbers of reflecting elements, i.e. $\left|L_i - L_j\right| \geq 0$, $\forall i \neq j$ and $i, j \in \{1, 2, \dots, N\}$, which may result in different geometric sizes. Let $\mathbf{\Phi}_n = \text{diag}([\kappa_{n1}e^{j\phi_{n1}}, \dots, \kappa_{nl}e^{j\phi_{nl}}, \dots, \kappa_{nL_n}e^{j\phi_{nL_n}}])$ represent the phase shift matrix of the nth IRS, where $\phi_{nl} \in [0, 2\pi)$ and $\kappa_{nl} \in (0, 1]$ represent the phase shift and its corresponding amplitude reflection coefficient, respectively, of the reflecting

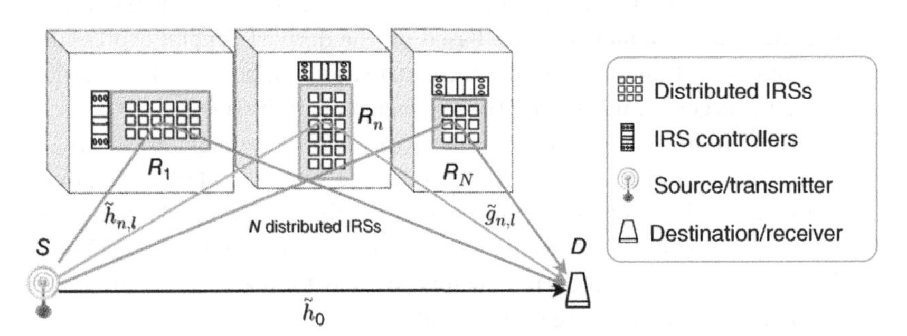

Figure 10.1 Schematic illustration of the considered distributed multi-IRS-assisted (DMI) wireless system.

element l on IRS n. We consider that all the wireless channels in the DMI system undergo Nakagami-m fading, which will be mathematically described next. Let $\tilde{h}_{nl}$ and $\tilde{g}_{nl}$ represent the *complex channel coefficients* of the incident channel from S to element l on IRS n, and the reflecting channel from that element to D, respectively, and let $\tilde{h}_0$ represent the $S \to D$ channel coefficient. The polar representations of the channels can be expressed as $\tilde{h}_0 = h_0 e^{j\angle\tilde{h}_0}$, $\tilde{h}_{nl} = h_{nl} e^{j\angle\tilde{h}_{nl}}$, $\tilde{g}_{nl} = g_{nl} e^{j\angle\tilde{g}_{nl}}$, where $h_0 \triangleq |\tilde{h}_0|$, $h_{nl} \triangleq |\tilde{h}_{nl}|$, and $g_{nl} \triangleq |\tilde{g}_{nl}|$ are *the real-valued channel magnitudes*, and $\angle\tilde{h}_0$, $\angle\tilde{h}_{nl}$, and $\angle\tilde{g}_{nl}$ are *the phases* of $\tilde{h}_0$, $\tilde{h}_{nl}$, and $\tilde{g}_{nl}$, respectively, with $\{\angle\tilde{h}_0, \angle\tilde{h}_{nl}, \angle\tilde{g}_{nl}\} \in [0, 2\pi]$.

In the DMI system, all the IRSs participate in assisting the $S \to D$ transmission, i.e. when S sends its signal x_S aiming for D, all N IRSs also receive this signal and are controlled to adjust their phase shifts to reflect the incident signal x_S towards D simultaneously over the same wireless channel. Herein, we assume that all IRSs steer their beams to D, and thus they cannot interfere with one another, as in [8, 10, 12]. As a result, D receives a superposition (combination) of the direct and reflecting signals. Let x_S, with $\mathbb{E}[|x_S|^2] = 1$, be the signal transmitted by S, using *coherent combination* [8], the received signal at D can be written as follows:

$$y^{\mathrm{DMI}} = \sqrt{P_S} x_S \left(\tilde{h}_0 + \sum_{n=1}^{N} \sum_{l=1}^{L_n} \tilde{g}_{nl} \kappa_{nl} e^{j\phi_{nl}} \tilde{h}_{nl} \right) + w_D \tag{10.1}$$

where P_S (dBm) denotes the transmit power and w_D denotes the circular complex Gaussian noise at D, i.e. $w_D idd \sim \mathcal{CN}(0, \sigma_D^2)$. Using the polar representations of the channels, the received SNR at D can be expressed as follows:

$$\begin{aligned} \mathrm{SNR}^{\mathrm{DMI}} &= \overline{\mathrm{SNR}} \left| h_0 e^{j\angle\tilde{h}_0} + \sum_{n=1}^{N} \sum_{l=1}^{L_n} g_{nl} \kappa_{nl} h_{nl} e^{j(\phi_{nl} + \angle\tilde{h}_{nl} + \angle\tilde{g}_{nl})} \right|^2 \\ &= \overline{\mathrm{SNR}} \left| h_0 + \sum_{n=1}^{N} \sum_{l=1}^{L_n} g_{nl} \kappa_{nl} h_{nl} e^{j(\phi_{nl} + \angle\tilde{h}_{nl} + \angle\tilde{g}_{nl} - \angle\tilde{h}_0)} \right|^2 \left| e^{j\angle\tilde{h}_0} \right|^2 \end{aligned} \tag{10.2}$$

where $\overline{\mathrm{SNR}} = P_S / \sigma_D^2$ denotes the *average transmit SNR* (dB).

To counteract the destructive effect of multi-path fading caused by the cascaded channels, the IRS phase shifts are reconfigured based on the optimal phase shift configuration, resulting in a constructive superposition of the direct and reflecting signals, which lead to the highest received SNR. Mathematically, the optimal phase shift configuration of reflecting element l on IRS n, denoted by ϕ_{nl}^*, satisfies

$$\mathrm{SNR}^{\mathrm{DMI}}(\phi_{nl}^*) = \sup_{\phi_{nl} \in [0, 2\pi)} \mathrm{SNR}^{\mathrm{DMI}}(\phi_{nl}), \forall l, \forall n \tag{10.3}$$

In (10.2), it is noted that $\left| e^{j\angle\tilde{h}_0} \right|^2 = 1, \forall \angle\tilde{h}_0$. Assuming that the channel state information (CSI) is available at S, by applying the triangle inequality for complex numbers to (10.2), we can figure out that the optimal solution for (10.3) can be obtained as follows:

$$\phi^*_{nl} = \angle\tilde{h}_0 - (\angle\tilde{h}_{nl} + \angle\tilde{g}_{nl}) \tag{10.4}$$

The solution in (10.4) yields a coherent combined signal at D in (10.1) having the largest amplitude. Knowing that $e^{j(\phi+2k\pi)} = e^{j\phi}, \phi^*_{nl} \in [0, 2\pi)$, the received SNR at D can be re-expressed as follows:

$$\mathrm{SNR}^{\mathrm{DMI}} = \overline{\mathrm{SNR}}\left|h_0 + \sum_{n=1}^{N}\sum_{l=1}^{L_n}\kappa_{nl}h_{nl}g_{nl}\right|^2 \tag{10.5}$$

In this section, we present in detail the mathematical description of the DMI system and its signal modelling. As can be observed from (10.5), the SNR is a function of $(2NL_n + 1)$ RVs, thus determining the exact distribution of the SNR is technically challenging. To deal with this problem, we propose a framework to obtain the statistical characterization of the e2e channel, as presented in Section 10.3.

10.3 Channel Characterization and Performance Analysis

First, we present some key distributions that will be used throughout the analysis. Let X be an RV that follows a Nakagami-m distribution, its cumulative distribution function (CDF) and probability density function (PDF) are given by [17]

$$F_X(x; m, \Theta) = \frac{\gamma\left(m, \frac{m}{\Theta}x^2\right)}{\Gamma(m)} \tag{10.6}$$

$$f_X(x; m, \Theta) = \frac{2m^m}{\Gamma(m)\Theta^m}x^{2m-1}e^{-\frac{m}{\Theta}x^2} \tag{10.7}$$

respectively, where $m > 0$ and $\Theta > 0$ are *shape* and *spread* parameters of the Nakagami-m distribution, respectively. Hereafter, we alternatively use the following representation to denote a Nakagami-m RV: $X \sim \mathrm{Nakagami}(m, \Theta)$. Thus, the magnitude distribution of each individual channel can be expressed as $h_0 \sim \mathrm{Nakagami}(m_0, \Theta_0)$, $h_{nl} \sim \mathrm{Nakagami}(m_{h_n}, \Theta_{h_n})$, and $g_{nl} \sim \mathrm{Nakagami}(m_{g_n}, \Theta_{g_n})$, where $l = 1, 2, \ldots, L_n$ and $n = 1, 2, \ldots, N$. Let Y be an RV that follows a Gamma distribution, its CDF and PDF are given by [17]

$$F_Y(y; \alpha, \varphi) = \frac{\gamma(\alpha, \varphi y)}{\Gamma(\alpha)}, y \geq 0 \tag{10.8}$$

$$f_Y(y; \alpha, \varphi) = \frac{\varphi^\alpha}{\Gamma(\alpha)}y^{\alpha-1}e^{-\varphi y}, y \geq 0 \tag{10.9}$$

respectively, where $\alpha > 0$ and $\varphi > 0$ denote the *shape* and *rate* parameters of the Gamma distribution, respectively. Hereafter, we alternatively use the following representation to denote a Gamma RV: $Y \sim \text{Gamma}(\alpha, \varphi)$. Let W be an RV that follows a log-normal distribution, its CDF and PDF are given by [17]

$$F_W(w; \nu, \eta) = \frac{1}{2} + \frac{1}{2}\,\text{erf}\left(\frac{\ln w - \nu}{\sqrt{2\eta^2}}\right) \tag{10.10}$$

$$f_W(w; \nu, \eta) = \frac{1}{w\sqrt{2\pi\eta^2}} e^{-\frac{(\ln w - \nu)^2}{2\eta^2}} \tag{10.11}$$

respectively, where ν and η^2, with $\eta > 0$, are the *mean* and *variance* of the log-normal distribution, respectively. Hereafter, we alternatively use the following representation to denote a log-normal RV: $W \sim \text{lognormal}(\nu, \eta)$.

For the sake of notational convenience, we define some RVs as follows: $U_{nl} \triangleq \kappa_{nl} h_{nl} g_{nl}$, $V_n \triangleq \sum_{l=1}^{L_n} U_{nl}$, $T \triangleq \sum_{n=1}^{N} V_n$, and $Z \triangleq h_0 + T$. Thus, from (10.1), the magnitude of the e2e channel of the DMI system can be represented by the RV Z. Because of the complicated structure of Z, deriving the true distribution of Z is infeasible. Therefore, to address this particular technical problem, we propose a framework for determining an approximation of the true distribution, which is presented as a three-step framework, as shown below:

- **Step 1**: We begin by determining which parametric distribution best approximates the true distribution. First, we simulate the true distribution using the data collected for the DMI system simulation. Through computer simulation, the true distribution is then heuristically and numerically matched to some candidate distributions, e.g. Gaussian, Burr, Gamma Weibull, log-normal distributions, merely to name a few, by using Matlab toolboxes, such as the fitdist function in the Statistics and Machine Learning Toolbox [18]. The best candidate distributions are then chosen, as they provide the highest goodness of fit, resulting in the most accurate numerical results for performance metrics of interest, e.g. EC and/or OP as will be shown in the following subsections.
- **Step 2**: In particular, we determine the probability distribution and associated parameters of the chosen distribution. To do so, we rely on the knowledge of the statistical characteristics of the true distribution of each component of Z, i.e. the kth moment of each individual RV. The main derivation is to find the estimators for the statistical characteristics, i.e. the parameters, of the chosen distribution. More specifically, to obtain the estimators, we match the unknown first kth moments of the chosen distribution with the population moments of the true distribution. Mathematically, we form a system of equations of unknown estimators and population moments and solve it to find the expressions of the estimators. For instance, if the chosen distribution to approximate the true distribution of the e2e channel magnitude is a parametric Gamma distribution with

the PDF given in (10.9), the challenging part is to determine the expressions of the estimators for α and φ of the Gamma distribution.

- **Step 3**: Once we obtain the statistical characteristics of the chosen distribution, we evaluate the accuracy of the obtained approximate distribution by examining the difference between the true and approximate distributions using the Kullback–Leibler divergence and the Kolmogorov–Smirnov test.

Next, we apply the proposed framework to determine the statistical characterization of Z, as presented in the following subsections.

10.3.1 Gamma Distribution-based Statistical Channel Characterization

Using the proposed framework, we prove that the Gamma distribution can accurately approximate the true distribution of Z, as stated in the following theorem.

Theorem 10.1 *The Gamma distribution with two parameters, α_Z and φ_Z, can be used to highly approximate the true distribution of Z, i.e. $Z \overset{\text{approx.}}{\sim} \text{Gamma}(\alpha_Z, \varphi_Z)$, where α_Z and φ_Z are determined as follows:*

$$\alpha_Z = \frac{(\mathbb{E}[Z])^2}{\text{VAR}[Z]} = \frac{[\overline{\mu}_Z(1)]^2}{\overline{\mu}_Z(2) - [\overline{\mu}_Z(1)]^2} \tag{10.12}$$

$$\varphi_Z = \frac{\mathbb{E}[Z]}{\text{VAR}[Z]} = \frac{\overline{\mu}_Z(1)}{\overline{\mu}_Z(2) - [\overline{\mu}_Z(1)]^2} \tag{10.13}$$

respectively, where $\overline{\mu}_Z(1)$ and $\overline{\mu}_Z(2)$ are derived as in (10.28) and (10.29), respectively. Plugging α_Z and φ_Z into (10.8) and (10.9), we obtained the approximate CDF and PDF of Z, i.e. $F_Z(z; \alpha_Z, \varphi_Z)$ and $f_Z(z; \alpha_Z, \varphi_Z)$, respectively.

Proof: Recalling that $h_0 \sim \text{Nakagami}(m_0, \Theta_0)$, the kth moment of h_0, which is defined as $\overline{\mu}_{h_0}(k) \triangleq \mathbb{E}[(h_0)^k]$, can be derived after several mathematical steps as follows:

$$\overline{\mu}_{h_0}(k) = \frac{\Gamma(m_0 + k/2)}{\Gamma(m_0)} \left(\frac{m_0}{\Theta_0} \right)^{-k/2} \tag{10.14}$$

We now shift our attention to the kth moment of U_{nl}. Recalling that, $U_{nl} = \kappa_{nl} h_{nl} g_{nl}$, where $U_{n1}, \dots, U_{nL_n}$ are i.i.d. RVs for a given n; however, h_{nl} and g_{nl} are i.n.i.d. RVs, $\forall n, \forall l$. Moreover, knowing that $f_{XY}(z) = \int_0^\infty \frac{1}{x} f_Y\left(\frac{z}{x}\right) f_X(x) dx$ for positive RVs X and Y, the PDF of U_{nl} can be expressed as follows:

$$f_{U_{nl}}(z) = \int_0^\infty \frac{1}{\kappa_{nl} x} f_{h_{nl}}\left(\frac{z}{\kappa_{nl} x}\right) f_{g_{nl}}(x) dx \tag{10.15}$$

As $h_{nl} \sim \text{Nakagami}(m_{h_n}, \Theta_{h_n})$ and $g_{nl} \sim \text{Nakagami}(m_{g_n}, \Theta_{g_n})$, from the PDF in (10.7), the PDF of U_{nl} can be further expressed as follows:

$$f_{U_{nl}}(z) = \frac{2(m_{h_n})^{m_{h_n}}}{\Gamma(m_{h_n})(\Theta_{h_n})^{m_{h_n}}} \frac{2(m_{g_n})^{m_{g_n}}}{\Gamma(m_{g_n})(\Theta_{g_n})^{m_{g_n}}} \frac{z^{2m_{h_n}-1}}{(\kappa_{nl})^{2m_{h_n}}} \times \int_0^\infty x^{2m_{g_n}-2m_{h_n}-1} \exp\left(-\frac{z^2}{(\kappa_{nl})^2} \frac{m_{h_n}}{\Theta_{h_n}} \frac{1}{x^2} - \frac{m_{g_n}}{\Theta_{g_n}} x^2\right) dx \quad (10.16)$$

Utilizing [[14], Eq. (3.478.4)], after some mathematical derivations, the exact PDF of U_{nl} is derived as follows:

$$f_{U_{nl}}(z) = \frac{4(\lambda_{nl})^{m_{h_n}+m_{g_n}}}{\Gamma(m_{h_n})\Gamma(m_{g_n})} x^{m_{h_n}+m_{g_n}-1} K_{m_{g_n}-m_{h_n}}(2\lambda_{nl} z) \quad (10.17)$$

where $\lambda_{nl} = \sqrt{\frac{1}{\kappa_{nl}^2} \frac{m_{h_n}}{\Theta_{h_n}} \frac{m_{g_n}}{\Theta_{g_n}}}$. Next, the kth moment of U_{nl}, which is defined as $\overline{\mu}_{U_{nl}}(k) \triangleq \mathbb{E}[(U_{nl})^k]$ can be derived after some mathematical steps as follows:

$$\begin{aligned} \overline{\mu}_{U_{nl}}(k) &= \int_0^\infty z^k f_{U_{nl}}(z) dz \\ &= (\lambda_{nl})^{-k} \frac{\Gamma(m_{h_n} + k/2)\Gamma(m_{g_n} + k/2)}{\Gamma(m_{h_n})\Gamma(m_{g_n})} \end{aligned} \quad (10.18)$$

It is noted that the exact PDF of U_{nl} in (10.17) can be used to obtain $\overline{\mu}_{U_{nl}}(k)$; however, if we use (10.17), the derivation of exact closed-form expressions of the $F_Z(z)$ and $f_Z(z)$ becomes intractable. To overcome this problem, utilizing (10.18), we match the true distribution of U_{nl} to a *Gamma* distribution. It is noted that U_{nl} are i.i.d. RVs with respect to $l = 1, 2, \ldots, L_n$ for a given n. Following Step 2 in our proposed framework, we point out that

$$U_{nl} \overset{\text{approx.}}{\sim} \text{Gamma}(\alpha_{U_n}, \varphi_{U_n}) \quad (10.19)$$

where the parameters of the Gamma distribution are given by

$$\alpha_{U_n} = \frac{(\mathbb{E}[U_{nl}])^2}{\text{VAR}[U_{nl}]} = \frac{[\overline{\mu}_{U_{nl}}(1)]^2}{\overline{\mu}_{U_{nl}}(2) - [\overline{\mu}_{U_{nl}}(1)]^2} \quad (10.20)$$

$$\varphi_{U_n} = \frac{\mathbb{E}[U_{nl}]}{\text{VAR}[U_{nl}]} = \frac{\overline{\mu}_{U_{nl}}(1)}{\overline{\mu}_{U_{nl}}(2) - [\overline{\mu}_{U_{nl}}(1)]^2} \quad (10.21)$$

and $f_{U_{nl}}(z; \alpha_{U_n}, \varphi_{U_n})$ is presented in (10.9). To simplify the analysis, we assume that amplitude reflection coefficients of all reflecting elements on the same IRS are identical, i.e. $\kappa_{nl} = \kappa_n, \forall l$, as in [19]. Next, the true distribution of V_n can be approximated by a Gamma distribution as follows:

$$V_n \overset{\text{approx.}}{\sim} \text{Gamma}(L_n \alpha_{U_n}, \varphi_{U_n}) \quad (10.22)$$

After some mathematical derivations, the closed-form expression of approximate CDF and PDF of V_n are obtained as follows:

$$F_{V_n}(z) \approx \frac{1}{\Gamma\left(L_n\alpha_{U_n}\right)}\gamma(L_n\alpha_{U_n}, \varphi_{U_n}z) \tag{10.23}$$

$$f_{V_n}(z) \approx f_{V_n}(z; L_n\alpha_{U_n}, \varphi_{U_n}) \tag{10.24}$$

where the expression of $f_{V_n}(z; L_n\alpha_{U_n}, \varphi_{U_n})$ has the formula presented in (10.9). Next, using the multinomial expansion formula [20], the kth moment of V_n, i.e. $\overline{\mu}_{V_n}(k) \triangleq \mathbb{E}[(V_n)^k]$, can be further written as follows:

$$\overline{\mu}_{V_n}(k) = \sum_{k_1=0}^{k}\sum_{k_2=0}^{k_1}\cdots\sum_{k_{L_n-1}=0}^{k_{L_n-2}}\binom{k}{k_1}\binom{k_1}{k_2}\cdots\binom{k_{L_n-2}}{k_{L_n-1}}$$
$$\times\overline{\mu}_{U_{n1}}(k-k_1)\overline{\mu}_{U_{n2}}(k_1-k_2)\ldots\overline{\mu}_{U_{nL_n}}(k_{L_n-1}) \tag{10.25}$$

We now shift our attention to the kth moment of T, which is defined as $\overline{\mu}_T(k) \triangleq \mathbb{E}[T^k]$. After a number of mathematical steps, the closed-form expression of $\overline{\mu}_T(k)$ can be obtained as follows:

$$\overline{\mu}_T(k) = \sum_{k_1=0}^{k}\sum_{k_2=0}^{k_1}\cdots\sum_{k_{N-1}=0}^{k_{N-2}}\binom{k}{k_1}\binom{k_1}{k_2}\cdots\binom{k_{N-2}}{k_{N-1}}$$
$$\times\overline{\mu}_{V_1}(k-k_1)\overline{\mu}_{V_2}(k_1-k_2)\ldots\overline{\mu}_{V_n}(k_{N-1}) \tag{10.26}$$

It is noted that h_0 and T are mutually independent RVs, thus, the kth moment of Z, which is defined as $\overline{\mu}_Z(k) \triangleq \mathbb{E}[Z^k]$, can be obtained based on the its summands, i.e. $\overline{\mu}_Z(k) = \mathbb{E}[(h_0+T)^k]$. Specifically, using the binomial theorem [20], and after several mathematical steps, the closed-form expression of $\overline{\mu}_Z(k)$ can be derived as

$$\begin{aligned}\overline{\mu}_Z(k) &= \mathbb{E}\left[\sum_{l=0}^{k}\binom{k}{l}(h_0)^l T^{k-l}\right]\\ &= \sum_{l=0}^{k}\binom{k}{l}\mathbb{E}\left[(h_0)^l T^{k-l}\right]\\ &= \sum_{l=0}^{k}\binom{k}{l}\mathbb{E}\left[(h_0)^l\right]\mathbb{E}\left[T^{k-l}\right]\\ &= \sum_{l=0}^{k}\binom{k}{l}\overline{\mu}_{h_0}(l)\overline{\mu}_T(k-l)\end{aligned} \tag{10.27}$$

As a result, from (10.27), the first and second moments of Z can be written as follows:

$$\overline{\mu}_Z(1) = \overline{\mu}_{h_0}(1) + \overline{\mu}_T(1) \tag{10.28}$$

$$\overline{\mu}_Z(2) = \overline{\mu}_{h_0}(2) + \overline{\mu}_T(2) + 2\overline{\mu}_{h_0}(1)\overline{\mu}_T(1) \tag{10.29}$$

respectively. From (10.14), the first and second moments of h_0 can be expressed as follows:

$$\overline{\mu}_{h_0}(1) = \frac{\Gamma(m_0 + 1/2)}{\Gamma(m_0)} \sqrt{\frac{\Theta_0}{m_0}} \tag{10.30}$$

$$\overline{\mu}_{h_0}(2) = \frac{\Gamma(m_0 + 1)}{\Gamma(m_0)} \frac{\Theta_0}{m_0} = \Theta_0 \tag{10.31}$$

respectively. Plugging (10.30) into (10.28), and combining (10.31), (10.26), and (10.25), we complete the proof of Theorem 10.1. ■

Furthermore, using the fact that for non-negative X and Y RVs, where $Y = cX^2$, we have $f_Y(y) = \frac{1}{2\sqrt{cy}} f_X(\sqrt{y/c})$ and $F_Y(y) = F_X(\sqrt{y/c})$, the approximate closed-form expressions of the CDF and PDF of Z^2 can be derived as follows:

$$f_{Z^2}(x) \approx \frac{{\varphi_Z}^{\alpha_Z}}{2\sqrt{x}\Gamma(\alpha_Z)} \left(\sqrt{x}\right)^{\alpha_Z - 1} e^{-\varphi_Z \sqrt{x}} \tag{10.32}$$

$$F_{Z^2}(x) \approx \frac{\gamma\left(\alpha_Z, \varphi_Z \sqrt{x}\right)}{\Gamma(\alpha_Z)} \tag{10.33}$$

respectively.

10.3.1.1 Gamma Distribution-based Ergodic Capacity Analysis

The EC of the DMI system is determined as the expected value of the instantaneous achievable rate over a number of channel realizations. Mathematically, the EC can be written as follows:

$$\begin{aligned} \overline{C}^{\text{DMI,Gam}} &= \mathbb{E}\left[\log_2\left(1 + \text{SNR}^{\text{DMI}}\right)\right] \\ &= \mathbb{E}\left[\log_2\left(1 + \overline{\text{SNR}}Z^2\right)\right] \end{aligned} \tag{10.34}$$

The EC in (10.34) can be further expressed as follows:

$$\begin{aligned} \overline{C}^{\text{DMI,Gam}} &= \int_0^\infty \log_2(1+z) f_{\overline{\text{SNR}}Z^2}(z) dz \\ &= \{\log_2(1+z)[F_{\overline{\text{SNR}}Z^2}(z) - 1]\}_0^\infty \\ &\quad - \frac{1}{\ln 2} \int_0^\infty \frac{1}{z+1} \left[F_{\overline{\text{SNR}}Z^2}(z) - 1\right] dz \\ &= \frac{1}{\ln 2} \int_0^\infty \frac{1}{z+1} \left[1 - F_Z\left(\sqrt{\frac{z}{\overline{\text{SNR}}}}\right)\right] dz \end{aligned} \tag{10.35}$$

where $\{g(x)\}_a^b = g(b) - g(a)$. Using the fact that $(1+x)^{-\xi} = \frac{1}{\Gamma(\xi)} G_{1,1}^{1,1}\left(x \middle| \begin{matrix} 1-\xi \\ 0 \end{matrix}\right)$ [[16], Eq. (8.4.2.5)], (10.35) can be further expressed as

$$\overline{C}^{\text{DMI,Gam}} \approx \frac{1}{\ln 2} \int_0^\infty G_{1,1}^{1,1}\left(z \middle| \begin{matrix} 0 \\ 0 \end{matrix}\right) \left[1 - \frac{\gamma\left(\alpha_Z, \varphi_Z \sqrt{\frac{z}{\overline{\text{SNR}}}}\right)}{\Gamma(\alpha_Z)}\right] dz \tag{10.36}$$

Using the following relationship of incomplete Gamma functions: $\Gamma(\alpha, x) + \gamma(\alpha, x) = \Gamma(\alpha)$ [14, Eq. (8.356.3)], we have

$$1 - \frac{\gamma\left(\alpha_Z, \varphi_Z \sqrt{\frac{z}{\overline{\text{SNR}}}}\right)}{\Gamma(\alpha_Z)} = \frac{\Gamma\left(\alpha_Z, \varphi_Z \sqrt{\frac{z}{\overline{\text{SNR}}}}\right)}{\Gamma(\alpha_Z)} \tag{10.37}$$

Using the incomplete Gamma function and the Meijer-G function's equivalence, i.e. $\Gamma\left(\alpha_Z, \varphi_Z \sqrt{\frac{z}{\overline{\text{SNR}}}}\right) = G_{1,2}^{2,0}\left(\varphi_Z \sqrt{\frac{z}{\overline{\text{SNR}}}} \middle| \begin{matrix} 1 \\ \alpha_Z, 0 \end{matrix}\right)$ [16, Eq. (8.4.16.2)], (10.37) can be further expressed as follows:

$$\overline{C}^{\text{DMI,Gam}} \approx \frac{1}{\Gamma(\alpha_Z) \ln 2} \int_0^\infty G_{1,1}^{1,1}\left(z \middle| \begin{matrix} 0 \\ 0 \end{matrix}\right) G_{1,2}^{2,0}\left(\varphi_Z \sqrt{\frac{z}{\overline{\text{SNR}}}} \middle| \begin{matrix} 1 \\ \alpha_Z, 0 \end{matrix}\right) dz \tag{10.38}$$

Next, the integral in (10.38) has the form of

$$\int_0^\infty x^{\alpha-1} G_{u,v}^{s,t}\left(\xi x \middle| \begin{matrix} (c_u) \\ (d_v) \end{matrix}\right) G_{p,q}^{m,n}\left(\omega x^{l/k} \middle| \begin{matrix} (a_p) \\ (b_q) \end{matrix}\right) dx$$

which can be expressed by another Meijer-G function using the identity [16, Eq. (2.24.1.1)]. After some mathematical derivations, an approximate closed-form expression of the EC is derived as follows:

$$\overline{C}^{\text{DMI,Gam}} \approx \frac{1}{\Gamma(\alpha_Z) \ln 2} \frac{2^{\alpha_Z - 1}}{\sqrt{\pi}} G_{3,5}^{5,1}\left(\frac{(\varphi_Z)^2}{4\overline{\text{SNR}}} \middle| \begin{matrix} 0, \frac{1}{2}, 1 \\ \frac{\alpha_Z}{2}, \frac{\alpha_Z+1}{2}, 0, \frac{1}{2}, 0 \end{matrix}\right) \tag{10.39}$$

10.3.1.2 Gamma Distribution-based Outage Probability Analysis

The OP is the probability of outage events that occur when the instantaneous achievable rate of the DMI system falls below a pre-defined spectral efficiency (SE) threshold R_{th} (b/s/Hz), i.e. $P_{\text{out}}^{\text{DMI}} = \Pr(\log_2(\text{SNR}^{\text{DMI}} + 1) < R_{\text{th}})$. From (10.5) and Theorem 10.1, an approximate closed-form expression of the OP is derived as

$$P_{\text{out}}^{\text{DMI,Gam}} = \Pr\left(\overline{\text{SNR}} Z^2 \leq \text{SNR}_{\text{th}}\right) \overset{(a)}{\approx} F_Z\left(\sqrt{\text{SNR}_{\text{th}} / \overline{\text{SNR}}}\right) \tag{10.40}$$

where $\mathrm{SNR}_{\mathrm{th}} = 2^{R_{\mathrm{th}}} - 1$ and step (a) in (10.40) is carried out as a result of the fact that $F_{X^2}(x) = F_X(\sqrt{x}), x > 0$.

10.3.2 Log-normal Distribution-based Statistical Channel Characterization

To gain a better understanding of the channel modelling and characterization of the DMI system, we show that the true distribution of the e2e channel magnitude can also be approximated by a log-normal distribution, which is stated in the following theorem.

Theorem 10.2 *The log-normal distribution with two parameters, ν_Z and η_Z, which are derived as in (10.41) and (10.42), can be used to approximate the true distribution of Z, i.e. $Z \overset{\text{approx.}}{\sim} \text{lognormal}(\nu_Z, \eta_Z)$. By plugging ν_Z and η_Z into (10.10) and (10.11), we obtained the approximate CDF and PDF of Z, i.e. $F_Z(z; \nu_Z, \eta_Z)$ and $f_Z(z; \nu_Z, \eta_Z)$, respectively.*

Proof: We follow Step 2 in our proposed framework to determine the two parameters ν_Z and η_Z of the approximate log-normal distribution of Z. Specifically, we consider an arbitrary log-normal RV X and match the first and second moments of Z with that of X. Mathematically, we formulate the following system of equations: $\mathbb{E}[Z] = \mathbb{E}[X]$ and $\mathbb{E}[Z^2] = \mathbb{E}[X^2]$. Following Fenton [21], after a number of mathematical steps, the estimators of ν_Z and η_Z can be determined as follows:

$$\nu_Z = \ln\left(\frac{(\mathbb{E}[Z])^2}{\sqrt{\mathbb{E}[(Z)^2]}}\right) = \ln\left(\frac{\left(\overline{\mu}_Z(1)\right)^2}{\sqrt{\overline{\mu}_Z(2)}}\right) \tag{10.41}$$

$$\eta_Z = \sqrt{\ln\left(\frac{\mathbb{E}[Z^2]}{(\mathbb{E}[Z])^2}\right)} = \sqrt{\ln\left(\frac{\overline{\mu}_Z(2)}{(\overline{\mu}_Z(1))^2}\right)} \tag{10.42}$$

Next, we plug $\overline{\mu}_Z(1)$ and $\overline{\mu}_Z(2)$, determined in (10.28) and (10.29), respectively, into (10.41) and (10.42) to obtain the values of ν_Z and η_Z, respectively. As a result, by plugging ν_Z and η_Z into (10.10) and (10.11), we obtain $F_Z(x; , \nu_Z, \eta_Z)$ and $f_Z(x; \nu_Z, \eta_Z)$, respectively. This concludes the proof of Theorem 10.2. ■

10.3.2.1 Log-normal Distribution-based Ergodic Capacity Analysis

Using the result stated in Theorem 10.2, the EC can be determined as $\overline{C}^{\mathrm{DMI,LN}} = \mathbb{E}\left[\log_2\left(1 + \overline{\mathrm{SNR}}Z^2\right)\right]$, which is derived as follows. We first introduce the following lemma to support the derivation of $\overline{C}^{\mathrm{DMI,LN}}$.

Lemma 10.1 *Given $X \sim \text{lognormal}(\nu_X, \eta_X)$, and let $\overline{C}_X = \mathbb{E}\left[\log_2\left(1 + \overline{\text{SNR}}X\right)\right]$, the approximate closed-form expression of $\overline{C}_X$ is $\Psi(\nu_X, \eta_X)$, where*

$$\Psi(\nu_X, \eta_X) \triangleq \frac{1}{\ln(2)}\left[\Xi\left(\frac{1}{\eta\sqrt{2}}, \frac{\ln(\overline{\text{SNR}}) + \nu}{\eta\sqrt{2}}\right) + \Xi\left(\frac{1}{\eta\sqrt{2}}, -\frac{\ln(\overline{\text{SNR}}) + \nu}{\eta\sqrt{2}}\right)\right.$$
$$\left. + \frac{\eta}{\sqrt{2\pi}} e^{-\frac{(\ln(\overline{\text{SNR}})+\nu)^2}{2\eta^2}} + \frac{\ln(\overline{\text{SNR}}) + \nu}{2}\,\text{erfc}\left(-\frac{\ln(\overline{\text{SNR}}) + \nu}{\eta\sqrt{2}}\right)\right] \tag{10.43}$$

where $\Xi(\cdot,\cdot)$ is defined in (10.47) and its closed-form expression is provided in (10.48).

Proof: From the definition of $\overline{C}_X$ in Lemma 10.1, $\overline{C}_X$ can be expressed as

$$\overline{C}_X = \int_0^{\infty} \log_2(1 + x)\frac{1}{\overline{\text{SNR}}} f_X\left(\frac{x}{\overline{\text{SNR}}}\right) dx \tag{10.44}$$

Substituting (10.11) into (10.44) yields

$$\overline{C}_X = \frac{1}{\overline{\text{SNR}}\eta\sqrt{2\pi}} \int_0^{\infty} \frac{\log_2(1 + x)}{x/\overline{\text{SNR}}} e^{-\frac{(\ln(x/\overline{\text{SNR}})-\nu)^2}{2\eta^2}} dx \tag{10.45}$$

Knowing the fact that $\log_2(1 + x) = \ln(1 + x)/\ln(2)$, $\overline{C}_X$ can be detailed as

$$\overline{C}_X = \frac{1}{\eta\sqrt{2\pi}\ln(2)} \int_0^{\infty} \frac{\ln(1 + x)}{x} e^{-\frac{[\ln(x)-(\ln(\overline{\text{SNR}})+\nu)]^2}{2\eta^2}} dx \tag{10.46}$$

Next, we define a function $\Xi(\cdot,\cdot)$ as follows

$$\Xi(a, b) \triangleq \frac{a}{\sqrt{\pi}} \int_0^1 e^{-[a\ln(x)-b]^2} \frac{\ln(1 + x)}{x} dx \tag{10.47}$$

To derive $\Xi(a, b)$, we first expand the term $\ln(1 + x)$ using the following identity: $\ln(1 + x) = \sum_{k=1}^{8} a_k x^k + \epsilon(x)$, $0 < x < 1$, provided by [22]; next, using the relationship in [14, Eq. (8.252.6)] and the identity $1 - \text{erf}(x) = \text{erfc}(x)$ [14, Eq. (8.250.4)], after some mathematical steps, the approximate closed-form expression of $\Xi(a, b)$ in (10.47) can be obtained as

$$\Xi(a, b) \approx \frac{e^{b^2}}{2}\sum_{k=1}^{8} a_k\,\text{erfc}\left(b + \frac{k}{2a}\right)\exp\left(\left(b + \frac{k}{2a}\right)^2\right) \tag{10.48}$$

With the expression of $\Xi(a, b)$ in (10.48), following the similar derivation steps as in [22, Subsection II-A], the approximate closed-form expression of the EC based on log-normal distribution, i.e. $\Psi(\nu_X, \eta_X)$, can be obtained as in (10.43). This concludes the proof of Lemma 10.1. ■

Invoking Theorem 10.2, i.e. $Z \overset{\text{approx.}}{\sim} \text{lognormal}(\nu_Z, \eta_Z)$, the PDF of Z^2 can be obtained; however, if such is the PDF of Z^2, we cannot obtain the closed-form expression of (10.44). To deal with this problem, we perform again Step 2 of the framework for the case ofZ^2, and after a number of mathematical steps, we obtain the distribution ofZ^2 as follows: $Z^2 \overset{approx.}{\sim}$ lognormal(ν_{Z^2}, η_{Z^2}), where $\nu_{Z^2} = \ln\left(\frac{(\overline{\mu}_Z(2))^2}{\sqrt{\overline{\mu}_Z(4)}}\right)$, $\eta_{Z^2} = \sqrt{\ln\left(\frac{\overline{\mu}_Z(4)}{(\overline{\mu}_Z(2))^2}\right)}$. Invoking (10.44) in Lemma 10.1, an approximate closed-form expression of $\overline{C}^{DMI,LN}$ can be obtained as follows:

$$\overline{C}^{\mathrm{DMI,LN}} \approx \Psi(\nu_{Z^2}, \eta_{Z^2}) \tag{10.49}$$

where $\Psi(\cdot,\cdot)$ is presented in (10.43).

10.3.2.2 Log-normal Distribution-based Outage Probability Analysis

Using the log-normal approximation of Z in Theorem 10.2, a new approximate closed-form expression of the OP can be obtained as follows:

$$P_{\mathrm{out}}^{\mathrm{DMI,LN}} = \Pr\left(\overline{\mathrm{SNR}}Z^2 \le \mathrm{SNR}_{\mathrm{th}}\right)$$

$$= F_Z\left(\sqrt{\mathrm{SNR}_{\mathrm{th}}/\overline{\mathrm{SNR}}}\right) \tag{10.50}$$

$$\approx \frac{1}{2} + \frac{1}{2}\,\mathrm{erf}\left(\frac{\ln\left(\sqrt{\mathrm{SNR}_{\mathrm{th}}/\overline{\mathrm{SNR}}}\right) - \nu_Z}{\sqrt{2\eta_Z^2}}\right) \tag{10.51}$$

where we use the CDF of Z in (10.10) to approximate (10.50) as (10.51).

In this section, we show that the e2e channel of the DMI system can be statically characterized by either Gamma distribution or log-normal distribution, as show in Sections 1.3.2 and 1.3.3, respectively. In addition, we use the obtained channel characterizations to evaluate the DMI system performance in terms of EC and OP. In the next section, we are going to verify the obtained analytical result through Monte-Carlo simulation and give more insights into the system performance through numerical results.

10.4 Numerical Results and Discussions

In this section, we present representative results of the approximate Gamma and log-normal distributions, OP, and EC, not only to validate the developed analysis but also to provide insights into the e2e performance. One of the technical contributions is that in our code, we use symbolic computations from the Symbolic Math Toolbox [18]. The reason for using symbolic functions is that the performance analysis of the DMI system over i.n.i.d. fading channels involves a huge number of RVs, especially when the number of reflecting elements on each IRS is large. When the floating-point arithmetic programming of Matlab handles such a large number of RVs, some non-symbolic operations are invalid, and the Matlab program returns not-a-number (NaN) values, which lead to inaccurate analytical results. Thus, to obtain accurate results, we need to use variable-precision arithmetic (vpa) functions, which can return values presented with 32 significant decimal digits of precision.

The simulation parameters are set to the values shown in Table 10.1 unless otherwise specified. Specifically, the equivalent noise power at D is calculated as $\sigma_D^2 = N_0 + 10\log(\mathrm{BW}) + \mathrm{NF}$ (dBm). To model the non-line-of-sight (NLoS) condition, we use the 3GPP Urban Micro (UMi) path-loss model [12, 19, 23]. It is noted that the path-loss is represented by the spread parameter of the Nakagami-m distribution, i.e. Θ_{XY} (dB) $= G_X + G_Y - 22.7 - 26\log(f_c) - 36.7\log(d_{XY}/d_0)$, where d_{XY} (m) is the distance between X and Y, where $X \in \{S, R_n\}$, $Y \in \{R_n, D\}$, and d_0 (m) stands for the reference distance, herein, $d_0 = 1$ m. To represent the number of elements installed at an individual IRS, we define the following vector: $\mathbf{R}_i = \{L_n^{\mathbf{R}_i} : n = 1, \dots, N\}$. To generate the numerical results, we consider two reflecting element settings, namely $\mathbf{R}_1$ and $\mathbf{R}_2$, where $L_n^{\mathbf{R}_1} = 20$ and $L_n^{\mathbf{R}_2} = 30$, respectively, with $n = 1, \dots, N$. The IRSs are uniformly deployed as $x_{R_n} \sim \mathcal{U}[5, 95]$, $y_{R_n} \sim \mathcal{U}[1, 9]$, whereas the coordinates of S and D are set as $(0, 0)$ and $(100, 0)$, respectively. In the M-staircase approximation, the number of steps is set as $M = 100$.

Table 10.1 Simulation parameters.

Parameters	Values
d_{SD} (m)	100
Height of IRS from the ground, H (m)	10
Number of distributed IRSs, N	5
Bandwidth, BW (MHz)	10 [19]
Carrier frequency, f_c (GHz)	3 [19]
Amplitude reflection coef., κ_{nl}	1 [19]
Nakagami shape parameter, m	$\sim \mathcal{U}[2, 3]$
Target SE, R_{th} (b/s/Hz)	1
Thermal noise power density (dBm/Hz)	−174 [19]
Transmit power, P_S (dBm)	[0, 30]
IRS's coordinate (m)	$x_{R_n} \sim \mathcal{U}[5, 95]$, $y_{R_n} \sim \mathcal{U}[1, 9]$
Antenna gain G_S, G_D, G_{R_n} (dB)	5 [19]
Noise figure, NF (dBm)	10 [19]

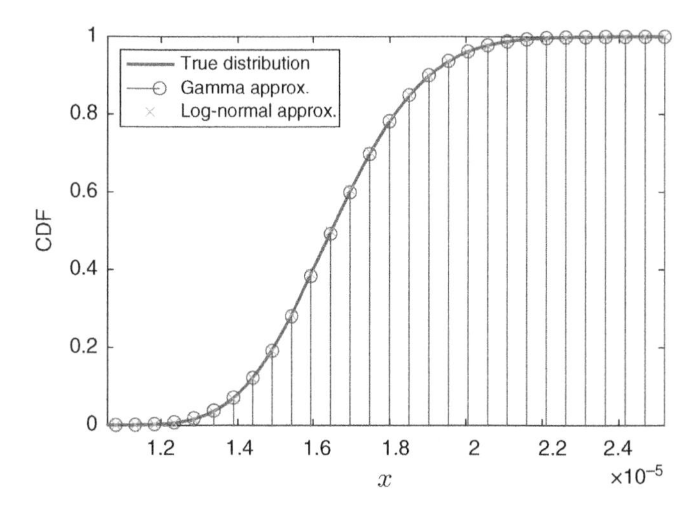

Figure 10.2 The simulation true CDF and the obtained analytical CDFs of Z in Theorems 10.1 and 10.2.

In Figures 10.2 and 10.3, we verify the correctness of our developed analysis in Theorems 10.1 and 10.2. In Figure 10.2, we plot the true, approximate Gamma and log-normal CDFs of Z. The true CDF and PDF are estimated from simulation data of the e2e channel, whereas the approximated Gamma and log-normal CDFs

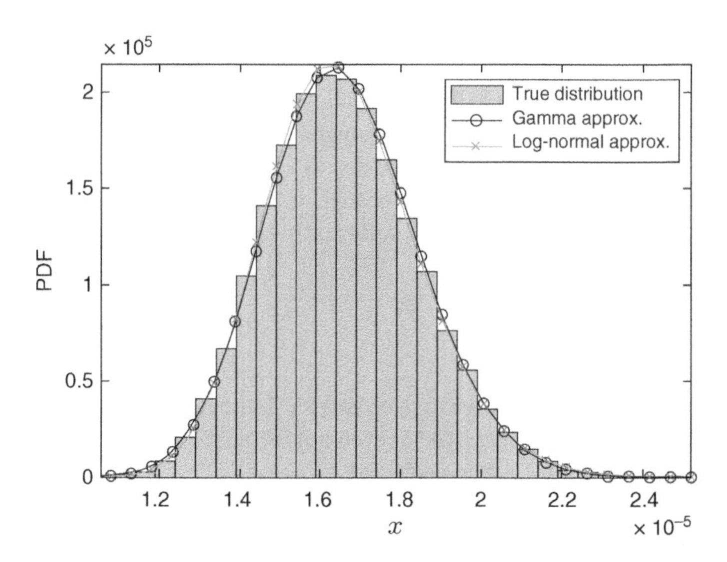

Figure 10.3 Graphical demonstration of the simulation true PDF and the obtained analytical PDFs of Z in Theorems 10.1 and 10.2.

are obtained from Theorems 10.1 and 10.2, respectively. The true, approximated Gamma and log-normal PDFs plotted in Figure 10.2 are obtained using the same method. We can observe from Figure 10.2 that the approximate CDF and PDF are well corroborated with the true CDF and PDF, which validates our analysis. Moreover, the approximate Gamma and log-normal distributions give similar results, which means that either the Gamma or log-normal distribution can be used to statistically model the e2e channel magnitude of the DMI system.

Next, we elaborate more on the Gamma and log-normal distributions to give more insights into the findings of our paper. Specifically, in our work, we have provided a key finding that both Gamma and log-normal distributions can be used to model the e2e channel of multi-IRS-assisted wireless systems. When comparing the accuracy between the Gamma and log-normal distributions, the difference occurs at the *left tail (the lower tail)* of the two distributions. To prove the above remark, we first theoretically compare the Gamma and log-normal distributions as follows.

Let U_i and V_i, $i = 1, 2, \dots$, be Gaussian RVs with zero mean and variance σ^2. Recalling that Y is the RV following a Gamma distribution with scale parameter α and rate parameter φ, mean $\mathbb{E}[Y] = \alpha/\varphi$, and variance $\mathrm{VAR}[Y] = \alpha/\varphi^2$, with the CDF and PDF presented in (10.9) and (10.8), respectively. Consequently, an RV R, where $R = \sqrt{Y}$, is a Nakagami-m RV. Considering that $\varphi \triangleq 1/2\sigma^2$ and α is integer, R can be re-expressed as

$$R = \sqrt{\sum_{i=1}^{\alpha} U_i^2 + \sum_{i=1}^{\alpha} V_i^2} \tag{10.52}$$

On the other hand, the right-hand side in (10.52) follows a Nakagami-m distribution with shape parameter $m = \alpha$ and spread parameter $\Theta = 2\alpha\sigma^2$.

From (10.52), we observe that Y can be expressed as a sum of 2α i.i.d. squared Gaussian RVs, each squared Gaussian RV has mean $\sigma^2 = \mathbb{E}[U_i^2] = \mathbb{E}[V_i^2]$ and variance $2\sigma^4 = \mathrm{VAR}[U_i^2] = \mathrm{VAR}[V_i^2]$. When α is sufficiently large, using the CLT, Y can be modelled as follows:

$$Y = \frac{\alpha}{\varphi}\left[1 + \frac{\mathcal{N}(0,1)}{\sqrt{\alpha}}\right], \quad \alpha \to \infty \tag{10.53}$$

where $\mathcal{N}(0, 1)$ denotes a standard Gaussian (Normal) RV. Although the shape α is previously considered to be an integer, (10.53) even holds true for α is real, since as $\alpha \to \infty$, the fractional part of α can be neglected.

Next, recall that W is the RV following a log-normal distribution with mean $\mathbb{E}[W] = e^{\nu+\eta^2/2}$, and variance $\mathrm{VAR}[W] = (e^{\eta^2} - 1)e^{2\nu+\eta^2}$, with the CDF and PDF being presented in (10.10) and (10.11), respectively. From the relationship between log-normal and Gaussian distributions, the RV W can be alternatively expressed as follows:

$$W = \exp(v + \eta \mathcal{N}(0,1)) \tag{10.54}$$

To show the relationship between the Gamma and log-normal distributions, following the method in [24], we match their first and second moments, i.e. means and variances, respectively, as follows:

$$\begin{cases} \exp\left(v + \dfrac{\eta^2}{2}\right) = \dfrac{\alpha}{\varphi} \\ \left(\exp\left(\eta^2\right) - 1\right)\exp\left(2v + \eta^2\right) = \dfrac{\alpha}{\varphi^2} \end{cases} \tag{10.55}$$

Solving (10.55) for η^2 and v yields

$$\eta^2 = \log\left(\frac{\alpha+1}{\alpha}\right), \quad v = \log\left(\frac{\alpha}{\varphi}\frac{\sqrt{\alpha}}{\sqrt{\alpha+1}}\right) \tag{10.56}$$

As α is sufficiently large, we have $\eta \to \frac{1}{\sqrt{\alpha}}$ and $v \to \log\left(\frac{\alpha}{\varphi}\right)$. Hence, by substituting the above result into (10.54), we have

$$W = \frac{\alpha}{\varphi}\left[1 + \frac{\mathcal{N}(0,1)}{\sqrt{\alpha}}\right], \quad \alpha \to \infty \tag{10.57}$$

This result is consistent with (10.53), which implies the similarity between the Gamma and log-normal distributions.

As shown in Figure 10.4, when α increases, the shape of the Gamma and log-normal distributions becomes more similar, especially at the lower tails of both distributions. It is noted that in our DMI system, the Nakagami shape parameter m is uniformly distributed in the range of $[2,3]$, consequently, the range of the shape parameter α_Z in (10.12) is $[77,120]$. Since the approximated Gamma and log-normal distribution give almost identical results with high α_Z, in next results, we use the approximate Gamma distribution to plot our results. Next, in Figure 10.5, we plot the EC and OP of the DMI system, i.e. $P_{\text{out}}^{\text{DMI,Gam}}$ and $\overline{C}^{\text{DMI,Gam}}$, presented in (10.40) and (10.39), respectively. As can be observed in Figures 10.5 and 10.6, the theoretical and simulation results are well corroborated, which validates our developed analysis.

In Figures 10.5 and 10.6, we consider different settings of reflecting elements, i.e. settings $\mathbf{R}_1$ and $\mathbf{R}_2$, to examine the impact of i.n.i.d. fading on the system performance. As can be seen in Figures 10.5 and 10.6, the OP decreases and the EC increases significantly as the number of passive reflecting elements installed at the IRSs increases. Because each IRS is subject to different fading severity under i.n.i.d. fading channels, as a result, the Nakagami-m distributions between different IRSs have different shape parameters and spread parameters. In addition, it is evidenced from Figures 10.5 and 10.6 that IRS-assisted systems

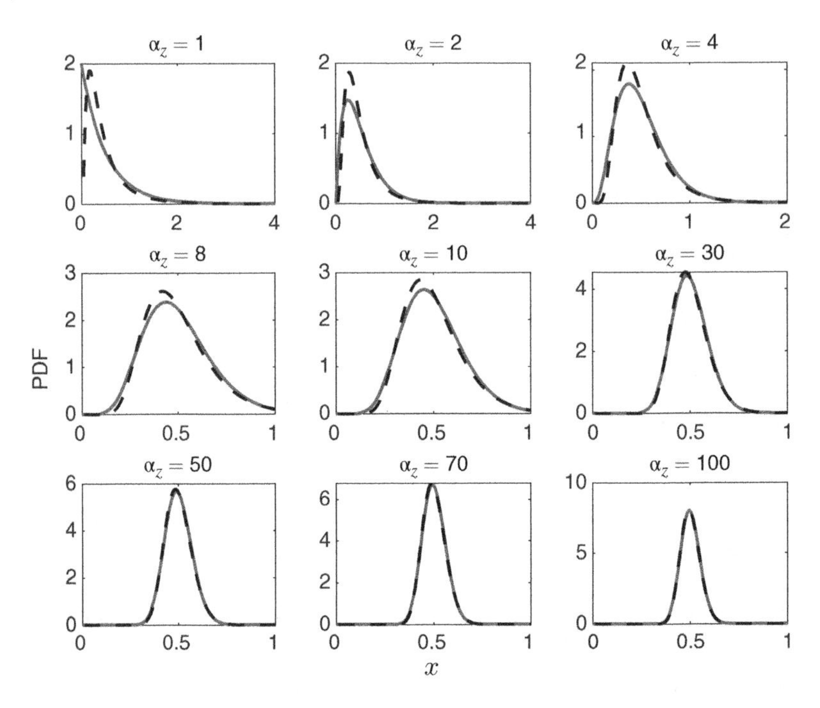

Figure 10.4 Comparison of the lower tails of the Gamma PDF (dashed curves) and the log-normal PDF (solid curves) with different values of α.

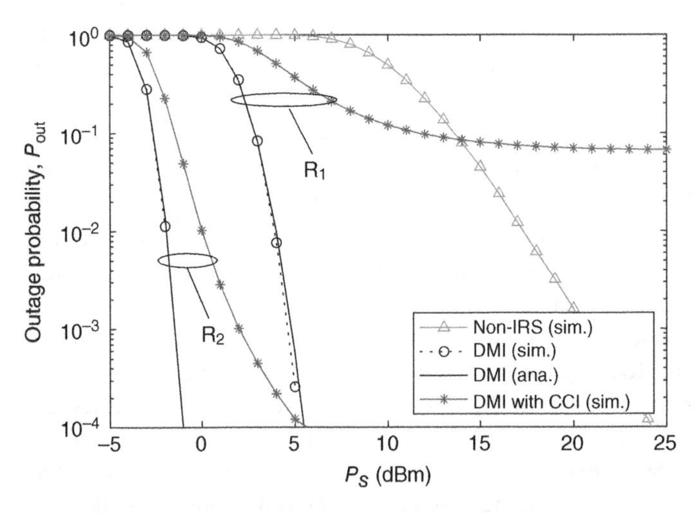

Figure 10.5 The OP as a function of transmit power P_S (dBm), where the co-channel interference is taken into account in the DMI system.

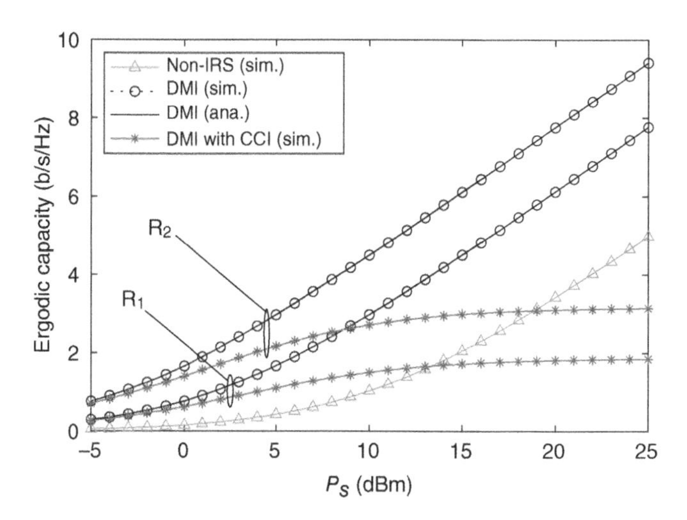

Figure 10.6 The EC as a function of transmit power P_S (dBm), where the co-channel interference is taken into account in the DMI system.

are superior to the non-IRS-assisted system. This is due to the fact that the more reflecting elements that are installed, the higher the e2e SNR can be achieved.

Moreover, in Figures (10.5) and (10.6), we investigate the impact of co-channel inter-cell interference on the performance of the considered multi-IRS-aided system in terms of the EC and OP, respectively. Specifically, we assume that two adjacent sources, located at $(-99, 0)$ and $(255, 0)$, respectively, are causing co-channel inter-cell interference to the DMI system. As can be observed in Figures 10.5 and 10.6, the co-channel inter-cell interference indeed decreases the performance of the multi-IRS-aided system, i.e. for a given transmit power, the OP increases and the EC decreases under the existing interference, and the stronger the interference is, the higher the performance degradation is.

As shown in this section, the developed analysis in Section 10.3 is validated through simulation results. Moreover, we present the in-depth discussion on the tail characteristics of Gamma and log-normal distributions. Furthermore, we clearly show the performance improvement achieved by the DMI system in terms of EC and OP over the conventional non-IRS systems.

10.5 Conclusions

In this chapter, we proposed a multi-IRS-assisted system, namely the DMI system. We focused on the statistical characterization and modelling of the system. To that end, we proposed a mathematical framework for determining the e2e fading channel distribution in the DMI system. The framework determined that the true distribution of the e2e channel magnitude can be approximated by either the Gamma or log-normal distributions. The approximate closed-form expressions of the EC and OP are then derived and used to evaluate system performance. Through our numerical results, the DMI system outperforms the conventional non-IRS-aided system in terms of EC and OP. In addition, we showed that the reflecting element setting and co-channel interference have a significant impact on system performance for a given total number of passive elements.

References

1 Di Renzo, M., Zappone, A., Debbah, M. et al. (2020). Smart radio environments empowered by reconfigurable intelligent surfaces: how it works, state of research, and the road ahead. *IEEE J. Sel. Areas Commun.* 38 (11): 2450–2525.

2 Wu, Q. and Zhang, R. (2019). Intelligent reflecting surface enhanced wireless network via joint active and passive beamforming. *IEEE Trans. Wireless Commun.* 18 (11): 5394–5409.

3 Liaskos, C., Nie, S., Tsioliaridou, A. et al. (2020). End-to-end wireless path deployment with intelligent surfaces using interpretable neural networks. *IEEE Trans. Commun.* 68 (11): 6792–6806.

4 Basar, E., Di Renzo, M., De Rosny, J. et al. (2019). Wireless communications through reconfigurable intelligent surfaces. *IEEE Access* 7: 116753–116773.

5 de Figueiredo, F.A.P., Facina, M.S.P., Ferreira, R.C. et al. (2021). Large intelligent surfaces with discrete set of phase-shifts communicating through double-Rayleigh fading channels. *IEEE Access* 9: 20768–20787.

6 Gan, X., Zhong, C., Zhu, Y., and Zhong, Z. (2021). User selection in reconfigurable intelligent surface assisted communication systems. *IEEE Commun. Lett.* 25 (4): 1353–1357.

7 Ibrahim, H., Tabassum, H., and Nguyen, U.T. (2021). Exact coverage analysis of intelligent reflecting surfaces with Nakagami-*M* channels. *IEEE Trans. Veh. Technol.* 70 (1): 1072–1076.

8 Galappaththige, D.L., Kudathanthirige, D., and Baduge, G.A.A. (2020). Performance analysis of distributed intelligent reflective surfaces for wireless communications. http://arxiv.org/abs/2010.12543.

9 Jung, M., Saad, W., Jang, Y. et al. (2020). Performance analysis of large intelligent surfaces (LISs): asymptotic data rate and channel hardening effects. *IEEE Trans. Wireless Commun.* 19 (3): 2052–2065.

10 Yang, L., Yang, Y., da Costa, D.B., and Trigui, I. (2021). Outage probability and capacity scaling law of multiple RIS-aided networks. *IEEE Wireless Commun. Lett.* 10 (2): 256–260.

11 Mei, W. and Zhang, R. (2021). Cooperative beam routing for multi-IRS aided communication. *IEEE Wireless Commun. Lett.* 10 (2): 426–430.

12 Yildirim, I., Uyrus, A., and Basar, E. (2021). Modeling and analysis of reconfigurable intelligent surfaces for indoor and outdoor applications in future wireless networks. *IEEE Trans. Commun.* 69 (2): 1290–1301.

13 Lyu, J. and Zhang, R. (2021). Hybrid active/passive wireless network aided by intelligent reflecting surface: system modeling and performance analysis. *IEEE Trans. Wireless Commun.* 20 (11): 7196–7212.

14 Gradshteyn, I.S. and Ryzhik, I.M. (2007). *Tables of Integrals, Series, and Products*, 7e. New York: Academic Press.

15 Abramowitz, M. and Stegun, I.A. (1965). *Handbook of Mathematical Functions: With Formulas, Graphs, and Mathematical Tables*. New York: Dover.

16 Prudnikov, A., Brychkov, Y., and Marichev, O. (1999). *Integrals and Series: More Special Functions*, vol. 3. Boca Raton, FL: CRC Press.

17 Peebles, P.Z. (2000). *Probability, Random Variables and Random Signal Principles*, 4e. New York: McGraw-Hill Science.

18 (2021). *MATLAB Version R2021a*. Natick, MA: The Mathworks, Inc.

19 Bjornson, E., Ozdogan, O., and Larsson, E.G. (2020). Intelligent reflecting surface versus decode-and-forward: how large surfaces are needed to beat relaying? *IEEE Wireless Commun. Lett.* 9 (2): 244–248.

20 da Costa, D.B., Ding, H., and Ge, J. (2011). Interference-limited relaying transmissions in dual-pop cooperative networks over Nakagami-*M* fading. *IEEE Commun. Lett.* 15 (5): 503–505.

21 Fenton, L. (1960). The sum of log-normal probability distributions in scatter transmission systems. *IEEE Trans. Commun.* 8 (1): 57–67.

22 Laourine, A., Stéphenne, A., and Affes, S. (2007). Estimating the ergodic capacity of log-normal channels. *IEEE Commun. Lett.* 11 (7): 568–570.

23 3GPP (2010). Evolved universal terrestrial radio access (E-UTRA); further advancements for E-UTRA physical layer aspects (release 9). *Document 3GPP-TR-36.814 V9.0.0*, March 2010.

24 Kosti, I.M. (2005). Analytical approach to performance analysis for channel subject to shadowing and fading. *IEE Proc. Commun.* 152 (6): 821.

11

RIS-Assisted UAV Communications

Mohammad O. Abualhauja'a[1], Shuja Ansari[1], Olaoluwa R. Popoola[1], Lina Mohjazi[1], Lina Bariah[2], Sami Muhaidat[3,4], Qammer H. Abbasi[1], and Muhammad Ali Imran[1]

[1]*James Watt School of Engineering, University of Glasgow, Glasgow, UK*
[2]*Technology Innovation Institute, Masdar City, Abu Dhabi, UAE*
[3]*Department of Electrical Engineering and Computer Science, KU Center for Cyber-Physical System, Khalifa University, Abu Dhabi, UAE*
[4]*Department of Systems and Computer Engineering, Carleton University, Ottawa, Canada*

11.1 Introduction

With the development of wireless technologies 5G, beyond 5G (B5G), and even sixth generation (6G), unmanned aerial vehicles (UAVs) are playing an increasingly significant role in enhancing the communication quality in various military and civilian applications. The role of UAVs in the future mobile networks has become attractable for both academics and industry with several potential applications in wireless communications due to their unique mobility and flexibility characteristics. UAVs may be utilized as airborne communication platforms to improve wireless network coverage, capacity, reliability, and energy efficiency. Due to the UAVs high mobility, they can be flexibly deployed to enhance the communication quality, while conventional terrestrial base stations (BSs) only serve the ground users in a fixed area. On the other hand, they can function as new aerial user equipment (UE) terminals in the cellular network serving various applications, ranging from data collection to item delivery [1, 2]. Moreover, future wireless networks are driven by explosive growing demand for high mobile data rates, reduced end-to-end latencies, and connectivity across a diversity of new applications such as the Internet of Things (IoT), Massive Machine Type Communication, etc. Hence, dense cell deployment and millimetre-wave communications are mandatory for serving these demands. However, dense cell networks face several challenges related to backhaul transmission and interference. In addition

Intelligent Reconfigurable Surfaces (IRS) for Prospective 6G Wireless Networks, First Edition.
Edited by Muhammad Ali Imran, Lina Mohjazi, Lina Bariah, Sami Muhaidat, Tie Jun Cui, and Qammer H. Abbasi.

to that, the millimetre-wave bands are affected by high propagation loss and sensitivity to blockage. UAVs can provide a low-cost and flexible solution for offloading the macro-cells and providing energy-efficient connectivity [2, 3]. Multiple UAVs can operate to support broadband services by forming multi-UAV flying Ad Hoc networks (FANETs). They can be also used in public safety scenarios in cases of emergency and fast service recovery for unexpected network drop [2, 3]. At the same time, they can operate with terrestrial BSs to enhance the performance of heterogeneous networks environment [2, 4]. However, the explosive growth in the number of deployed UAV nodes implies the need to rethink traditional beamforming mechanisms to address several challenges, such as dynamic placement, high mobility, and other challenges related to the existing terrestrial networks planning. To this effect, reconfigurable intelligent surfaces (RISs) may be deployed to assist in UAV communication and enhance the overall performance of the network in a spectrum-and energy- and cost-effective manner. Beamforming design and optimization can enable a highly efficient self-reconfiguration of RISs and adaptable UAV implementations. RIS-aided UAV communication shows an improved performance in terms of power efficiency, coverage, capacity, and reliability. This opens the door for autonomous UAV-assisted networks, which are highly scalable, adaptable, and robust to environmental changes [5–12].

Reliable and improved communications performance and coverage are required to meet the needs of the growing UAV market with significant attention being paid to UAV-enabled wireless communication networks. Standardization for non-terrestrial networks is already an on-going effort by third-Generation Partnership Project (3GPP), in the framework of 5G [13]. Nevertheless, further enhancements are expected. Leveraging intelligent beamforming solutions enables UAV networks, or FANETs, to adaptively optimize their actions and efficiently manage their resources while meeting a diverse set of constraints. This will open the door for new application use cases, such as UAV-based remote sensing, UAV-based real-time multimedia streaming networks, UAV-enabled intelligent transportation systems, UAV-based intelligent navigation and monitoring systems, UAV-assisted communications, etc. To reap the benefits of future UAV-based applications, it is essential to address several challenges related to the unique features of the UAVs and their channel characteristics such as air-to-ground channel modelling, optimal placement, trajectory planning, resource management, energy efficiency, and other challenges related to network planning [2, 14]. For instance, UAV BS or relay has a smaller coverage area due to the energy limitations and blockage caused by buildings and other objects. At the same time, they suffer from severe interference due to their high line of sight (LOS) probability with ground BSs [6]. Moreover, the existing terrestrial networks are planned to serve ground users, and hence UAVs, which usually fly at a relatively high altitude, receive weaker signal.

In the light of the foregoing, due to the complex urban environment, the links between the UAV and ground BSs or ground users may suffer high-penetration loss, multi-path fading, or signal blockage, resulting severe channel quality degradation. By deploying RISs, the interaction of RISs with impinging electromagnetic (EM) waves can be intentionally controlled through connected passive elements to enhance the performance of the UAV network while also improving the coverage.

Recently, noticeable research efforts proposed RIS-UAV joint deployment as a promising solution to extend wireless networks coverage and enhance their performance [5–12]. These works considered optimizing RIS-assisted UAV networks to utilize joint RIS-UAV deployment in wireless networks. The work in [15] analysed the performance of RIS-assisted UAV communication system, where UAV acts as a relay solution and RIS is deployed to support the link from BS to the UAV. Their results show that RIS deployment can enhance system performance in terms of reliability, average bit error rate (BER), and average capacity. In [5], authors proposed an algorithm to maximize the achievable rate of a UAV BS serving a ground user by jointly optimizing the RIS passive beamforming matrix and the UAV trajectory. In [6], authors proposed system considered multiple RIS deployment supporting a UAV BS to maximize the received power by a ground user. In [8], authors proposed a UAV-enabled solution, where the UAV is equipped with an RIS and acts as a passive relay between a ground BS and ground UEs to maximize the secure energy efficiency against eavesdropper by optimizing UAV trajectory, RIS passive beamforming, user association, and transmitted power. It is worth noting that in order to reap the benefit of RIS-UAV joint deployment, the proposed optimization algorithm must consider optimizing both RIS passive beamforming and UAV trajectory.

11.2 Background

The use of UAVs, commonly known as drones, has recently experienced tremendous development, particularly in non-military applications, such as public safety, traffic monitoring, photography, delivery services, and communication networks [2, 3, 14]. UAV applications have a market value of approximately $127 billion [16]. Due to their various and diverse uses, UAVs come in a variety of shapes and sizes in practise. While there is no universal standard for UAV classification, UAVs can be classified based on a variety of factors including purpose, weight or payload, size, endurance, wing design, control techniques, cruising range, flying altitude, and energy feeding mechanism [17, 18].

Their fast and dynamic deployment, high probability of LOS links, and affordably low maintenance and infrastructure costs compared to the terrestrial BSs are the key attributes that make UAVs considered as a promising solution for future mobile networks [2, 3, 14]. They can operate in the network as aerial UEs for various applications, or they can be used as aerial BSs and relays to provide capacity and coverage solutions, especially for IoT applications that transmit the signal over a low range [2, 3, 19]. They can be used to provide broadband services to the remote areas, where building a terrestrial network need costly infrastructure. Moreover, as the 6G mobile networks propose a three-dimensional (3D) heterogeneous architecture, UAVs can operate with the terrestrial BSs to support broadband services and to enhance the performance of heterogeneous networks environment by forming multi-UAV FANETs. They can be also used in cases of emergencies and disasters to provide a fast service recovery for unexpected network drop [2, 3, 18]. At the same time, they can cooperate with the terrestrial BSs to enhance the performance of heterogeneous networks environment [2, 4].

11.3 The Role of UAVs in the Future Mobile Networks and Their Unique Characteristics

To create a clear image of how UAVs may be utilized in the wireless networks, this section will go through their potential uses and list their unique characteristics compared to the traditional terrestrial networks. Furthermore, these characteristics will be investigated to point out opportunities and challenges. It is worth noting that this part places emphasis on potential uses and opportunities, leaving the challenges to be discussed in greater depth in Section 11.4.

11.3.1 UAV Characteristics

From a research viewpoint, to provide the full horizon of UAVs applications and benefits, it is essential to highlight the fundamental differences between UAV communications and conventional terrestrial communications. After all, researchers have studied terrestrial communication for decades, so what is genuinely new about UAV communications? The best answer comes from examining some characteristics of UAVs. On the one hand, most UAVs fly at a high altitude, resulting in a wider ground coverage when they are deployed as aerial BSs or relays. On the other hand, if the UAV is considered as a UE, it becomes a challenge because the existing terrestrial networks are planned to serve ground users, so the UAV which usually fly at a relatively high altitude receives a weaker signal due to the BS antennas down tilting [20].

Compared to the conventional terrestrial communications, UAV communications provide several compelling benefits, which are detailed below:

- **Dynamic deployment**: Due to their flexible placement and trajectory (3D mobility), UAVs are ideal for on-demand deployment at specific locations, creating a cost-effective and fast solution appropriate for unexpected or emergency cases. At the same time, as shown in Figure 11.1, their trajectory can be planned to enhance quality of service (QoS) according to real-time requirements by providing shorter link distance and less signal obstruction [2, 18, 21, 22]. Moreover, dynamic deployment of UAVs as BSs can be utilized to maximize the lifetime of mobile terminals, minimize the average latency, and maximize the number of covered users under energy and spectrum constraints [22–26].
- **High LOS probability**: The features of the air-to-ground (A2G) channel tend to differ from those of other terrestrial communication channels. For instance, as illustrated in Figure 11.2. UAV communications have an additional high LOS probability benefit over the terrestrial communications [18, 27]. In fact, LOS probability mainly depends on three factors: the UAV altitude, elevation angle, and type of the propagation environment [2, 13, 18]. High LOS probability allows deploying stronger and more reliable communication links with relatively larger macro-diversity gains [3, 14].
- **Easy deployment and affordably low-infrastructure cost**: In contrast to terrestrial BSs, UAVs are easy to deploy and do not require costly infrastructure. This allows extending network access to remote regions that are difficult to reach. Google's Loon project and Facebook Aquila are good examples of a contemporary initiatives to build massive UAV networks to deliver broadband access to rural areas.

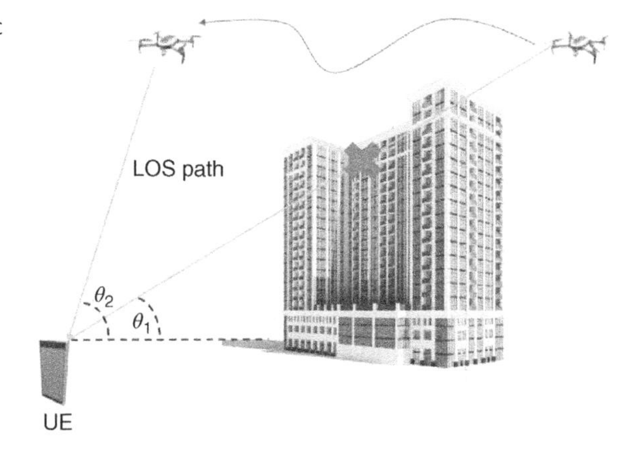

Figure 11.1 UAV dynamic deployment.

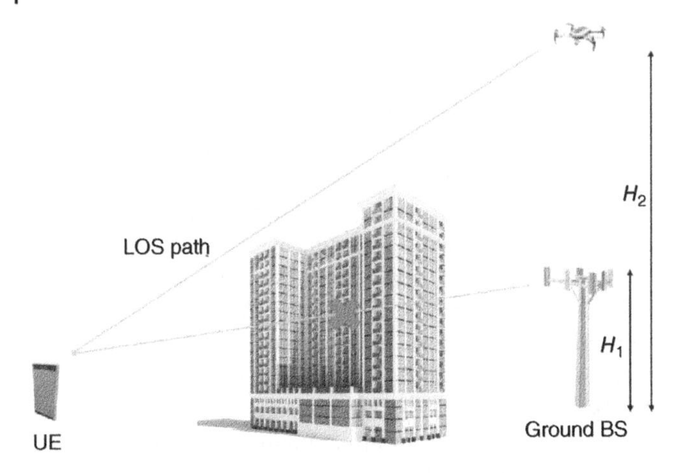

Figure 11.2 LOS probability.

11.4 Challenges of UAV Communications

Due to their unique characteristics and physical attributes, UAVs will play a fundamental role in realizing the development of future mobile networks. However, in order to integrate a reliable UAV communications system, several challenges and limitations must be addressed. In this section, we provide an overview on the key technical challenges that arise in UAV communications for different deployment paradigms.

11.4.1 Air-to-Ground (3D) Channel Modelling

An accurate channel model is essential for the proper capacity dimensioning and coverage planning of communication systems. When employing UAVs in applications such as UAV-assisted communications, cellular-connected UAVs, and IoT communications, an accurate A2G channel modelling is critical to predicting the performance of UAV wireless communications in terms of coverage and capacity [2, 27, 28]. The characteristics of A2G channels differ substantially from those of traditional terrestrial communication channels. Specifically, due to their unique attributes, UAVs experience higher LOS probability. In fact, the A2G channel depends on multiple factors such as UAV height, elevation angle, and the wireless environment [2, 13, 27, 28]. Moreover, because of UAVs dynamic mobility patterns, A2G channels can experience greater spatial and temporal variations in non-stationary channels than traditional terrestrial communication channels. Thus, Doppler shifts and spread must also be considered for certain communication applications [13, 27–29]. Finally, noise caused by the UAV motors and circuitry, UAV body shadowing, and antenna characteristics must be also considered [27].

11.4.2 Three-dimensional Deployment of UAVs

Optimal 3D placement is one most critical aspects of UAV-assisted communications. Energy-efficient maximal coverage can be achieved by utilizing the flexible UAVs deployment as aerial communications platforms [2, 23]. However, due to their high mobility, it is a challenging task to achieve optimal UAVs 3D deployment, especially in multi-UAVs deployment scenarios. In fact, there are multiple factors that must be considered such as inter-cell interference, energy and spectrum constraints, ground users' geographical distribution, traffic demands, and the varying A2G channel characteristics (e.g. path loss and LOS probability depend on UAV altitude and elevation angle as show in Figure 11.3) [2, 24].

11.4.3 Optimal Trajectory Planning

Optimizing the UAV trajectory allows maintaining the quality of communication links by providing shorter and more reliable connections which results in higher data rates and lower latencies. However, planning the optimal UAV path brings new challenges since it requires considering several factors such as inter-cell interference, backhauling, propagation environment, users traffic demands, and other physical constraints, i.e. UAV dynamics, flying speed, and acceleration [2, 30].

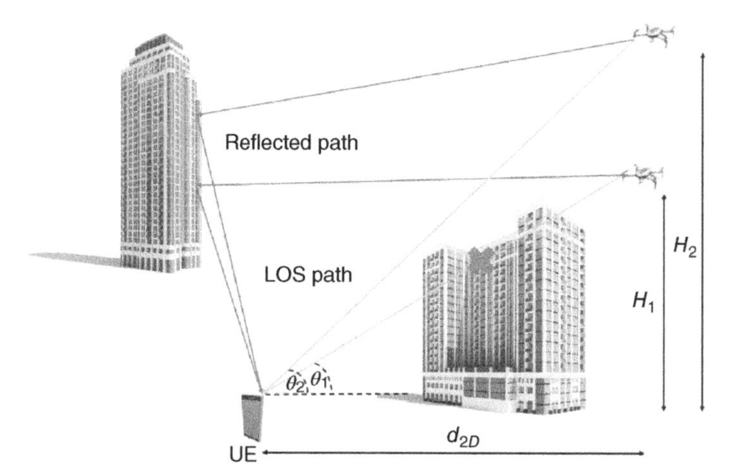

Figure 11.3 Relationship between path loss and LOS probability to the UAV altitude and elevation angle.

11.4.4 Network Planning for Cellular-connected UAV Applications

The existing cellular networks are planned to support conventional ground users, which have different characteristics from the aerial users. For instance, BSs antennas are usually down tilted to provide coverage for the ground users and limit inter-cell interference. At the same time, UAVs which fly at relatively high altitude can be served by antennas side-lobes. However, studies reveal that as the UAV altitude increases, the received signal becomes significantly weaker [14, 20]. Hence, further enhancements and new technologies are needed to reshape the traditional terrestrial networks to provide sufficient coverage and signal quality to support the cellular-connected UAV applications [2, 14, 20]. The 3GPPP measurements show that number of radio link failures (RLF) increases significantly when UAV altitude exceeds 120 m. Moreover, due to their fast movement, cellular-connected UAVs introduce new challenges in terms of mobility management and the essential need to ultra-reliable low-latency communications (URLLCs) navigation and control links [2, 13].

11.4.5 Interference Caused by Ground BSs

Due to their relatively high altitude, UAVs will receive LOS signals from multiple terrestrial BSs. Field trials show that the number of detectable cells increases as the UAV altitude increases. Hence, the interference in cellular-connected UAV systems significantly increases at high UAV altitude causing low signal-to-interference-and-noise-ratio (SINR). Moreover, it can be noticed in that as the range of the detectable BSs increases, the risk of detecting cells that carry the same physical-cell-identity (PCI) value (PCI conflict) also increases.

11.5 RIS-assisted UAV Communications: Integration Paradigms and Use Cases

Current communication systems are being designed following the assumption that the wireless propagation environment is an uncontrollable random process which can be only modified by the processes at the transmit and reception ends, such as adaptive modulation, coding, beamforming, and power control [31, 32]. Anyway, this assumption is not always valid. For instance, modern relaying solutions, typically classified into amplify-and-forward (AF) and decode-and-forward (DF) relaying, have the capability to enhance the wireless channel [32].

Future mobile 6G networks will be able to offer heterogeneous services, such as sensing and localization, jointly with ultra-high throughput, ultra-low latency, and ultra-reliable communications. The controlled wireless environment is becoming

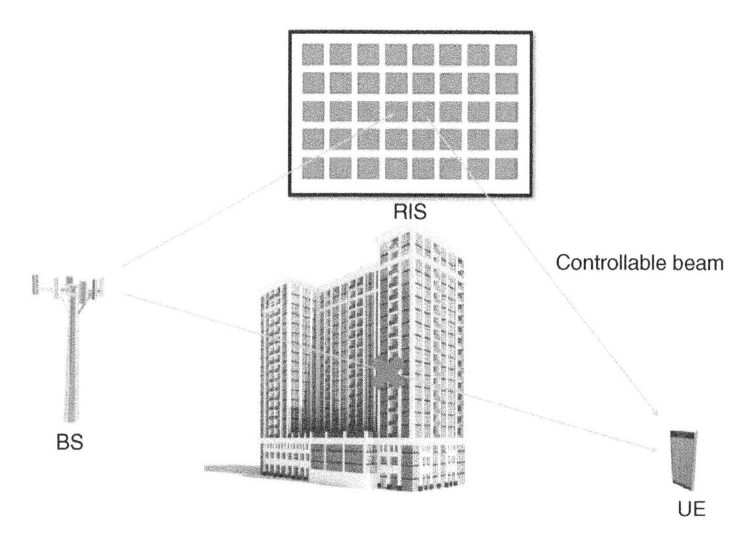

Figure 11.4 RIS-assisted transmission.

a key factor to realize the development of these networks and achieve the flexibility required at the new architectural platform of 6G networks, where the wireless channels can be controlled and customized in real-time. The term smart radio environment (SRE) refers to the process of turning the wireless environment into a controllable variable [31–34].

As seen in Figure 11.4, the RIS is a massive array of low-cost reflecting or scattering passive elements. These elements can be configured using a simple controller to change the incident wave phase, amplitude, frequency, or polarization. It coherently adds the signal reflections and focuses them where they are most needed, and at the same time, eliminates any source of noise or interference [35, 36]. It can be easily deployed on different surfaces such as indoor walls, buildings, and even aerial platforms providing the opportunity to control various wireless environments [34]. In addition to its ease of deployment, RIS is lower cost and more energy-efficient compared to the active relaying solutions, where it creates virtual LOS links between the transmitter and the receiver via passive beamforming, without the need for any transmit RF chains [34, 37]. Moreover, the usage of a huge number of active devices such as small cells, relays, and remote radio heads (RRHs) causes an aggregative interference, while RIS operates as a passive array without employing any transmit RF chains, which makes it easier to be deployed without causing additional interference [38]. On top of that, RIS is compatible with existing technologies and can operate in wideband full duplex without the need for self-interference cancellation techniques [35, 37, 38].

Massive multiple-input multiple-output (M-MIMO) antennas and beamforming are key-enabler technologies for improving communication systems performance with respect to capacity, speed, spectral efficiency, and massive connectivity. However, M-MIMO technology introduces design challenges related to complexity, energy consumption, and cost. Hence, it is a critical research area to develop low-complexity and low-cost solution. RIS can be easily integrated into the wireless communication networks and smartly tune the random radio environment, to improve coverage, throughput, energy consumption, and communication security [35, 39]. RIS can be also jointly deployed with M-MIMO systems to improve network performance using both passive and active beamforming techniques [38]. In addition to that, RIS can actively tune the wireless channels; hence, it is a key element to realize the notion of the SRE. Moreover, the usage of large numbers of active devices such as small cells, relays, and RRHs causes an aggregative interference. While the RIS operates as a passive array without employing any transmit RF chains, which make it easier to be deployed without causing additional interference [38]. On top of that, RIS is compatible with existing technologies and can operate in wide-band full duplex without the need for self-interference cancellation techniques [37, 38, 40].

Based on their functionality, there are two major paradigms for integrating UAVs into cellular networks. On the one hand, under the cellular-connected category, UAVs operate as new aerial users terminals in a cellular network to serve different applications. On the other hand, UAV-aided communications, in which UAVs serve as aerial communication platforms (BSs, access points [APs], or relays). Hence, there are multiple scenarios for integrating RIS into UAV cellular networks [2, 3, 18].

11.5.1 RIS to Support UAV-assisted Communications Air-to-Ground (A2G) Links

High mobility of the UAVs allows them to act as a reliable and responsive communication platform (aerial BS/relay) where UAVs can provide a low-cost and flexible solution for offloading the macro-cells and providing energy-efficient connectivity. UAVs can be developed as cost-effective aerial communication platforms to extend network coverage to remote regions that are difficult to access or adding extra capacity to the existing networks. Moreover, in the events of large-scale natural disasters and emergencies, the existing terrestrial networks may suffer from huge damage. UAV-assisted communication is a promising technology that can provide fast, flexible, and reliable solution to overcome network failure in similar scenarios. Furthermore, by providing the essential coverage, UAVs can play a significant role in public safety scenarios during natural disasters to support search-and-rescue operations and disaster response.

In addition to that, as the number of IoT devices is growing exponentially. In the coming few years, there will be billions of devices connected to the mobile networks. These devices are integrated everywhere, at homes, vehicles, and even as wearable devices. Due to their limited energy and blockage caused by the environment, IoT devices are usually unable to communicate over long distances. At the same time, the massive amounts of data transmitted through these widely deployed devices require a real-time processing in most scenarios. Otherwise, it will become useless. Hence, it is essential to create reliable connectivity solutions that overcome both time and energy constraints. Their flexibility, low cost, and dynamic deployment make UAV-assisted communications an ideal candidate to overcome these challenges.

RISs can be implemented to assist A2G links between the aerial platforms and ground users to maximize UAV coverage area and increase transmit power efficiency. By incorporating RISs into UAV-assisted wireless networks, a virtual LOS links between UAVs and ground users can be established via passive beamforming, resulting in a wider coverage area, more efficient transmissions, and reduced UAV mobility.

11.5.2 RIS to Support Cellular-Connected UAV Ground-to-Air (G2A) Links

UAVs can operate as new aerial users/terminals in a cellular network to serve different applications such as remote sensing, edge computing, public safety, and delivery services. The existing cellular networks are planned to support conventional ground users, which have different characteristics from the aerial users. For instance, BSs antennas are usually down tilted to provide coverage for the ground users and limit inter-cell interference. At the same time, UAVs which fly at relatively high altitude can be served by antennas side-lobes. However, studies reveal that as the UAV altitude increases, the received signal becomes significantly weaker. Hence, further enhancements and new technologies are needed to reshape the traditional terrestrial networks to provide a sufficient coverage and signal quality to support the cellular-connected UAV applications. The 3GPPP measurements show that number of RLF increases significantly as the UAV altitude increases. Moreover, due to their fast movement, cellular-connected UAVs introduce new challenges in terms of mobility management and the essential need to URLLC navigation and control links.

As RISs are deployed for local coverage only, their operating ranges are usually much shorter than those of active BSs/relays, which makes it easier to practically deploy RISs without interfering each other. Hence, RIS can be deployed to improve ground-to-air (G2A) connections and overcome the signal-to-noise ratio (SNR) degradation caused by the severe interference when UAV flies at relatively high altitude.

11.5.3 RIS-equipped Aerial Platforms RIS to Support Air-to-Air (A2A) Links

In this scenario, UAV will be equipped with an RIS to create LOS links between BSs and users at coverage holes or blind spots. UAVs equipped with RISs can also support communication between aerial platforms to benefit from the reliability of air-to-air (A2A) links. Moreover, future wireless networks are driven by explosive growing demand for high mobile data rates, reduced end-to-end latencies, and connectivity across a diversity of new applications such as the IoT, Massive Machine Type Communication. Hence, dense cell deployment and millimetre-wave communications are mandatory for serving these demands. However, dense cell networks face several challenges related to backhaul transmission and interference. In addition to that, the millimetre-wave bands affected by high propagation loss and sensitivity to blockage, this implies the use of reliable LOS links. In this context, UAVs equipped with RISs, can dynamically support these technologies by creating LOS wherever they are needed.

Due to large size of RIS and limited payload of UAVs, this scenario is less practical than other deployment scenarios. However, the work in [41] proposed aerial UAV swarm-enabled aerial RISs. By introducing a cooperative transmission scheme to benefit from the flexibility of UAVs and at the same time to overcome their limited payload and battery capacity issue. Suggested solution can serve multiple applications by supporting links between ground BSs and UAVs, aerial BSs and ground UEs, and UAV to UAV links (Figure 11.5).

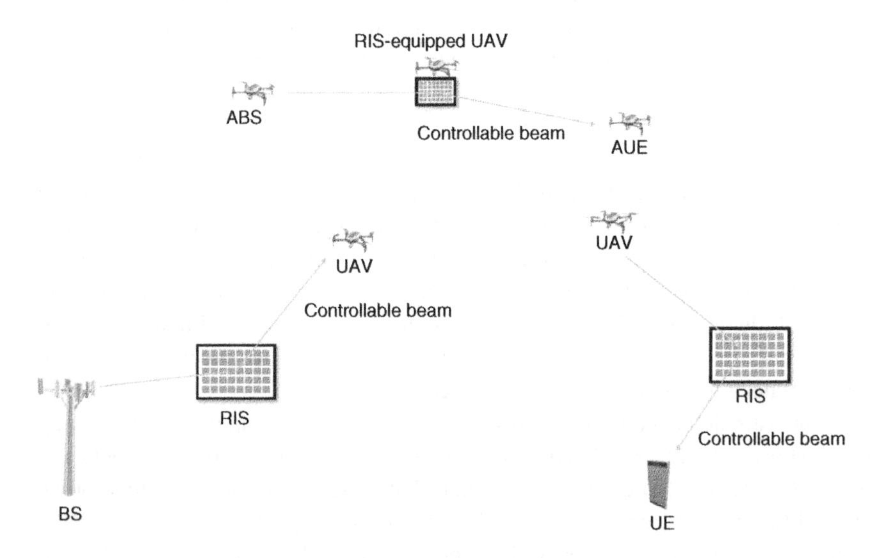

Figure 11.5 RIS-assisted UAV implementation scenarios.

11.6 Preliminary Investigations

This section investigates the performance of RIS-assisted UAV communication systems, by showing a comparison between RIS and DF relaying.

11.6.1 RIS versus Relay

The RIS can be easily integrated into wireless communication networks to smartly control the random radio environment, so as to improve the coverage, throughput, and energy efficiency [35]. Relays are also widely recognized as a promising solution for wireless network coverage extension.

Similar to the RIS, relay-supported links experience better channel propagation conditions compared to the direct transmission links in case of weak or blocked direct paths. Relays can be classified depending on the relaying protocol into AF and DF relaying. Although the AF relays are less complex, they also amplify the signal noise [42], whereas the DF relays show better performance in terms of SNR and achievable rate.

The authors in [42] proposed a comparison between RIS and DF relaying-supported transmissions. Their results show that RIS should be equipped with a large number of elements to outperform the relay. In this work, we compare the performance of RIS and relay-assisted UAV communications, using channel gains modelled by 3GPP for aerial vehicles, in terms of achievable rate and power requirements. In addition, the effect of varying the UAV height on the system performance is also investigated.

11.6.1.1 System Model

We consider a downlink transmission system consisting of a single antenna terrestrial BS and a UAV that acts as an aerial UE. The transmission is either supported by a DF relay or an RIS, as shown in Figure 11.6. We assume that the relay and the UAV are equipped with a single omni-directional antenna. The RIS is equipped

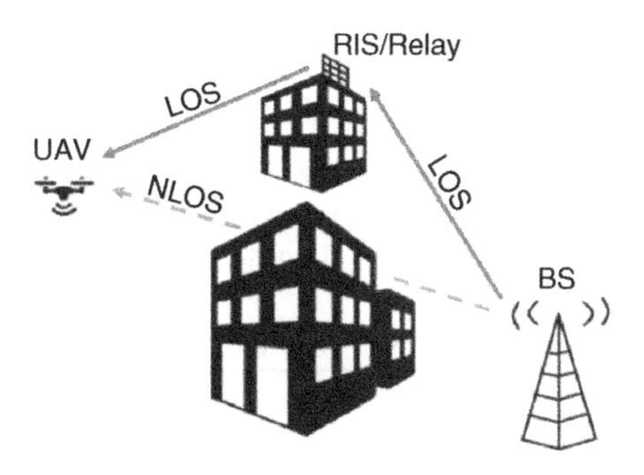

Figure 11.6 RIS/relay-assisted UAV communication system.

with a uniform linear array (ULA) of M reflecting elements. The links from the BS to the RIS, and from the RIS to the UAV are assumed to be LOS channels. We assume a non-line-of-sight (NLOS) channel model for the link from the BS to the UAV. Three different modes of data transmission are considered: (i) direct or single-input single-output (SISO) transmission, (ii) RIS-supported transmission, (iii) relay-supported transmission.

11.6.1.2 Direct Transmission

Let x be the transmitted signal and $h_{SU} \in \mathbb{C}$ represents the channel gain between the BS and UAV, accordingly, the received signal at the UAV can be written as follows:

$$y = h_{\mathrm{SU}}\, x + n \tag{11.1}$$

where $n \sim \mathcal{CN}(0, \sigma)$ is the additive white Gaussian noise (AWGN) with zero mean and σ variance.

Using (11.1), the rate at the UAV is given by:

$$R_{\mathrm{SISO}} = \log_2\left(1 + \frac{p|h_{\mathrm{SU}}|^2}{\sigma^2}\right) \tag{11.2}$$

where p is the power of the transmitted signal.

11.6.1.3 RIS-supported Transmission

In this setup, the RIS reflects the incident signal in the direction of the UAV. Let $\Theta = \left\{e^{j\theta_1}, e^{j\theta_2}, \dots, e^{j\theta_M}\right\}$ be the RIS diagonal phase-shift matrix, where $\theta_i \in [0, 2\pi)$, $i \in \{1, \dots, M\}$ is the phase shift of the ith reflecting element, and $\alpha \in [0, 1)$ is the RIS reflection coefficient. Then the received signal at the UAV is

$$y = \left(h_{\mathrm{SU}} + \alpha h_{\mathrm{SR}}^T \Theta h_{\mathrm{RU}}\right) x + n \tag{11.3}$$

where h_{SR} and $h_{\mathrm{RU}} \in \mathbb{C}$ are the channel gains from the BS to the RIS and from the RIS to the UAV, respectively.

Based on (11.3), the SNR at the UAV can be written as follows:

$$\gamma = \frac{p\left|h_{\mathrm{SU}} + \alpha h_{\mathrm{SR}}^T \Theta h_{\mathrm{RU}}\right|^2}{\sigma^2} \tag{11.4}$$

To minimize the transmit power, the RIS elements phase shifts are selected to coherently combine the signals from different paths. Hence, the maximum instantaneous SNR at the UAV is written as follows:

$$\gamma = \frac{p\left|h_{\mathrm{SU}} + h_{\mathrm{SR}}^T h_{\mathrm{RU}}\right|^2}{\sigma^2} \tag{11.5}$$

Using (11.3) and (11.5), the maximum rate at the UAV can be written as follows:

$$R_{\text{RIS}} = \log_2\left(1 + \frac{p\left|h_{\text{SU}} + h_{\text{SR}}^T h_{\text{RU}}\right|^2}{\sigma^2}\right) \tag{11.6}$$

11.6.1.4 Relay-supported Transmission

The relaying system transmission is divided into two stages: in the first stage, the BS sends the signal to the relay, and the signal received by the relay can be written as follows:

$$y_1 = h_{\text{SR}}\, x_1 + n_1 \tag{11.7}$$

where $h_{\text{SR}} \in \mathbb{C}$ is the channel gain from the BS to the relay, x_1 is the transmitted signal from the BS, and $n_1 \sim \mathcal{CN}(0, \sigma)$ is the AWGN at the relay. In the second stage, the relay decodes the received signal and forwards it to the UAV. Therefore, the signal received by the UAV can be written as follows:

$$y_2 = h_{\text{RU}}\, x_2 + n2 \tag{11.8}$$

where $h_{\text{RU}} \in \mathbb{C}$ is the channel gain from the relay to the UAV, x_2 is the transmitted signal from the relay, and $n_2 \sim \mathcal{CN}(0, \sigma)$ is the AWGN of the second stage.

The UAV receiver performs selection combining to obtain the desired signal. Utilizing (11.7) and (11.8), the SNR of the DF relay-supported transmission can be expressed as [42]:

$$\gamma = \min\left(\frac{p_1\left|h_{\text{SR}}\right|^2}{\sigma^2}, \frac{p_1\left|h_{\text{SU}}\right|^2}{\sigma^2} + \frac{p_2\left|h_{\text{RU}}\right|^2}{\sigma^2}\right) \tag{11.9}$$

where p_1 and p_2 are the powers of the transmitted signals from the BS and the relay, respectively.

Using (11.9), the rate at the UAV is

$$R_{\text{RIS}} = \frac{1}{2}\log_2\left(1 + \min\left(\frac{p_1\left|h_{\text{SR}}\right|^2}{\sigma^2}, \frac{p_1\left|h_{\text{SU}}\right|^2}{\sigma^2} + \frac{p_2\left|h_{\text{RU}}\right|^2}{\sigma^2}\right)\right) \tag{11.10}$$

11.6.1.5 Results and Discussion

In this section, simulation results are represented to evaluate each transmission mode. The channel gains are modelled using the 3GPP Urban Micro (UMi) for aerial vehicles 'from [13], Table B-1' with a carrier frequency of 3 GHz. We extended the setup in [42] to a three-dimensional setup to fit the used channel models as illustrated in Figure 11.7.

Figure 11.8 shows the achievable rate for SISO, RIS with varying number of elements, and DF relaying. It is observed that RIS needs more than 100 elements to improve upon the DF relay performance. Figure 11.9 shows the transmit

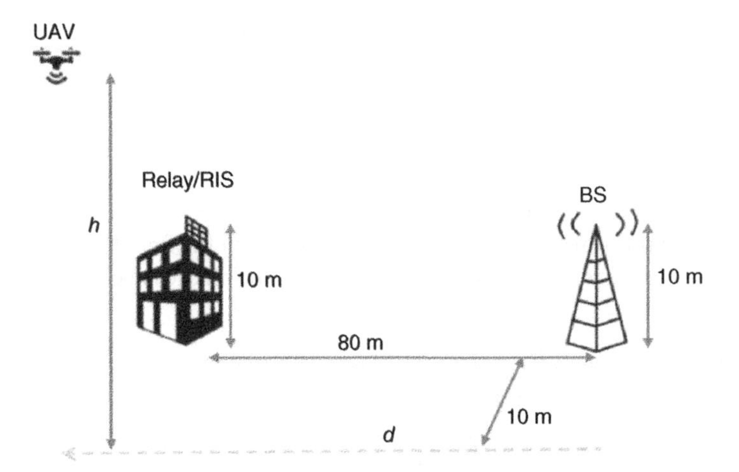

Figure 11.7 The simulation setup for RIS/relay-assisted UAV communication system.

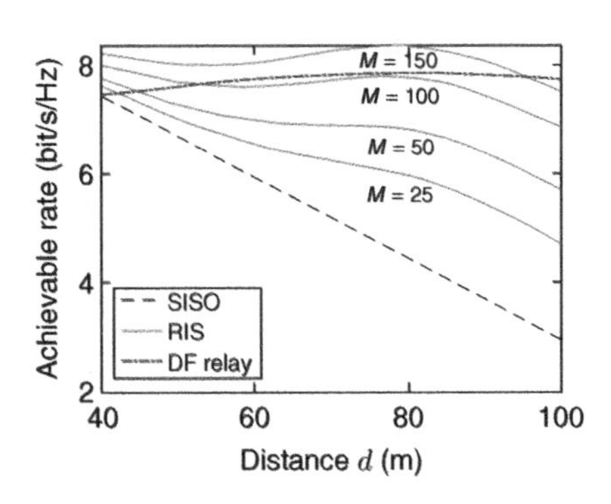

Figure 11.8 The achievable rate for different transmission modes as a function of distance d, with a fixed UAV height $h = 25$ m.

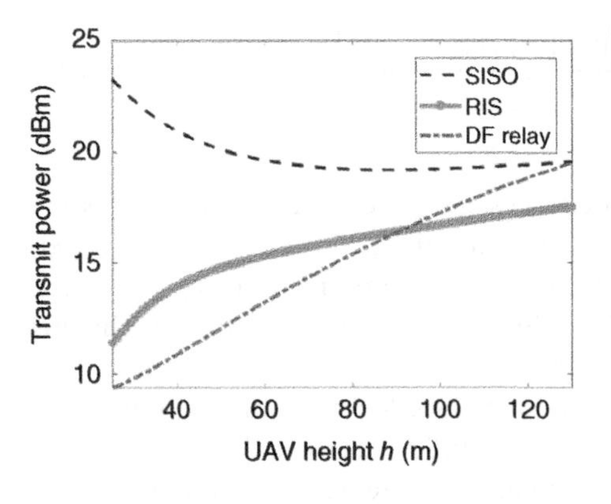

Figure 11.9 The transmit power needed to achieve a rate of $R = 6$ bit/s/Hz as a function of UAV height, with $d = 70$ m.

power that is needed to achieve a rate of $R = 6$ bit/s/Hz for different transmission modes and different UAV heights with ($M = 150$). The SISO scenario requires the highest power and the DF relay-assisted transmission requires the least power. It can be also noticed that the gap between RIS and DF relaying decreases as the UAV height increases, and RIS outperforms DF relaying for altitudes greater than 100 m. The reason is that for higher UAV altitudes, the channel gain for NLOS link slightly improves. On the other hand, it becomes worse for the LOS link, and the relay-supported links experience better channel gain compared to the direct transmission in case of weak direct links.

11.7 Conclusions

Up to date, the potential application area of these RIS-assisted UAV communications has not been explored in the real-time wireless propagation environment. It is desired to develop solutions to meet the QoS and energy requirements in highly dynamic and non-uniform environments of RIS-assisted UAV networks. To this effect, we presented the fundamental characteristics of the UAVs, major paradigms for integrating the UAVs into the wireless networks, and their possible applications, as well as addressing open problems and challenges in the UAV communications and shed light on the possible RIS-assisted UAV systems scenarios. Furthermore, we analysed the performance of the RIS-assisted UAV systems that is also presented. Results show that the RIS should be equipped with more than one hundred elements to match up to a single DF relay performance in terms of achievable rate. It was also observed that an RIS with sufficiently high number of elements attains a better performance, in terms of power requirements, in the case of high-UAV altitudes.

To significantly boost system performance, it's critical for the future research to develop algorithms that optimize the transmit precoders, passive beamforming matrices, and resource allocation, as well as the trajectory of the UAV. It is worth noting that although the availability of channel state information (CSI) is crucial for effective beamforming and optimization, its acquisition is a non-trivial issue in RIS-assisted wireless communications since RIS cannot actively transmit/receive signals as BS or user due to the lack of RF chains.

References

1 Dao, N.N., Pham, Q.V., Tu, N.H. et al. (2021). Survey on aerial radio access networks: toward a comprehensive 6G access infrastructure. *IEEE Commun. Surv. Tutorials* 23 (2): 1193–1225.

2 Mozaffari, M., Saad, W., Bennis, M. et al. (2019). A tutorial on UAVs for wireless networks: applications, challenges, and open problems. *IEEE Commun. Surv. Tutorials* 21 (3): 2334–2360.

3 Li, B., Fei, Z., and Zhang, Y. (2019). UAV communications for 5G and beyond: recent advances and future trends. *IEEE Internet Things J.* 6 (2): 2241–2263.

4 Cicek, C.T., Gultekin, H., Tavli, B. et al. (2019). UAV base station location optimization for next generation wireless networks: overview and future research directions. *1st International Conference on Unmanned Vehicle Systems-Oman (UVS), 2019 February*. http://dx.doi.org/10.1109/UVS.2019.8658363.

5 Li, S., Duo, B., Yuan, X. et al. (2020). Reconfigurable intelligent surface assisted UAV communication: joint trajectory design and passive beamforming. *IEEE Wireless Commun. Lett.* 9 (5): 716–720.

6 Ge, L., Dong, P., Zhang, H. et al. (2020). Joint beamforming and trajectory optimization for intelligent reflecting surfaces-assisted UAV communications. *IEEE Access* 8: 78702–78712.

7 Wang, L., Wang, K., Pan, C. et al. (2020). Joint trajectory and passive beamforming design for intelligent reflecting surface-aided uav communications: a deep reinforcement learning approach.

8 Long, H., Chen, M., Yang, Z. et al. (2020). Joint trajectory and passive beamforming design for secure UAV networks with RIS. *2020 IEEE Globecom Workshops (GC Wkshps)*, pp. 1–6.

9 Liu, X., Liu, Y., and Chen, Y. (2021). Machine learning empowered trajectory and passive beamforming design in UAV-RIS wireless networks. *IEEE J. Sel. Areas Commun.* 39 (7): 2042–2055.

10 Shafique, T., Tabassum, H., and Hossain, E. (2021). Optimization of wireless relaying with flexible UAV-borne reflecting surfaces. *IEEE Trans. Commun.* 69 (1): 309–325.

11 Ranjha, A. and Kaddoum, G. (2021). URLLC facilitated by mobile UAV relay and RIS: a joint design of passive beamforming, blocklength, and UAV positioning. *IEEE Internet Things J.* 8 (6): 4618–4627.

12 Li, S., Duo, B., Di Renzo, M. et al. (2021). Robust secure UAV communications with the aid of reconfigurable intelligent surfaces. *IEEE Trans. Wireless Commun.* 20 (10): 6402–6417.

13 3GPP (2018). Enhanced LTE Support for Aerial Vehicles. *Technical report (TR) 36777*. 3rd Generation Partnership Project (3GPP).

14 Zeng, Y., Lyu, J., and Zhang, R. (2019). Cellular-connected UAV: potential, challenges, and promising technologies. *IEEE Wireless Commun.* 26 (1): 120–127.

15 Yang, L., Meng, F., Zhang, J. et al. (2020). On the performance of RIS-assisted dual-hop UAV communication systems. *IEEE Trans. Veh. Technol.* 69 (9): 10385–10390.

16 Li, Y., He, L., Ye, X. et al. (2016). Geometric correction algorithm of UAV remote sensing image for the emergency disaster. *2016 IEEE International Geoscience and Remote Sensing Symposium (IGARSS)*, pp. 6691–6694.

17 Fotouhi, A., Qiang, H., Ding, M. et al. (2019). Survey on UAV cellular communications: practical aspects, standardization advancements, regulation, and security challenges. *IEEE Commun. Surv. Tutorials* 21 (4): 3417–3442.

18 Zeng, Y., Wu, Q., and Zhang, R. (2019). Accessing from the sky: a tutorial on UAV communications for 5G and beyond. Proceedings of the IEEE, vol. 107, no. 12, pp. 2327–2375.

19 Liu, Y., Dai, H.N., Wang, Q. et al. (2020). Unmanned aerial vehicle for internet of everything: opportunities and challenges. *Comput. Commun.* 155: 66–83. http://dx.doi.org/10.1016/j.comcom.2020.03.017.

20 Ma, D., Ding, M., Hassan, M. (2020). Enhancing cellular communications for uavs via intelligent reflective surface. *2020 IEEE Wireless Communications and Networking Conference (WCNC)*, pp. 1–6.

21 Zeng, Y., Zhang, R., and Lim, T.J. (2016). Wireless communications with unmanned aerial vehicles: opportunities and challenges. *IEEE Commun. Mag.* 54 (5): 36–42.

22 Lai, C.C., Chen, C.T., and Wang, L.C. (2019). On-demand density-aware UAV base station 3D placement for arbitrarily distributed users with guaranteed data rates. *IEEE Wireless Commun. Lett.* 8 (3): 913–916.

23 Alzenad, M., El-Keyi, A., Lagum, F. et al. (2017). 3-D placement of an unmanned aerial vehicle base station (UAV-BS) for energy-efficient maximal coverage. *IEEE Wireless Commun. Lett.* 6 (4): 434–437.

24 Lakew, D.S., Masood, A., and Cho, S. (2020). 3D UAV placement and trajectory optimization in UAV assisted wireless networks. *2020 International Conference on Information Networking (ICOIN)*, pp. 80–82.

25 Shakhatreh, H. and Khreishah, A. (2018). Optimal placement of a UAV to maximize the lifetime of wireless devices. *2018 14th International Wireless Communications Mobile Computing Conference (IWCMC)*, pp. 1225–1230.

26 Cherif, N., Jaafar, W., Yanikomeroglu, H. et al. (2020). On the optimal 3D placement of a UAV base station for maximal coverage of UAV users. GLOBECOM 2020 - 2020 IEEE Global Communications Conference, 2020, pp. 1–6.

27 Khawaja, W., Guvenc, I., Matolak, D.W. et al. (2019). A survey of air-to-ground propagation channel modeling for unmanned aerial vehicles. *IEEE Commun. Surv. Tutorials* 21 (3): 2361–2391.

28 Khuwaja, A.A., Chen, Y., Zhao, N. et al. (2018). A survey of channel modeling for UAV communications. *IEEE Commun. Surv. Tutorials* 20 (4): 2804–2821.

29 Azari, M.M., Rosas, F., Chiumento, A. et al. (2017). Coexistence of terrestrial and aerial users in cellular networks. *2017 IEEE Globecom Workshops (GC Wkshps)*, pp. 1–6.

30 Fouda, A., Ibrahim, A.S., Güvenç, I. et al. (2019). Interference management in UAV-assisted integrated access and backhaul cellular networks. *IEEE Access* 7: 104553–104566.

31 Renzo, M.D., Zappone, A., Debbah, M. et al. (2020). Smart radio environments empowered by reconfigurable intelligent surfaces: how it works, state of research, and road ahead. *IEEE J. Selected Areas Commun.* 38 (11): 2450–2525

32 Bjornson, E., Ozdogan, O., and Larsson, E.G. (2020). Reconfigurable intelligent surfaces: three myths and two critical questions. *IEEE Commun. Mag.* 58 (12): 90–96. http://dx.doi.org/10.1109/MCOM.001.2000407.

33 Wu, Q., Zhang, S., Zheng, B. et al. (2021). Intelligent reflecting surface-aided wireless communications: a tutorial. *IEEE Trans. Commun.* 69 (5): 3313–3351.

34 Liu, Y., Liu, X., Mu, X. et al. (2021). Reconfigurable intelligent surfaces: principles and opportunities. *IEEE Commun. Surv. Tutorials* 23 (3): 1546–1577.

35 Zhao, J. (2019). A survey of intelligent reflecting surfaces (IRSs): towards 6G wireless communication networks.

36 Jung, M., Saad, W., Debbah, M. et al. (2021). On the optimality of reconfigurable intelligent surfaces (RISs): passive beamforming, modulation, and resource allocation. *IEEE Trans. Wireless Commun.* 20 (7): 4347–4363.

37 Zheng, B. and Zhang, R. (2020). Intelligent reflecting surface-enhanced OFDM: channel estimation and reflection optimization. *IEEE Wireless Commun. Lett.* 9 (4): 518–522.

38 Wu, Q. and Zhang, R. (2019). Intelligent reflecting surface enhanced wireless network via joint active and passive beamforming. *IEEE Trans. Wireless Commun.* 18 (11): 5394–5409.

39 Gong, S., Lu, X.,Hoang, D.T. et al. (2020). Toward smart wireless communications via intelligent reflecting surfaces: a contemporary survey. *IEEE Commun. Surv. Tutorials* 22 (4): 2283–2314. http://dx.doi.org/10.1109/COMST.2020.3004197.

40 Abdalla, A.S., Rahman, T.F., and Marojevic, V. (2020). UAVs with reconfigurable intelligent surfaces: applications, challenges, and opportunities.

41 Shang, B., Shafin, R., and Liu, L. (2021). UAV swarm-enabled aerial reconfigurable intelligent surface. *IEEE Wireless Commun.* 28 (5): 156–163.

42 Björnson, E., Özdogan, O., and Larsson, E.G. (2019). Intelligent reflecting surface versus decode-and-forward: how large surfaces are needed to beat relaying? *IEEE Wireless Commun. Lett.* 9 (2): 244–248.

12

Optical Wireless Communications Using Intelligent Walls*

Anil Yesilkaya[1], Hanaa Abumarshoud[2], and Harald Haas[1]

[1] *LiFi Research and Development Centre (LRDC), Department of Electronic & Electrical Engineering, The University of Strathclyde, Glasgow, UK*
[2] *James Watt School of Engineering, The University of Glasgow, UK*

12.1 Introduction

The current surge in the number of mobile devices and the emerging internet-of-things (IoT) and internet-of-everything (IoE) applications are posing an unprecedented demand on wireless connectivity [1]. Mobile device density in service areas is approaching the region of more than one wireless device per square metre, which indicates that enhancing the capacity of wireless networks will become even more crucial in the foreseeable future. The use of conventional radio frequency (RF) small cells in such dense deployments proves challenging due to severe inter-cell co-channel interference (CCI). This is mainly attributed to the fact that RF signals travel beyond the cell boundaries. Moreover, there is a general consensus that the limited RF spectrum resources are not sufficient to future-proof wireless communications. Therefore, other parts of the electromagnetic (EM) spectrum have been explored for wireless connectivity. Particularly, optical wireless communications (OWC) presents a promising solution to satisfy the demands of future wireless networks. The main advantages offered by OWC are the freely available resources in the optical band of the EM spectrum which is in terahertz, with the visible light (VL) band alone being almost 2600 times larger than the full RF spectrum [2]. Utilizing the entire optical spectrum is not practically feasible due to the limitations of the transceivers front-ends

*This research has been supported in part by EPSRC under Established Career Fellowship Grant EP/R007101/1, Wolfson Foundation and European Commission's Horizon 2020 research and innovation program under grant agreement 871428, 5G-CLARITY project.

Intelligent Reconfigurable Surfaces (IRS) for Prospective 6G Wireless Networks, First Edition.
Edited by Muhammad Ali Imran, Lina Mohjazi, Lina Bariah, Sami Muhaidat,
Tie Jun Cui, and Qammer H. Abbasi.

and the necessary electro-optical (EO) and opto-electrical (OE) conversions. However, advancements in solid-state lighting and semi-conductor devices are continuously leading to better capabilities, which means higher electrical bandwidth and narrower spectral emissions.

OWC technologies include free space optical (FSO) communication, ultraviolet (UV) communication, optical camera communication (OCC), visible light communications (VLC), and light-fidelity (LiFi). FSO communication systems provide point-to-point links over relatively large transmission distances and are mainly used for enabling low-cost wireless backhaul links indoors and outdoors. UV communication systems provide long-distance non-line-of-sight (NLoS) wireless connectivity through atmospheric scattering. UV links are robust to link blockage and, thus, do not require perfect alignment between the transmitter and the receiver. OCC utilizes cameras as receivers and operates in the VL spectrum. OCC provides connectivity for low-rate, short-range line-of-sight (LoS) applications. VLC also utilizes VL for wireless connectivity by employing a photo-diode (PD) as a receiver and offers high-speed, short-range LoS connectivity. The term LiFi refers to a network solution that utilizes wireless optical links, mainly VLC and infra-red (IR), to provide bidirectional connectivity with mobility support and seamless coverage.

LiFi arises as a promising solution to add a new dimension to spectrum heterogeneity in future wireless networks [3, 4]. A LiFi network constitutes multiple interconnected short-range optical access points (APs), referred to as attocells. Each attocell serves a small number of users within a coverage area of a few square meters. The high directionality of light signals in LiFi networks allow for extreme cell densification and, hence, provide an effective solution to wireless coverage, and thus ultra-high capacity, in extremely dense deployments.

LiFi is envisioned to play a significant role in future cellular networks and IoT applications, including smart healthcare provisioning, smart infrastructure management, high-precision manufacturing, self-driven vehicles, and remote robots operations, to count a few. These applications rely on smart autonomous operations and reliable high-speed connectivity that facilitate real-time interactions between different entities. Until recently, enhancing the communication capabilities of LiFi systems was primarily focused on the development of transmission and reception capabilities in the face of undesirable uncontrollable channel conditions. However, breakthrough advancements in programmable meta-surfaces resulted in a paradigm shift in the way wireless signal propagation is dealt with: from fully uncontrollable to tunable and customized wireless channel engineering. The recent rise of revolutionary intelligent reconfigurable surfaces (IRS) technology means that the environment itself can be programmed in order to enhance the performance of wireless communication systems. This enhancement comes in terms of spectral efficiency, energy efficiency, link reliability, and security. Specifically, an intelligent surface constitutes a number of meta-surfaces that are artificially engineered in order to allow full manipulation of the incident EM waves. The EM response of these meta-surfaces can be altered on a macroscopic level to control the amplitude and phase of the incident beams. Based on this, it is possible to effectively control wireless signals to achieve the desired performance gains.

This chapter discusses the use of IRS or 'intelligent walls' in the context of LiFi systems. Specifically, we discuss the background of this emerging technology and present some of its applications in Section 12.2. A case study for high-performance IRS-enabled LiFi systems is presented in Section 12.3, while Section 12.4 discusses some of the related challenges and future research directions.

12.2 Optical IRS: Background and Applications

In this section, we provide a comprehensive literature review of the integration of IRSs in OWC and shed light on the different applications of this emerging concept.

12.2.1 IRS from the Physics Perspective

The EM response of a surface is typically determined by the material it is made of as well as its geometry. Surface reflection behaviour can be classified into one of three responses: specular, diffuse, or glossy. Perfectly smooth surfaces act as mirrors and reflect light in a specular manner according to Snell's law of reflection. Rough surfaces, on the other hand, scatter the incident light in all directions. Surfaces with a glossy nature reflect light in a way that contains both specular and diffuse components.

The term IRS refers to a surface containing an array of periodically arranged metasurface elements that are engineered to produce a controlled response to impinging light signals. The EM response of each element of the IRS array can be adjusted by tuning the surface impedance through electrical voltage stimulation, which can be controlled via field-programmable gate array (FPGA) chips. Various IRS designs have been presented in the literature using different numbers of layers as well as different materials, including liquid crystal, meta-lenses with artificial muscles, doped semi-conductors, and electromechanical switches. The fundamental architecture of an IRS is depicted in Figure 12.1. Since the physical-layer characteristics of IRSs can be controlled by software, they are also termed software-defined surfaces (SDSs). Although the use of metasurfaces in OWC have only recently became an active area of research, their capabilities in light manipulation have already been tested and developed in the field of flat optics [6, 7].

Reconfigurable metasurfaces are two-dimensional, artificial periodic structures that consist of programmable space-variant, sub-wavelength metallic or dielectric elements known as 'meta-atoms'. The most distinct feature of metasurfaces

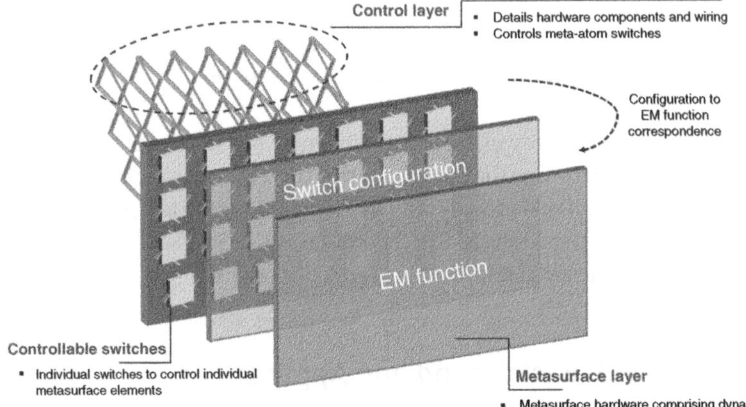

Figure 12.1 General architecture of an IRS [5].

is that their EM characteristics can be reconfigured in order to introduce an engineered response to the incident wave-front by manipulating the outgoing photons. Based on this, the electric and magnetic properties of these structures can be engineered in order to effectively control the key properties of EM waves, namely, amplitude, phase, and polarization. Based on this, two categories of phenomena can be observed, namely, direction-related phenomena and intensity-related phenomena. Direction-related phenomena steer the propagation of the rays through scattering, reflection, and refraction, while intensity-related phenomena affect the signal power in the form of amplification, attenuation, or absorption. In the following, we list four of the most important light manipulation functionalities that can be realized by metasurfaces:

- **Refractive index tuning**: the refractive index of meta-surfaces can be tuned to change the behaviour of light rays as they pass through the material. Controlling the imaginary part of the refractive index affects how the material amplifies or absorbs the light. The real part can be positive or negative. With a negative refractive index, it is possible to reverse the phase velocity of the light-wave and bend it in a direction that is impossible with a positive index [8]. It is also possible to realize near-zero index materials which allow perfect wave transmission in one direction [9].
- **Anomalous reflection**: meta-surfaces make it possible to break the law of reflection and reflect the light with an angle that is different to the angle of incidence. Based on this, the light rays impinging on a meta-surface element can be steered into desired directions [10]. This is done by employing different reflecting phases at different meta-elements in the surface to tune the

phase distribution over the meta-surface so that the reflected waves interfere constructively in the desired direction.

- **Signal amplification and attenuation**: meta-surfaces can be tuned so as to provide light amplification, attenuation, or even complete absorbance [11]. The amplitude control can be achieved by varying the conversion efficiency of each of the meta-elements via structurally birefringent meta-atoms.
- **Wavelength decoupling**: meta-surfaces offer the possibility of engineering a wavelength-specific EM response to decouple and independently control different wavelengths [12]. Based on this, it is possible to allow signals within a specific wavelength range to be reflected in a certain direction while absorbing signals with other wavelength values as in the multiplexer/de-multiplexer metasurface in [12] which decomposes the light signal into multiple channels.

Light manipulation functionalities are illustrated in Figure 12.2. Although the use of meta-surfaces in OWC have only recently become an active area of research, their capabilities to realize light manipulation have already been tested and developed in the field of flat optics.

Most of the research efforts in metasurface fabrication focus on engineering the phase profile of a wave-front since it is a critically important design feature for many applications. An example of meta-surface-enabled optical phase control is the beam deflector presented in [13]. The beam deflector consists of a super-cell with 15 nano-disks, and the diameter of each of the nano-disks can be changed to produce a phase shift from 0 to 2π with $\pi/7$ phase increments. The proposed

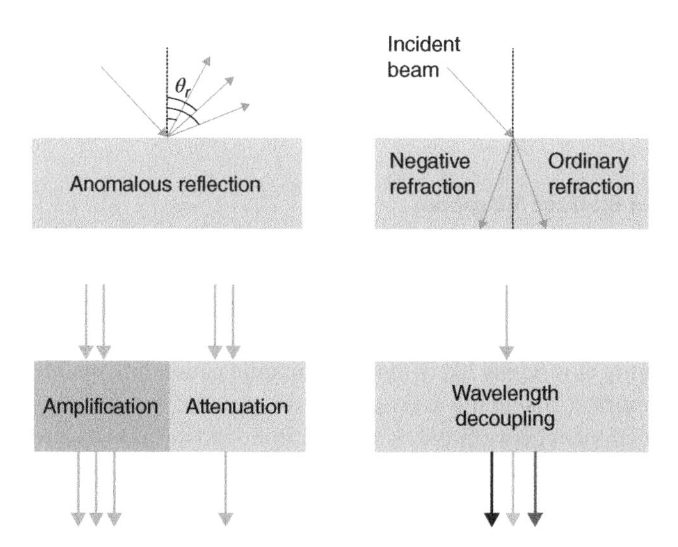

Figure 12.2 Meta-surfaces light manipulation functionalities.

design offers full 2π-phase control and a near-unity transmission efficiency at an operating wavelength of 715 nm.

Widening the scope of metasurfaces applications requires the control of more than just the phase of the waveform. For this reason, research efforts have pushed towards achieving simultaneous control of multiple signal parameters. For example, a metasurface that can control both the amplitude and phase of light signals was proposed in [14]. The amplitude is controlled by varying the conversion efficiency of the meta-atoms, while the phase is controlled by altering the in-plane orientation of the meta-atoms. These two degrees of freedom enable complete and independent control of the amplitude and phase of the optical signal.

Similarly, the work in [15] showed the possibility of achieving independent phase, and polarization control with sub-wavelength spatial resolution and an experimentally measured efficiency of up to 97%.

Another case of combining multiple functions in one metasurface is the multiplexer/de-multiplexer demonstrated in [12] in which wavelength, polarization, and space multiplexing capabilities were integrated into one single ultra-thin metasurface in order to decompose the light signal into multiple channels.

The afore-mentioned examples demonstrate that it is possible to effectively manoeuvre the physico-chemical characteristics of metasurfaces in the VL spectrum, which opens the door for many IRS applications in OWC systems, as will be discussed in the following sections.

12.2.2 IRS Applications in OWC

In the following, we discuss some potential applications for the integration of IRS in OWC systems.

12.2.2.1 Reflection for Blockage Mitigation

Optical signals do not penetrate through solid objects and they typically exhibit increased attenuation and less diffraction compared to RF signals. As a result, the optical link quality primarily relies on the existence of a LoS path between the transmitter and the receiver. If the LoS path is obstructed, i.e. due to the existence of a blocking entity, it is highly likely that the communication link quality will significantly deteriorate. Due to the dependency on LoS, OWC systems typically require perfect alignment between the transceiver front-ends so as to ensure a reliable transmission. Ensuring such alignment is easier in cases with no or low mobility, i.e. in VLC systems with static users. However, the case is more complicated in LiFi networks which are required to fully support high-user mobility and ubiquitous connectivity. One of the key challenges in such networks is that the user terminal randomly changes its position and orientation which could lead to obstruction of the LoS link between the user and the LiFi AP.

Along with the LoS component, receivers also perceive multi-path components that result from the diffusion and reflection of light signals throughout the surrounding environment. Since both the direct LoS signal and the reflected signals carry the same data, they can be added at the receiver to maximize the total received power. In a typical indoor environment, however, NLoS signals are totally uncontrolled and nearly isotropic in their spatial distribution and are, therefore, weak at their average intensity. As a result, the effect of multi-path reflections on the received signal intensity is typically negligible [16]. The use of IRS mounted on the walls of indoor spaces with LiFi coverage means that the reflection of the light signals could possibly be controlled so as to add significantly to the received signal at the mobile device and compensate for the loss of the LoS component in the event of a blockage. Moreover, using IRS in indoor LiFi systems can enhance the coverage of the LiFi APs by mitigating the dead-zone problem for cell-edge users. Since the coverage of an indoor LiFi AP is typically confined within the beam width of the transmitted light, non-uniform coverage distribution is typically exhibited. The reduced link reliability at the cell edge means that users moving in the proximity of the LiFi APs will need to perform frequent horizontal handovers, i.e. between neighbouring LiFi APs, or vertical handovers, between the LiFi network and the RF network. With the use of wall-mounted IRSs, users close to the wall can receive strong reflected signals that allow for meaningful signal-to-noise-ratio (SNR) [17]. This implies that a cell-edge user in the vicinity of the IRS can still connect to the LiFi AP despite having a weak LoS path.

The concept of using wall-mounted IRS in indoor OWC systems was investigated in [18]. Two types of IRSs, namely, intelligent meta-surface reflectors and intelligent mirror arrays reflectors, were studied. For the meta-surface reflectors, a macroscopic model for the metasurface elements is adopted to abstract them as anomalous reflective rectangular blocks. It is assumed that the phase gradient is kept constant over each metasurface element and that the phase discontinuity of each metasurface element can be tuned independently of the other elements to reflect the incident light in a specific direction, i.e. the detector's centre. For mirror array reflectors, identical rectangular mirrors with two rotational degrees of freedom are used. In the aforementioned work, it was shown that the focusing and steering capabilities of IRS are proportional to the number of reflecting elements up to a specific number of elements. Furthermore, it was shown that using a 25 cm × 15 cm reflector on the wall can enhance the received power by up to five times compared to the direct LoS link. The presented results imply that IRS technology provide a promising solution to compensate for LoS blockage in LiFi systems, particularly when the communication link is susceptible to blockage because of the movement of the user device or other objects. In other words, by leveraging IRS, we can enable a high-performance operation for mobile devices under mobility conditions and link blockages as well as random device orientations, making LiFi resilient to these probabilistic factors.

12.2.2.2 Enhanced Optical MIMO

Multiple-input-multiple-output (MIMO) configurations are used to provide wireless communication systems with multiple transmit/receive paths so as to enable the simultaneous transmission of multiple streams of data signals. With the application of appropriate signal processing, MIMO systems can enhance spectral efficiency and received signal quality. The application of MIMO systems in OWC is an attractive solution due to the existence of large numbers of light emitting diodes (LEDs) in a single luminaire and many luminaires in various indoor settings. In the context of OWC, MIMO systems can be achieved by deploying multiple LEDs on the AP side and/or multiple PDs on the user side [19].

While MIMO configurations have the potential to improve the performance of OWC systems, the achieved performance enhancement is highly dependant on the condition of the MIMO channel matrix. If the MIMO channel has a strong spatial correlation, the rank of the channel matrix will be low. In a multiplexing MIMO system, this means that only a few MIMO sub-channels are actually usable while the remaining sub-channels are deemed unreliable for transmitting and receiving data. In a diversity MIMO system, the high-channel correlation implies that the characteristics of different diversity sub-channels are very similar. Consequently, if one of them is unreliable, the remaining sub-channels are likely to be unreliable as well, which diminishes the diversity gain. In a spatial modulation (SM) system, part of the information signal is carried on the index of the activated LED [20, 21]. Therefore, high-channel correlation makes it hard to distinguish the difference between various spatial symbols, which results in high bit error ratios (BERs) [22, 23]. Therefore, it is clear that establishing a well-conditioned MIMO channel matrix in OWC systems is essential to ensure that various MIMO configurations can be utilized to achieve the desired advantages.

The MIMO channel conditions in OWC are different to RF. The RF wireless channels comprise multiple NLoS paths with small-scale fading, which results in MIMO channels with weak spatial correlation. On the other hand, OWC channels do not exhibit fading characteristics and rely mainly on front-end characteristics and positions. In the case of similar or symmetrical positions of multiple PDs with respect to (w.r.t.) the AP, the corresponding MIMO sub-channels will have a strong spatial correlation, which forms a major challenge in OWC MIMO systems [24, 25]. One of the possible solutions to tackle the high-channel correlation in optical MIMO is to carefully align the locations of the LEDs and the PDs such that they create uncorrelated channel paths as was proposed in [26]. While this solution could be feasible in fixed point-to-point communications, it is not always possible in LiFi systems that support high-user mobility.

IRS technology provides an effective solution to improve poorly conditioned optical MIMO systems by creating controllable multi-path channels. The reflection coefficients of the IRS elements can be optimized such that the rank of the MIMO channel matrix is enhanced and the observed sub-channels are sufficiently distinguishable. The higher spatial decorrelation will enable higher spectral efficiency and boost the achievable data rate of optical MIMO systems [17]. The rank improvement capabilities of IRSs were studied in [27], where a high multiplexing gain was achieved even when the LoS MIMO channel is rank-deficient. It is noted that enhancing the MIMO channel rank requires joint optimization of the reflection coefficients of the IRS elements and the transmit precoding matrix, which is a non-convex optimization problem [28].

12.2.2.3 Media-Based Modulation

The concept of media-based modulation (MBM) can be considered as a special case of index modulation (IM) [29]. While conventional IM systems employ multiple transmitters to carry distinct data symbols on the indices of the activated transmitters, MBM can be realized with a single transmitter aided with IRS. In MBM, an IRS array can be used to provide an additional dimension for data modulation by controlling the EM response of each of its reflecting elements. MBM offers some advantages compared to conventional IM including lower cost and higher flexibility [30, 31].

The basic idea of MBM is explained in the following section. The transmission system comprises of the transmitting AP and an IRS equipped with multiple reflecting elements. Each reflecting element receives the data signal from the AP and reflects it to the intended receiver with a specific reflection coefficient. Different reflection coefficients correspond to different observed channel realizations at the receiving PD, so we can carry data bits on the state of the reflecting element. At the receiver, recovering the intended data symbols involves decoding the M-ary modulated symbol as well as estimating the channel state of each IRS reflecting element. This necessitates the knowledge of the channel state information (CSI) of all the reflecting elements which could be obtained using passive pilot-based channel estimation in which each IRS passively reflects the pilot sequences sent by the user to the AP to estimate its channel coefficients [32].

In theory, the spectral efficiency of MBM can be enhanced by increasing the number of IRS reflecting elements, as well as the number of possible reflection states. However, the enhancement is practically limited by the degree of decorrelation between the sub-channels created by different IRS elements. This is because the Euclidean distance between the MBM constellation points depends on the values of the possible channel realizations. As a result, the desired performance enhancement can be only realized if these combinations are clearly distinguishable at the receiving terminal.

12.2.2.4 Enhanced Optical NOMA

Non-orthogonal multiple access (NOMA) is a spectrally efficient multi-user access technique that has been widely investigated for use in OWC systems [33–35]. In NOMA, signals of different users are multiplexed in the power domain by assigning distinct power levels to different users' signals. This process is referred to as superposition coding and the power allocation coefficients are usually determined based on the users' channel conditions. The basic principle is that users with more favourable channel conditions are allocated lower-power levels than those with weaker channel conditions. Successive interference cancellation (SIC) is then performed at the receiver side to decode and subtract the signals with higher-power levels first until the desired signal is extracted. NOMA has been shown to provide promising performance gain in OWC due to the high SNR and the somewhat deterministic nature of the optical channel, which facilitates the acquisition of CSI of the users with less overhead compared to RF fading channels [36].

The operating principle of NOMA is based on having distinct channels for different users. However, this is not always guaranteed in OWC due to the high correlation between the optical channels. In fact, the characteristics of the optical channel imply that it is likely that multiple users experience similar channel conditions, hindering the feasibility of NOMA. IRS technology offers the possibility to overcome this limitation and create distinct channels for different users by dynamically altering the perceived channel gains. Moreover, by carefully designing the reflection coefficients of the IRS elements, it is possible to create better conditions for effective power allocation which, in turn, leads to successful SIC and high reliability. Moreover, IRS-enabled NOMA offers enhanced user fairness compared to conventional NOMA systems. This is because, in conventional NOMA, the decoding order of a user depends on its channel gain value compared to the rest of the users. To this effect, users with a lower decoding order, i.e. a lower channel gain, must always decode their signals with the existence of interference from users with a higher decoding order. Dynamic IRS configuration means that it would be possible to change the users decoding order despite their locations, leading to enhanced fairness [37]. The work in [38] proposed a framework for the joint design of the NOMA decoding order, power allocation, and the reflection coefficients in an IRS-assisted NOMA VLC system. The NP-hard multi-dimensional optimization problem was solved by a heuristic technique with the objective of enhancing the bit error ratio BER performance. The presented results showed that optimized IRS leads to higher-link reliability, particularly when the links are subject to the adverse effects of link blockage and random device orientation.

12.2.2.5 Enhanced PLS

The fact that light signals cannot penetrate through opaque objects makes OWC systems particularly secure in confined spaces, provided that eavesdroppers are located outside the room [39–43]. This means that OWC systems provide inherent physical layer security (PLS) compared to RF systems where the signals can easily be intercepted by eavesdroppers from behind a wall. Nevertheless, OWC links are susceptible to eavesdropping by malicious users that exist within the coverage area of the AP of interest.

IRS technology offers a solution to enhance the security of such systems by implementing IRS-assisted PLS techniques. Several adaptive methods can be utilized to tune the optical properties of the environment so as to enhance the link reliability for legitimate users while degrading the reception quality for potential eavesdroppers. Moreover, the IRS elements can be tuned to produce friendly jamming signals by creating randomized multi-path reflections to produce artificial noise at the eavesdropper without affecting the transmission of the legitimate user. Another possibility to use IRS to boost the PLS is to employ secure beamforming in MIMO systems. Multiple reflecting elements can be configured to produce a precoding matrix so that the signal can be decoded only at the intended receiver location to reduce the probability of signal interception.

In order to realize IRS-based PLS, the secrecy performance of such systems, i.e. secrecy capacity and secrecy outage probability, must be analysed in relation to the locations and capabilities of IRS. The work in [44] proposed a framework for maximizing the data security of an IRS-assisted VLC system. An array of wall-mounted mirrors was utilized and the orientation of each mirror was adjusted so as to maximize the achievable secrecy capacity. It was found that the IRS channel gain is highly affected by any change in the mirrors' orientation, which makes it hard to find the optimal combination of orientation angles. In order to overcome this issue, the orientation optimization problem was converted into a reflected spot position finding problem. As a result, the secrecy capacity can be maximized by finding the optimal position of the reflected spots for each mirror, which greatly reduces the complexity.

12.3 Case Study: High Performance IRS-Aided Indoor LiFi

12.3.1 Channel Modelling

In this section, the adopted methodology for obtaining the channel properties of an indoor LiFi network, where the system performance is enhanced via the deployment of IRSs, will be presented. Then, the effect of IRSs on the optical link budget, channel impulse response (CIR), channel frequency response (CFR), and other important channel parameters will also be detailed.

Unlike conventional RF-based wireless communication systems, the operation wavelengths are in the nm region in LiFi, which makes the wavelengths comparable with atoms and molecules that comprise the surrounding materials. The penetration characteristics of the EM propagation in the VL and IR spectra become weak due to this relative comparability. As a result, the optical channel becomes highly dependent on the surface geometry as well as the coating characteristics. In order to capture the optical channel in a realistic manner, we propose a non-sequential Monte Carlo ray-tracing (MCRT)-based channel modelling technique, which is able to capture the realistic transmitter (TX), receiver (RX), environment geometry, and coating material characteristics. Note that the simplest implementation of an indoor LiFi system is to utilize off-the-shelf LEDs and PDs at the TX and RX, respectively. Thus, the information will be transmitted from the TX to RX incoherently by using the subtle changes encoded onto instantaneous light intensity, which could be detected at the RX. Consequently, the measurement of incoherent irradiance for given scenarios will be the main target in our simulations.

The proposed MCRT simulation toolbox, which is depicted in Figure 12.3, amalgamates the geometrical non-sequential ray tracing (NSRT) capabilities of Zemax OpticStudio version 21.3.2 [45] with our custom MATLAB-based computation and post-processing libraries. The accuracy of the ray-tracing in Zemax Opticstudio for only LoS and LoS-plus-NLoS cases have been reported to match the real-world measurements closely by [46–49]. Accordingly, the mean squared error (MSE) between the Zemax OpticStudio simulations and real-world

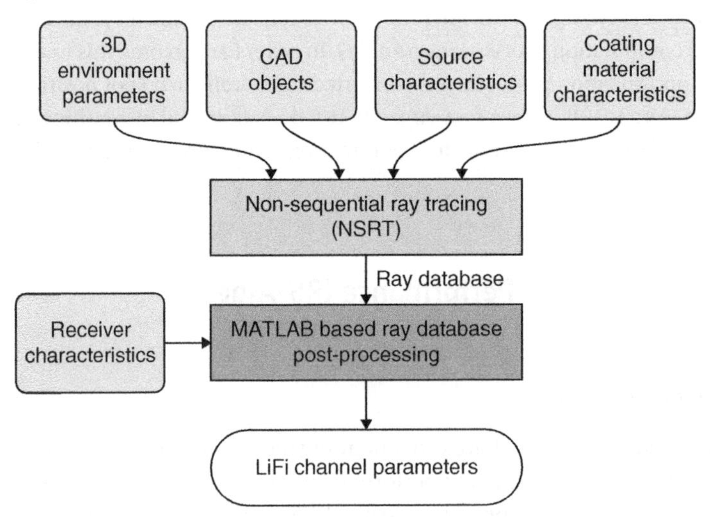

Figure 12.3 Block diagram for MCRT-based LiFi channel modelling environment.

measurements of 0.7%, 1%, and 1.5% are achieved for LoS, only first-order NLoS and higher-order NLoS reflection scenarios, respectively [46]. Therefore, a flexible and highly accurate optical channel modelling compared to the recursive method [50, 51]-based calculations is obtained. It is important to note that the recursive method lacks the ability to model complex geometrical shapes, sources, receivers, and coating materials, which is crucial in our meta-material based IRS application. As can be seen from Figure 12.3, the environment parameters, computer-aided-design (CAD) object models, as well as source and coating material characteristics are inputted into our simulation environment. Since the main purpose of LiFi is to achieve broadband information transmission and illumination by using non-imaging off-the-shelf components simultaneously, the NSRT method is adopted in our simulations.

12.3.1.1 Generation of the Indoor Environment

A typical indoor LiFi application scenario within a cuboid-shaped room, which represents a typical indoor room structure for homes, offices, hospitals, etc., with dimensions of 5 m × 5 m × 3 m is adopted in this work. In Figure 12.4,

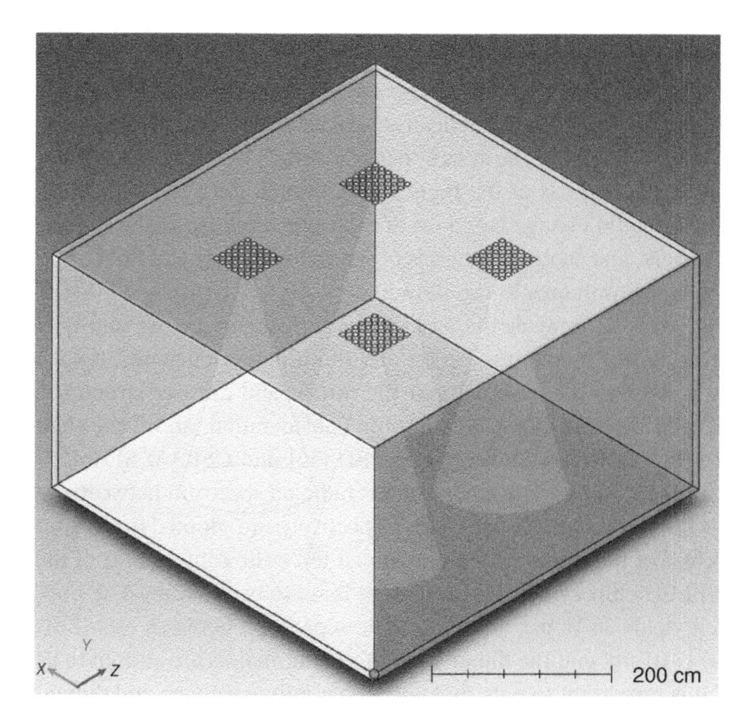

Figure 12.4 Isometric view of the considered scenario. The global origin point of the simulation environment is indicated by the dark grey circle located in the corner of the room.

the scenario under investigation is depicted, where four luminaires serve as optical APs. Each optical AP is placed at the corners of a square, with sides measuring 2 m, that is located at the centre of the ceiling. To harvest the benefits of the IRS functionality, the side walls are assumed to be covered with meta-material based wallpaper, which is depicted only for a single wall in Figure 12.4 for the sake of simplicity. Thus, the reflection characteristics of the side walls will be altered dynamically by using IRSs to enhance the achievable SNR as well as the average error performance of a mobile user equipment (UE).

The global positions and orientations of the transmit LED array based luminaires, receive PDs and IRSs are defined by 3×1 vectors, which are defined w.r.t. the global origin point $O(0, 0, 0)$. Note that the global origin of the system is depicted by the dark grey circle in Figure 12.4. The location vector of each element could be given in the format of $\mathbf{v} = (v_x, v_y.v_z)$, where x, y, and z axis components of the vector are denoted by v_x, v_y, and v_z, respectively. Similarly, the orientation vectors of each object are given by $\mathbf{o} = (o_x, o_y, o_z)$, where the rotation w.r.t. x, y, and z axis, in other words, within the yz, xy planes, are given by o_x, o_y, and o_z, respectively.

12.3.1.2 Source Characterization

The application of metasurfaces in LiFi requires the development of optical models of the transmit LEDs, receive PDs, and reflective surfaces. These models need to be as realistic as possible to capture the system performance. The accurate selection of the operation wavelengths is of the highest importance since the achievable system performance could change based on the spectral response profiles of the transmitters, receivers, and coating materials of the walls, ceiling, and floor.

The generic indoor luminaires in the illumination market typically consist of a concave mirror, an LED array as the base, and a diffuser. Since our aim is to use the luminaires both for communication and illumination purposes, a 6×6 LED chip array is used as the base without the mirror and diffuser structures. To take both VL and IR band characteristics into consideration, an off-the-shelf OSRAM GW QSSPA1.EM High-Power White LED [52] and OSRAM SFH 4253 High-Power IR LED [53] chips with a continuous radiated spectrum between the limits of 0.383–0.780 μm and 0.770–0.920 μm, respectively, are adopted in our simulations. The origin of the luminaires are assumed to be the centre point of the LED array. Each LED chip within the luminaires is set to be radiating 1 W optical power, which yields 36 W per luminaire. The separation between each LED chip in the array is 10 cm, and the dimensions of the luminaires are chosen to be 60 cm $\times$ 60 cm. It is important to note that both the nominal wattage and dimensions of the luminaires are chosen to replicate generic commercial LED flat panel products [54].

The realistic radiometric spectral characteristics of the sources are defined by using the spectrum files (.spcd) in Zemax OpticStudio within our simulation environment. Thus, the related spectral distribution values for the adopted sources are obtained after processing the data-sheet values, which report the real-world measurements results. Then, the spcd files are obtained accordingly and fed into the simulation environment, which contain the relative spectral distribution coefficients w.r.t. the corresponding wavelength. The coefficients in the spcd file determine the density of a ray with the given wavelength among all the other rays with various wavelengths in MCRT. Thus, the optical power dedicated to the given wavelength will become directly proportional to the measured relative coefficients in ray-tracing simulations, which will replicate the real-source spectral characteristics in the simulation environment. The relative radiometric colour spectrum plots for the adopted VL and IR sources are given in Figure 12.5a,b, respectively. As can be seen from the figures, the spectrum distribution function of the white LED, $f_w(\cdot)$, which is plotted against the wavelength, λ, has two peaks at $\lambda_1 = 0.450$ μm and $\lambda_2 = 0.604$ μm. This is primarily due to the manufacturing process of the LEDs, where a yellow colour-converting phosphor coating (λ_2) is superimposed onto the blue source (λ_1) to achieve white light emission. The correlated colour temperature (CCT) of our adopted white LED model becomes 3000 K, which sits on the warmer side of the scale.

Similarly, the relative radiometric distribution function for the IR LED, $f_i(\cdot)$, is also depicted against the wavelength in Figure 12.5b, where there is a single peak value at the $\lambda_3 = 0.860$ μm. Another important parameter is the spectral spread of the adopted sources, where the bandwidth value for the IR source of 0.150 μm shows intrinsic dominant monochromatic characteristics, where most of the power is concentrated within a relatively small range of wavelengths. On the contrary, the white source with the bandwidth value of 0.398 μm presents

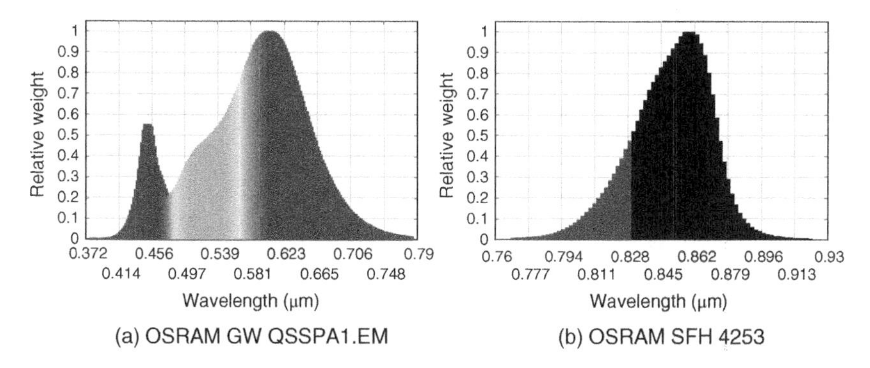

Figure 12.5 Relative radiometric colour spectrum of; (a) VL [0.382 0.780] μm and (b) IR [0.770 0.920] μm band LED chips that used in the MCRT simulations.

a widespread poly-chromatic profile. The importance of the source chromacity for the optical link budget will be clearer in the following subsections when the coating material reflectances are presented. In our simulation environment, each source profile is represented by 200 data points, which is the maximum value that is permitted by the simulation environment.

The realistic spatio-angular profile of the sources are also obtained from the measurement results provided by the manufacturer. The source model files with a large number of ray recordings, which are obtained by real-world goniometer measurements, are utilized to represent the adopted source spatio-angular characteristics. The main advantage of this method is the ability to model complex sources without requiring knowledge of inner opto-electrical and quantum effects dominated working principles. The radiation patterns of the VL and IR band LEDs are given by the directivity plots in Figure 12.6a,b, respectively. In these polar plots, the normalized radiant intensity distribution of the sources are plotted against the polar angle w.r.t. a source located in the $-y$ direction, refer to Figure 12.4. Furthermore, the different colours in the plots represent the various azimuth angle scans, 0°, 45°, 90°, and 135°, for the spherical coordinate system. It can be inferred from the figure that both sources have a Lambertian-like emission pattern with a strong y-axis symmetry, which corresponds to the semi-angle of half power of $\Phi_h = 60°$. Compared to the ideal diffuse (Lambertian distributed) emitter model in [55] with point source and receiver assumptions, the spatial domain characteristics for sources, receivers, and coating materials are considered in our simulations. It is also important to determine the number of rays that will be traced in the MCRT-based channel modelling approach, since it will determine the resolution of the simulations. Moreover, the statistical significance of the transmitter, receiver, and reflection models needs to be maintained in MCRT by the utilization of the law of large numbers. Therefore, as a rule of thumb, the number of rays that are traced in MCRT-based channel modelling applications are generally chosen to be in the order of millions [56, 57]. To trace sufficient number of

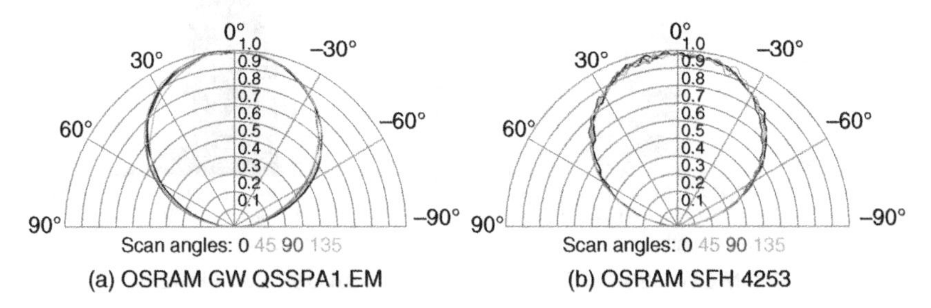

Figure 12.6 Source directivity plots of the VL (a) and IR (b) band LED chips for the azimuthal angles [0 45 90 135].

rays and realistically model the adopted sources, the spatio–angular profile of the IR band source is represented by 5×10^6 measured rays in our simulations. For the VL band, blue, and yellow spectra have been represented by 5×10^6 measured rays each, which yields 10^7 measured rays in total. In the following subsection, the optical modelling for the meta-surfaces as well as the coating materials will be detailed.

12.3.1.3 IRS and Coating Material Characterization

The coating surface characteristics for the IRS aided LiFi applications is of crucial importance since the achievable system performance will directly be affected by the total optical power that is transferred from the TX to the RX. Furthermore, the system reliability is also dictated by higher-order reflections if the direct LoS link is not available. As depicted in Figure 12.4, the meta-material based IRSs, which are envisaged to be implanted on the wallpaper, have 'ON' and 'OFF' states. Please note that the 'ON' and 'OFF' states of the meta-materials represent two indoor application scenarios: (i) with IRS and (ii) without IRS deployment, respectively. When the IRSs are in the 'ON' state, the electrical/mechanical/electro-mechanical meta-surfaces are activated and they act as micro-mirrors, which means that the reflectance of the wall increases. Since the manufacturing process of the meta-materials is complex and costly, the small grid of meta-surfaces are envisaged to be utilized on the wallpaper instead of being produced as a large single sheet of meta-surface. Therefore, depending on the density of the meta-materials, the average reflectance value of the wallpaper when the IRSs are activated could vary.

To model the IRS implanted wallpaper as realistically as possible, the relative reflectance of the side walls is assumed to be 90% in our simulations when the IRSs are 'ON'. The remaining 10% of the incident optical power is assumed to be absorbed by the coating material of the meta-surfaces to take real-life imperfections into consideration. Furthermore, the reflection characteristics of the IRSs are chosen to follow the Phong reflection model with 75% specular and 25% diffuse reflection components, since the wallpaper and meta-material mixture contains both rough and shiny surface structures. Note that the diffuse reflections are scattered into v rays with Lambertian distribution. In other words, the bi-directional scatter distribution function (BSDF) and the resultant intensity could be given by $1/\pi$ and $\cos(\theta_{s\text{-}s})$, for each scattered ray, respectively. The parameter $\theta_{s\text{-}s}$ denotes the angle between the specular component and the scattered rays. In every reflection (bounce) of a light ray, 25% of the 90% of the incident optical power is equally divided among all the scattered rays due to the adopted fractions of reflection and scattering coefficients. Similarly, 75% of the 90% of the incident optical power is allocated to the specular component. Thus, the number of the total rays in the system that must be traced after the κth reflection becomes $n_R = (\nu + 1)^{\kappa}$, where $0 \leq \kappa \leq \kappa_{max}$ represents the arbitrary number of reflections in the system. The parameter κ_{max} is the maximum number of reflections that are considered in the MCRT simulations. On the contrary, diffuse reflection characteristics with no specular components, which stem from the rough nature of the surfaces [56–61], are adopted as the reflection characteristics of the wallpapers when the IRSs are 'OFF'. Accordingly, the incident light rays become scattered into ν rays, which follow a Lambertian distribution. Lastly, the reflection profile of the wallpaper is directly dependent on the adopted surface coating material properties of the side walls.

In order to capture the realistic surface coating characteristics of the walls when the IRSs are OFF, the measurement-based spectral reflectance data had to be included in our simulations. Accordingly, the wall coating material which introduces relatively low reflectance in both the VL and IR spectra, 'Cobalt Green Paint Pigment', was chosen as the main coating for the side walls as well as the ceiling. The main reason behind the selection of this material is to have a relatively small reflection contributions when IRSs are not deployed. Hence, we can simulate and report the achievable rate performance difference between IRS 'OFF' and 'ON' states, which correspond to the worst and best-case scenarios, respectively. Furthermore, the floor of the room is assumed to be covered with a 'Black Polyester Pile Carpet', in order to mimic a realistic indoor scenario. The measured relative reflectivity weight values for the chosen materials w.r.t. the measurement spectra are obtained from the United States Geological Survey (USGS) High Resolution Spectral Library Version 7 [62].

The visual representations of the adopted coating materials are given in Figure 12.7. Furthermore, the measured relative reflectivities of the adopted coating materials against the measurement spectrum, 0.35–2.5 μm, are depicted for 2151 data points in Figure 12.8. In Figures 12.9 (a), (b) the portion of the reflectivity spectrum that corresponds to the adopted source characteristics, $f_v(\lambda)$ and $f_i(\lambda)$, is given for the chosen materials, respectively. Accordingly, the effective source spectral distribution after the multiplication of the source characteristics

Figure 12.7 The material that is used in our simulations [62]: Cobalt green paint pigment.

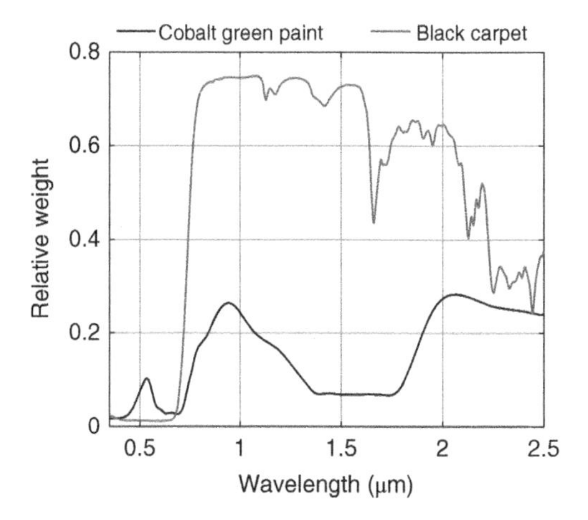

Figure 12.8 Relative spectral reflectivity values of the coating materials, which are adopted for IRS-aided LiFi channel modelling simulations.

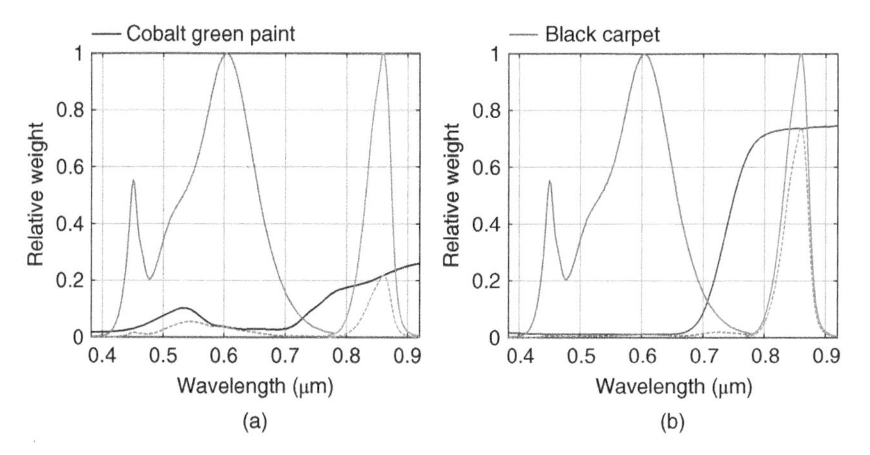

Figure 12.9 Relative spectral reflectivity values of the chosen materials with respect to the adopted VL and IR band source characteristics. The $f_v(\lambda)$ and $f_i(\lambda)$ are depicted as light grey solid lines on the left and right hand side of each plot, respectively. The resultant distribution of the sources after multiplication with the material characteristics are given by dotted lines under the respective curves.

and the coating material reflectivity values are plotted by dotted lines under the respective source plot. It can be seen from Figures 12.9 (a), (b) that the reflectivity characteristics of the black carpet are approximately 0.01 and 0.75 for the VL and IR bands, respectively. Similarly, the cobalt green paint has a relative reflectance value of 0.05 and 0.2 in the VL and IR bands, respectively. It is important to note that the reflectivity values of the coating material could significantly change within the spectra of each source as shown by Figure 12.9 (a), (b). Thus, representing such fluctuations with an average value, as implemented by the recursive method-based techniques [50, 51], would introduce a significant error in optical channel characterization. In our simulation environment, each coating material is represented by 200 spectral data points, as this is the maximum number that is allowed by our simulation environment. In the next subsection, the receiver characterization will be detailed.

12.3.1.4 Receiver Characterization

Although other MCRT-based toolkits are also proposed in the literature, the techniques in [56, 57, 63] are not capable of capturing the optical channel completely since they are unable to reflect the receiver spatio-angular and spatio-spectral characteristics within their calculations. Similar to the source modelling procedure, spatial, angular, and spectral parameterization is needed for realistic detector modelling.

It is important to emphasize that two different source spectra: VL and IR band emission characteristics are adopted in this work. Therefore, the spectral responsivity curves at the RX must match the intended sources. Accordingly, two silicone PIN PDs: (i) OSRAM SFH 2716 with peak sensitivity at $\lambda_v = 0.62$ µm, and (ii) OSRAM SFH 2704 with the peak sensitivity at $\lambda_i = 0.9$ µm are adopted as the VL and IR band receivers, respectively. Hence, non-imaging bare PD models, without any front-end optics, that are rectangular shaped with 1 cm^2 active area are generated in our simulation environment. The relative responsivity functions of the SFH 2716, $g_v(\cdot)$, and SFH 2704, $g_i(\cdot)$, plotted against the wavelength, λ, are given in Figure 12.10 (a), (b), respectively. Accordingly, the solid dark grey lines depict the relative spectral response weights of each detector, where the left and right hand side light grey colours are $f_v(\lambda)$ and $f_i(\lambda)$, respectively. Moreover, the overall spectral response for VL and IR bands, which are calculated by $f_v(\lambda)g_v(\lambda)$ and $f_i(\lambda)g_i(\lambda)$, are plotted as dotted lines under the VL and IR source spectral distribution plots, respectively. As can be seen from Figure 12.10a,b each detector closely matches their intended sources and severely attenuates the out-of-band signals. For instance, the OSRAM SFH 2716 detector responsivity curve matches the OSRAM GW QSSPA1.EM emission spectra with an average responsivity of approximately 0.7. However, the same detector filters more than 50% of the optical power emitted from OSRAM SFH 4253. In a similar manner, the OSRAM SFH 2704 detector introduces a responsivity value of approximately 0.95 for the OSRAM SFH 4253 source, where it filters more than 40% of the optical power from the OSRAM GW QSSPA1.EM source.

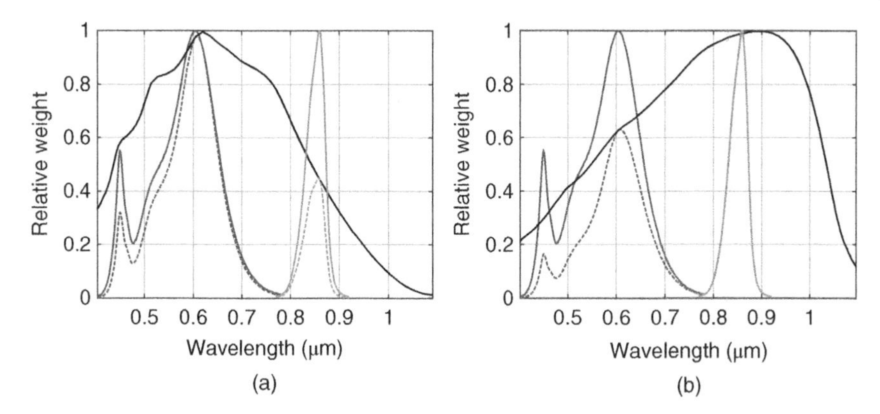

Figure 12.10 The relative spectral response curves for the adopted detectors, $g_v(\lambda)$ and $g_i(\lambda)$ (solid dark grey) with the relative spectral distributions for the OSRAM GW QSSPA1.EM, $f_v(\lambda)$ (light grey on the left hand side in each figure) and OSRAM SFH 4253, $f_i(\lambda)$ (light grey on the right hand side in each figure). The overall spectral response for VL and IR bands are plotted as dotted lines under the respective source spectral distribution plots. (a) OSRAM SFH 2716 with $\lambda_v = 0.62$ μm and (b) OSRAM SFH 2704 with $\lambda_i = 0.9$ μm.

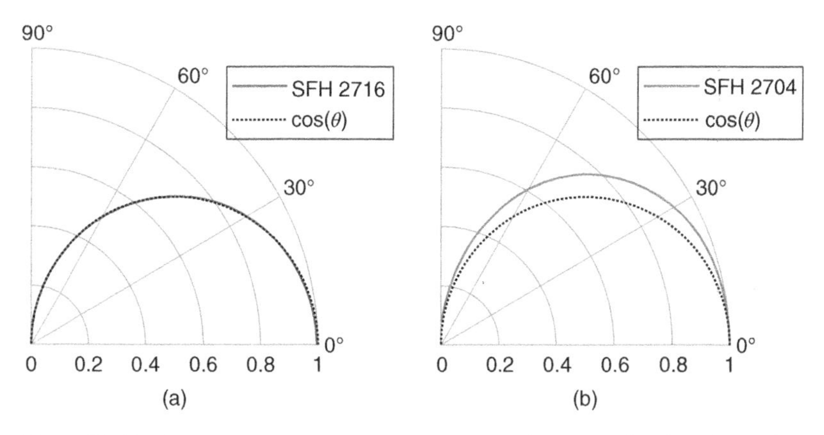

Figure 12.11 Relative angular responsivity characteristic plots for the adopted VL (a) and IR (b) band detectors. The ideal cosine responsivity curve is given by black dotted line as a benchmark. (a) OSRAM SFH 2716 and (b) OSRAM SFH 2704.

The relative angular responsivity profile of the chosen detectors, which are obtained via goniometer measurements conducted and reported by the manufacturer, are fed into our simulation environment after multiple processing stages. The relative angular responsivity curves of the OSRAM SFH 2716 and OSRAM SFH 2704 detectors are given in Figure 12.11a,b, respectively. Furthermore, the ideal cosine angular profile for the receivers, $\cos(\theta)$, which is depicted by a dotted dark grey line is also given as a benchmark. The parameter θ denotes the angle of incidence of a ray that strikes the detector. It can be seen from the angular profile of the OSRAM SFH 2716 VL band photo-detector that it closely follows the ideal cosine detection responsivity. However, this is not the case for the IR band SFH 2704 photo-detector, due to the real-world geometrical imperfections introduced in the manufacturing process of the receiver micro-chips. Please note that the curves in Figure 12.11 (a), (b) are obtained by the real world measurements. Hence, any real-world imperfections and/or manufacturing errors are captured by the measurement results, unlike the ideal cosine curve. It is also important to emphasize that the non-ideal angular characteristic will play a significant role in the IR band CIRs as the received optical power values will be multiplied with a modified cosine profile of the receiver. In the next sub-section, the channel models, which are obtained via MCRT simulations by considering a realistic source, IRS, coating and receiver characteristics, will be presented.

12.3.2 Obtaining the Channel Models

The multipath CIR between source S and receiver R could be expressed by our MCRT simulation results as follows:

$$h(t; S, R) = \sum_{i=1}^{i_{\text{hit}}} P_i \delta(t - t_i) \tag{12.1}$$

where the parameters P_i, i_{hit}, and t_i denote the received incoherent irradiance, total number of rays that hit the receiver and the elapsed time for the ith ray to reach the receiver, respectively. Note that incoming rays that strike the detector surface introduce various irradiance and time-of-flight values due to the different ray paths, even when there is no NLoS path in the system. The main reason behind this is the spatial dispersion of the rays, which emerges due to the realistic LED and PD geometrical models in our simulations. Unlike the point source and receiver assumptions in the analytical models, the source and receiver models are accurately shaped as their actual micro-chip form factor, which effects the generation and capture of the traced rays. Therefore, to reduce the temporal fluctuations caused by spatial dispersion and ensure the statistical significance, data binning also known as clustering, on $h(t; S, R)$ is applied, which yields the discrete-time optical CIR by

$$h[n; S, R] = \sum_{n=0}^{N_{\text{b}}-1} \tilde{P}_n \delta(n - t_n), \quad \text{for } n \in \{0, 1, \ldots, N_{\text{b}} - 1\} \tag{12.2}$$

Accordingly, the number of bins is calculated by

$$N_{\text{b}} = \left\lceil \frac{t_{\text{L}} - t_1}{\Delta w} \right\rceil \tag{12.3}$$

where the time of arrival for the first and last rays are denoted by t_1 and t_L, respectively. Moreover, bin widths are also given by Δw. The bin edge for the nth bin could also be calculated by $t_n = t_1 + n\Delta w$. The cumulative irradiance within the given bin interval is calculated by $\tilde{P}_n = \sum_{\forall i} P_i, \forall i \in [t_n\ t_{n+1}]$, if $n = N_b - 1$ and $\forall i \in [t_n\ t_{n+1})$, otherwise. Note from (12.2) and (12.3) that the temporal domain accuracy is directly related to the bin width, Δw, where the resulting discrete-time CIR closely approximates the actual channel when the bin width approaches zero, $\lim_{\Delta w \to 0} h[n; S, R] \approx h(t; S, R)$.

Other important channel characterization parameters could also be devised by using the CIR expression obtained in (12.2). Accordingly, the optical CFR is described in terms of the discrete-time CIR as follows:

$$H(f; S, R) = \int_{-\infty}^{\infty} h(t; S, R) e^{-j2\pi ft}\, dt \approx \sum_{n=0}^{N_b-1} h[n; S, R] e^{-j\frac{2\pi kn}{N}},$$
$$\text{for } k \in \frac{\Delta f}{N} \circ \left\{ -\frac{N}{2}, -\frac{N}{2} + 1, \dots, \frac{N}{2} - 1 \right\} \tag{12.4}$$

where the element-wise multiplication operation is given by $\circ$. It is important to note from the above expression that the continuous time fast Fourier transform (FFT) and discrete Fourier transform (DFT) of the channel will closely approximate each other as the bin width approaches zero. The sampling frequency is calculated as $\Delta f = 1/\Delta w$. Moreover, the number of subcarriers in the DFT operation is also calculated by $N = 2^{\lceil \log_2(N_b) \rceil}$. Another important parameter is the direct-current (DC) channel gain or total optical power of the impulse response, which can also be calculated by using (12.4),

$$H(0; S, R) = \int_{-\infty}^{\infty} h(t; S, R)\, dt$$
$$\approx \sum_{n=-\infty}^{\infty} h[n; S, R] = \sum_{n=0}^{N_b-1} \sum_{\kappa=0}^{\kappa_{max}} h^{(\kappa)}[n; S, R] \tag{12.5}$$

The average transmitted and received optical powers could be linked by using the above DC channel gain as follows: $P_R = H(0; S, R) P_S$. Similarly, by using the previous expression, the path loss (PL) in decibels could be given by

$$\text{PL} = -10 \log_{10} H(0; S, R) \tag{12.6}$$

The root-mean-squared (RMS) delay spread and mean delay are two important measures to define the multipath richness of the channel, which also indicates the impact of inter-symbol interference (ISI) on the system performance. Hence, the RMS delay spread can be calculated by using the second and zeroth central moments of $h(t; S, R)$ as follows:

$$\tau_{\text{RMS}} = \sqrt{\frac{\mu_2(\tau_0)}{\mu_0(\tau_0)}} = \sqrt{\frac{\int_{-\infty}^{\infty} (t - \tau_0)^2 h^2(t; S, R)\, dt}{\int_{-\infty}^{\infty} h^2(t; S, R)\, dt}} = \sqrt{\frac{\sum_{n=0}^{N_b-1} (n - \tau_0)^2 h^2[n; S, R]}{\sum_{n=0}^{N_b-1} h^2[n; S, R]}} \tag{12.7}$$

where the mean delay is given in terms of the zeroth and first raw moments of $h[n; S, R]$ by

$$\tau_0 = \frac{\mu_1(0)}{\mu_0(0)} = \frac{\mu_1(0)}{\mu_0(\tau_0)} = \frac{\int_{-\infty}^{\infty} t h^2(t; S, R)\, dt}{\int_{-\infty}^{\infty} h^2(t; S, R)\, dt} = \frac{\sum_{n=0}^{N_b-1} n h^2[n; S, R]}{\sum_{n=0}^{N_b-1} h^2[n; S, R]} \tag{12.8}$$

Lastly, the relative power of the LoS component compared to the total received optical power is also an important factor to determine the dominance of the LoS path. The dominance of the LoS component directly indicates higher-system reliability and link quality in cases where the direct link is broken. Accordingly, the 'flatness factor' of the optical channel is calculated by using the Rician K-factor as follows:

$$\rho = \frac{K}{K+1} = \frac{P_{\text{LoS}}}{P_{\text{LoS}} + P_{\text{NLoS}}} = \frac{\int_{-\infty}^{\infty} h^{(0)}(t; S, R)\, dt}{\sum_{\kappa=0}^{\kappa_{\max}} \int_{-\infty}^{\infty} h^{(\kappa)}(t; S, R)\, dt} = \frac{\sum_{n=0}^{N_b-1} h^{(0)}[n; S, R]}{H(0; S, R)} \tag{12.9}$$

where the Rician K-factor is defined as $K = P_{\text{LoS}}/P_{\text{NLoS}}$ [64]. The MCRT-based channel characterization results for IRS aided LiFi will be provided for both VL and IR bands and will be presented in the following sub-section.

12.3.2.1 MCRT Channel Characterization Results

In this subsection, the optical channels obtained with the proposed MCRT toolkit will be presented for both VL and IR bands when IRSs are ON and OFF. Moreover, the effect of user mobility on the channel parameters will also be investigated. Three mobile UE locations; L1, L2, and L3, as depicted in Figure 12.12, are chosen to generalize the UE mobility. Accordingly, the point L1 is located near the corner and UE will receive reflections from two side walls. Similarly, point L2 is located near the side wall aligned with the centre of the room. The UE will primarily receive reflections from a single wall. Lastly, point L3 is located at the centre of the room, which will yield a significant LoS path but negligible side wall

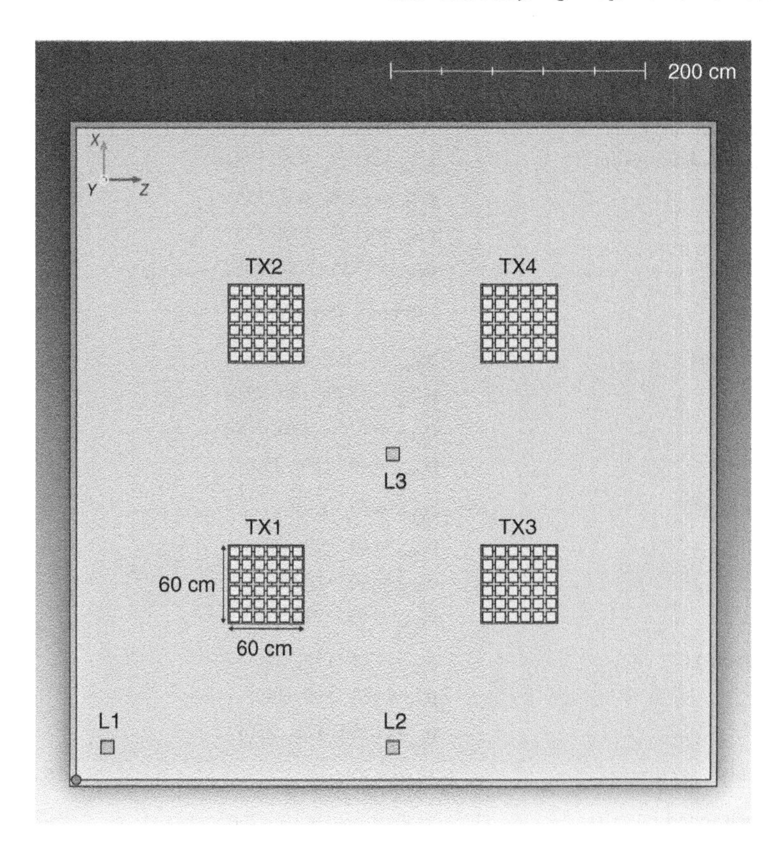

Figure 12.12 Top view of the considered scenario with transmitter (TX1, TX2, TX3, and TX4) and receiver locations (L1, L2, and L3). The global origin point of the simulation environment is indicated by the dark grey circle located at the bottom left hand side of the figure.

reflections. Note that the average UE height is taken as $H_{\text{UE}} = 0.8H_{\text{human}}$, which is reported in [23]. The parameter H_{human} represents the average of the mean female and male height values (168.75 cm) for England obtained in 2016 [65]. Hence, the average height of the mobile user becomes 135 cm. The complete set of parameters used in the MCRT-based optical channel simulations could be found in Table 12.1. It is important to note that the interior surfaces of the room are designed to introduce spectrum dependent reflection characteristics. When the IRSs are active, the side walls introduce mostly specular reflections, which corresponds to 75% of all reflections Thus, 25% of the all reflections becomes diffuse when IRSs are ON. The diffuse reflections are designed to consist of $\nu = 5$ scattered rays for every surface in the simulation environment. Note that the trade-off between the time complexity and reflection accuracy in our simulations is directly controlled by the number of scattered rays, ν. Hence,

Table 12.1 Details of the parameters used in the MCRT simulations.

Room dimensions	5 m × 5 m × 3 m
LED luminaire positions (cm)	$\mathbf{p}_{TX1} = (135,\ 300,\ 135)$ $\mathbf{p}_{TX2} = (335,\ 300,\ 135)$ $\mathbf{p}_{TX3} = (135,\ 300,\ 335)$ $\mathbf{p}_{TX4} = (335,\ 300,\ 335)$
IRS dimensions	250 cm × 150 cm
IRS positions (cm)	$\mathbf{p}_{IRS1} = (250,\ 150,\ 0)$ $\mathbf{p}_{IRS2} = (500,\ 150,\ 250)$ $\mathbf{p}_{IRS3} = (250,\ 150,\ 500)$ $\mathbf{p}_{IRS4} = (0,\ 150,\ 250)$
IRS orientations (°)	$\mathbf{o}_{IRS1} = (0,\ 0,\ 0)$ $\mathbf{o}_{IRS2} = (0, -90,\ 0)$ $\mathbf{o}_{IRS3} = (0, -180,\ 0)$ $\mathbf{o}_{IRS4} = (0, -270,\ 0)$
PD positions (cm)	$\mathbf{p}_{L1} = (25,\ 135,\ 25)$ $\mathbf{p}_{L2} = (25,\ 135,\ 250)$ $\mathbf{p}_{L3} = (250,\ 135,\ 250)$
Number of chips per luminaire	36 (6 × 6)
Number of generated rays per LED chip | luminaire	VL band: 10×10^6 |360×10^6 IR Band: 5×10^6 |180×10^6
Power per luminaire (P_S)	36 W
Model of the LED chips	VL band: OSRAM GW QSSPA1.EM IR band: OSRAM SFH 4253
FWHM of the LED Chips	120°
Model of the PDs	VL band: OSRAM SFH 2716 IR band: OSRAM SFH 2704
Effective area of the PDs	1 cm^2
FWHM of the PDs	OSRAM SFH 2716: 120° OSRAM SFH 2704: 132°
Coating materials	Cobalt green paint, $\nu = 5$ Black carpet, $\nu = 5$
Time resolution (bin width, Δw)	0.2 ns

$v = 5$ is determined to be a sufficient value to model the reflection characteristics accurately without saturating the computational resources. The floor and ceiling always introduce diffuse reflections, similar to the side walls, when IRSs are inactive. The parameter 'minimum relative ray intensity' in our MCRT simulations, which decides when to terminate the trace of a single ray is chosen to be 10^{-5} and 10^{-4} for VL and IR band simulations, respectively. Similar to [56, 57], a trace of a single ray is terminated when the optical power of the light ray is decreases to 0.001% and 0.01% of the initial intensity in VL and IR band simulations, respectively. Note that the overall reflectivity of the coating materials in the VL band is significantly lower compared the IR band, which is compensated via smaller 'minimum relative ray intensity' value. The number of rays generated per LED chip is chosen to be five million in our simulation environment. Therefore, a total of 180 million rays per luminaire are generated in IR band simulations. However, since the white LED chips consists of yellow and blue components, a total of 360 million rays per luminaire are generated in our VL band simulations. It is also important to note that the receive PDs are assumed to capture the light rays that strike the front face of the detector surface. In our simulations, both the transmit luminaires and receive PDs are assumed to be orientated towards $-y$ and $+y$ axes directions, respectively. Consequently, the front face of the PDs becomes the face that looks towards $+y$ direction, whereas the other face is assumed to be insensitive to the incoming light to model realistic receiver characteristics. Lastly, all four luminaires are assumed to be transmitting the same information in our channel measurement simulations without loss of generality.

12.3.2.2 VL Band Results

The CIR plots of the VL channels when the IRSs were in both the ON and OFF states are given for UE locations L1, L2, and L3 in a 2×3 matrix formation in Figure 12.13. The rows of the 2×3 subplot matrix represents the location of the UE, and the columns show the states of the IRSs. Furthermore, the previously mentioned channel parameters for the same configuration are also given in Table 12.2.

For the UE location L1, the CIR results are depicted by Figure 12.13a,b. As can be seen from the figures, the magnitudes of the channel taps are significantly higher in the IRS ON state compared to when the IRS is OFF, as expected. More specifically, the peak CIR and DC channel gain in the IRS-ON state are approximately 48.56% and 141.6% higher compared to the IRS-OFF state for point L1, respectively. The main reason behind this significant difference is the enhanced wall reflectivity coefficients when the IRSs are activated. This effect can also be observed by the channel dispersion, where the RMS delay spread is approximately 41.97% larger when IRSs are active compared to the IRS-OFF state for UE location L1. It can also be seen from the figures that the contribution of higher-order

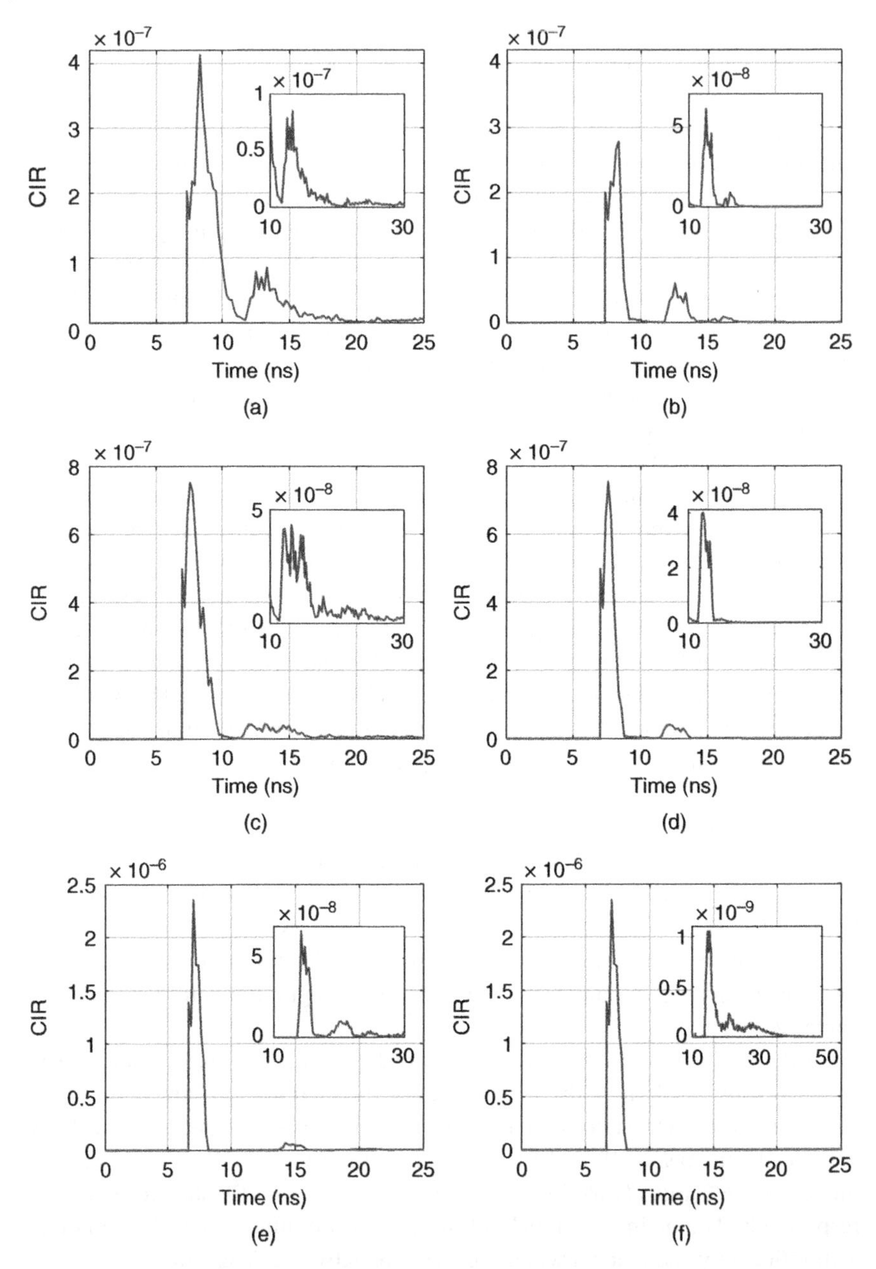

Figure 12.13 IRS-aided indoor LiFi CIR simulation results for VL band. The results are obtained by the proposed MCRT-based simulation technique. Each row represents the UE's location, whereas the columns are the state of the IRSs. (a) IRS ON $h(t; \forall S, \mathrm{L1})$, (b) IRS OFF $h(t; \forall S, \mathrm{L1})$, (c) IRS ON $h(t; \forall S, \mathrm{L2})$, (d) IRS OFF $h(t; \forall S, \mathrm{L2})$, (e) IRS ON $h(t; \forall S, \mathrm{L3})$, and (f) IRS OFF $h(t; \forall S, \mathrm{L3})$.

Table 12.2 Details of the VL band optical channel parameters when IRSs are ON and OFF.

IRS ON (VL band)					
R	κ_{max}	$H[0; \forall S, R]$	τ_0 (ns)	τ_{RMS} (ns)	ρ
L1	33	$4.856E^{-6}$	8.762	1.414	0.402
L2	38	$6.707E^{-6}$	7.815	0.780	0.615
L3	38	$1.123E^{-5}$	7.161	0.428	0.934
IRS OFF (VL Band)					
R	κ_{max}	$H[0; S, R]$	τ_0 (ns)	τ_{RMS} (ns)	ρ
L1	4	$2.010E^{-6}$	8.178	0.996	0.967
L2	4	$4.183E^{-6}$	7.574	0.449	0.983
L3	4	$1.052E^{-5}$	7.151	0.312	0.998

reflections from the side walls, ceiling, and floor surfaces yield a very high LoS spike (between 6–10 ns) as well as multiple NLoS spikes (between 11–30 ns) in IRS-ON case. The maximum number of reflections captured for the given configuration becomes $\kappa_{max} = 33$ when the IRSs are ON, unlike IRS-OFF case, where $\kappa_{max} = 4$. Furthermore, the contribution of the NLoS channel power compared to the whole CIR is approximately 59.8% and 3.3% when the IRSs are ON and OFF, respectively. It is important to emphasize that this significant difference between the NLoS channels prove that the deployment of the IRSs substantially increases system reliability in cases where the direct LoS channel is blocked.

In Figure 12.13 (c), (d) the CIR results are depicted for UE location L2. Accordingly, the peak CIR and channel DC gain when the IRSs are ON becomes 0.11% lower and 60.34% higher, respectively, compared to the case where IRSs are OFF. The reason behind the close match among the peak values of the CIR consists of the significantly higher LoS component that come from all four sources. Compared to location L1, the DC channel gain has increased by approximately 32.57% and 106.44% at point L2 when the IRSs are ON and OFF, respectively. Similarly, the time dispersion of the channel becomes 73.72% higher in the case where the IRSs are ON compared to the case where the IRSs are OFF for L2. The RMS delay spread has decreased 44.84% and 121.83% from point L1 to L2 when IRSs are ON and OFF, respectively. The maximum number of reflections in point L2 also becomes 38 and 4 when the IRSs are ON and OFF, respectively. The contribution of the NLoS path compared to the whole CIR is 38.5% and 1.7% for the IRS-ON and

IRS-OFF states, respectively. As can be seen from Figure 12.13c, the activation of IRSs creates the first and second tier of reflections, which are attached to the LoS component between 7 and 10 ns. Moreover, much higher-order reflections induced by the IRSs, could also be observed in between 13 and 30 ns.

In Figure 12.13 (e), (f), the CIR plots are provided for the mobile UE location L3. As can be clearly seen from these figures, the CIRs (especially the LoS components) are closely matching, as expected. The only difference between two cases is the larger magnitude of the higher-order reflections when IRSs are active, which could be observed between 12 and 25 ns region in both figures. However, the maximum number of reflections captured in our simulations are 38 and 4 for the cases IRS-ON and IRS-OFF, respectively. The reason behind this phenomena is the resolution and accuracy of our MCRT simulations, which is able to capture very small irradiance fluctuations even if they have no significance for communication purposes. The reflections captured from the side walls, ceiling, and floor becomes significantly lower for location L3 even when the IRSs are active, which is due to the large physical distance between UE and the side walls. The peak value of the CIR and DC channel gain are only 0.11% and 14.23% higher when IRSs are ON and OFF, respectively. The channel RMS delay spread is also increased 125.4% when IRSs are activated for point L3. The NLoS channel component comprises the 6.6% and 0.2% of the whole CIR for IRS-ON and IRS-OFF cases in location L3, respectively.

12.3.2.3 IR Band Results

The IR band CIR plots and related channel parameters are also given in Figure 12.14 and Table 12.3, respectively.

In Figure 12.14 (a), (b), the IR band CIRs plots for the mobile UE location L1 when the IRSs are ON and OFF, respectively, are depicted. Similar to the VL band results, the channel conditions significantly improve with the utilization of the IRSs when the mobile user is close to the highly reflective IRSs and far from the sources. As can be seen from Figure 12.14 (a), a very high peak emerged by the first- and second-order reflections, depicted between 8 and 12 ns, which is merged with the LoS component. Furthermore, the third and further order reflections, depicted between 13 and 50 ns, are also significantly enhanced with the aid of the IRSs. The peak value of the CIR and DC channel gain values increased approximately 42.3% and 161.5% when IRSs are activated, respectively. Similarly, an increase of 184.4% in the RMS delay spread can also be observed with the employment of the IRSs compared to the case when IRSs are OFF. The number of captured bounces from the side walls, ceiling, and the floor becomes 22 and 3 when IRS-ON and IRS-OFF, respectively. Please note that the difference between the VL and IR bands stems from the different spectral reflection profiles of the materials as well as the 'minimum relative ray intensity' value in our simulations. The fraction of the NLoS

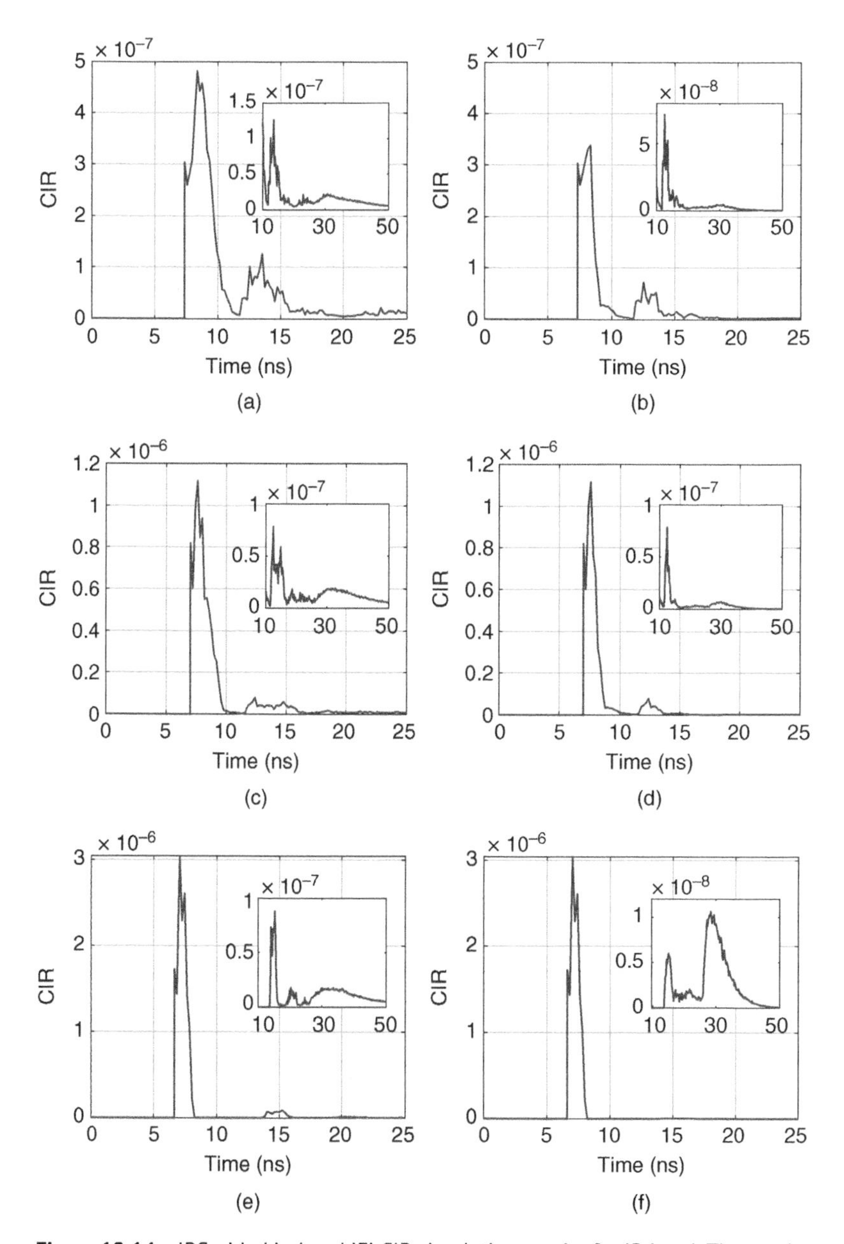

Figure 12.14 IRS-aided indoor LiFi CIR simulation results for IR band. The results are obtained by the proposed MCRT-based simulation technique. Each row represents the UE's location, whereas the columns are the state of the IRSs. (a) IRS ON $h(t; \forall S, \text{L1})$, (b) IRS OFF $h(t; \forall S, \text{L1})$, (c) IRS ON $h(t; \forall S, \text{L2})$, (d) IRS OFF $h(t; \forall S, \text{L2})$, (e) IRS ON $h(t; \forall S, \text{L3})$, and (f) IRS OFF $h(t; \forall S, \text{L3})$.

Table 12.3 Details of the IR band optical channel parameters when IRSs are ON and OFF.

IRS ON (IR band)					
R	κ_{max}	$H[0; \forall S, R]$	τ_0 (ns)	τ_{RMS} (ns)	ρ
L1	22	8.486E^{-6}	9.165	3.433	0.308
L2	25	1.125E^{-5}	7.875	1.838	0.529
L3	28	1.630E^{-5}	7.201	0.843	0.845
IRS OFF (IR band)					
R	κ_{max}	$H[0; S, R]$	τ_0 (ns)	τ_{RMS} (ns)	ρ
L1	3	3.245E^{-6}	8.146	1.207	0.805
L2	3	6.699E^{-6}	7.547	0.601	0.888
L3	3	1.427E^{-5}	7.175	0.374	0.965

component compared to the whole IR band channel becomes 69.2% and 19.5% for IRS-ON and IRS-OFF cases, respectively. The reason behind the higher NLoS component in the IR band compared to its VL band counterpart is the higher reflectivity values of the coating materials in the IR band (Figure 12.8). Lastly, the maximum number of bounces considered in location L1 also becomes 22 and 3 for the IRS-ON and IRS-OFF cases, respectively. The IR band CIR plots and the related parameters are also provided for UE location L2 in Figure 12.14c,d and Table 12.3, respectively. Similar to the VL band results, the impact of the IRSs could be seen from Figure 12.14c in the first- and second-order reflections (7–10 ns) as well as the higher-order reflections (12–50 ns). However, the effect of the IRSs are not as dominant as in location L1 due to the dominant LoS link (0th order reflection). The peak value of CIR and DC channel gain when IRS-ON are 0.12% lower and 67.94% higher compared to IRS-OFF case, respectively. Again, the reason behind this is the effect of the IRSs on the higher-order reflections, which also indicate that the system capacity will be higher when IRSs are activated if the direct LoS link is blocked. The RMS delay spread also increases around 205.82% when the IRSs are activated, which shows the significance of the higher-order reflections. Compared to location L1, the DC channel gain increases by 32.57%, whereas the RMS delay spread decreases by 46.46% in location L2 when the IRSs are ON. For cases where the IRSs are OFF, the DC channel gain increase is approximately 106.44% and the RMS delay spread decrease is around 50.21% from point L1 to L2, respectively. Moreover, only the NLoS path contributions compared to the whole channel becomes 47.1% and 11.2% for IRS-ON and IRS-OFF cases, respectively, when the

UE is located at L2. The number of bounces encountered in our MCRT simulations becomes 25 and 3 for IRS-ON and IRS-OFF cases, respectively, when the UE is at L2.

In Figure 12.14e,f, the CIR plots for mobile UE location L3 when IRSs are ON and OFF, respectively, are depicted. The peak value of the CIR and DC channel gain are increased 0.16% and 14.24% when the IRSs are activated for UE location L3, respectively. As can be seen from the channel parameters as well as Figure 12.14e,f, the effect of the IRSs on the main lobe, between 6 and 8 ns, is minimal due to the dominant LoS power. However, the optical power emerges due to the higher-order reflections, between 14 and 50 ns, is significantly higher in IRS-ON case compared to IRS-OFF. Furthermore, the RMS delay spread of the channel increases by more than 125.4% when the IRSs are activated at location L3. The maximum number of bounces captured in our simulations for the case when the IRSs are ON and OFF becomes 28 and 3, respectively, when the UE is at L3. The NLoS contribution against the whole channel optical power also increased from 3.5% to 15.5% when the IRSs are activated for a mobile user located at L3.

12.3.3 The Achievable Rates for IRS-aided LiFi

In this subsection, the achievable rate curves for the IRS-aided indoor LiFi application scenario will be provided. The maximum achievable system capacity for LiFi systems could simply be calculated as follows [2, 55]:

$$C(R) = \frac{1}{2}B\log_2\left(1 + \frac{4P_S^2 H^2(0; \forall S, R)}{\sigma_w^2}\right) \tag{12.10}$$

where the average transmitted optical power, DC channel gain, the effective noise power in the electrical domain and communication network's effective bandwidth are denoted by P_S, $H(0; S, R)$, σ_w^2, and B, respectively. Note that the effective noise term at the UE consists of the addition of shot and thermal noises, where the shot noise emerges as a result of ambient light sources and the information bearing signal itself. In cases where high ambient light power at the PD is significantly larger than the transmit signal power, the shot noise becomes signal independent. Therefore, the high-intensity shot noise at the RX could be modelled as a summation of independent low-power Poisson processes, which could be approximated as a zero mean Gaussian distribution. Consequently, the effective noise could be modelled as additive white Gaussian noise (AWGN), $w \sim \mathcal{N}\left(0, \sigma_w^2\right)$, where $\sigma_w^2 = \sigma_{\text{shot}}^2 + \sigma_{\text{thermal}}^2$. A real Gaussian distribution with mean μ and variance σ^2 is denoted by $\mathcal{N}(\mu, \sigma^2)$.

By using Tables 12.2 and 12.3, the DC channel gain values for both the VL and IR band could be obtained for the cases where IRSs are ON and OFF w.r.t. UE locations L1, L2, and L3. Also note from Table 12.1, the transmitted optical power is chosen to be $P_S = 36$ W per luminaire. The effective channel bandwidth, B, for the considered system could be calculated by using the expression,

$$B = \frac{\Delta f(N-1)}{N} = \frac{\left(2^{\lceil \log_2(N_b) \rceil} - 1\right)}{2^{\lceil \log_2(N_b) \rceil} \Delta w} \approx \Delta f \quad \text{if } N \gg 64 \tag{12.11}$$

Please note from (12.10) and (12.11) that typical LiFi networks are not limited by channel bandwidth, but by the electrical domain bandwidth of the front-end opto-electronic devices such as LEDs and PDs. In this work, the optical channel bandwidth is considered as the main limiting factor for the system as it inherently emerges. On the contrary, the electrical channel bandwidth, caused by the font-end, opto-electronic frequency response, is strictly dependent on the manufacturing technology behind the transmit LEDs and PDs. Therefore, the channel impairments introduced by the electrical domain components could be avoided as technology advances. Moreover, in our simulations, the number of DFT subcarriers and the effective optical channel bandwidth could be calculated by using the values in Table 12.1 as $N = 1024$ and $B = 5$ GHz.

In Figure 12.15, the achievable capacity plots for IRS-ON and IRS-OFF cases when the UE is located at L1, L2, and L3. Accordingly, the VL band plots are given for the [30 50] dB SNR-per-bit ($E_{b,\,\text{elec}}/N_0$) region in Figure 12.15a. As can be seen from figure that the maximum achievable capacity increases exponentially within the low SNR region. Please note from Figures 12.13 and 12.14 that the channel magnitudes are in the 10^{-6} and 10^{-7} region, thus, the electrical domain path loss at the RX becomes about −120 and −140 dB, respectively [66]. Therefore, the SNR measured at the TX side within the interval of 30–50 dB will be classified as the low SNR regime for this application. For UE location L1, the achievable data rate becomes approximately 500 and 90 Mbits/s when the IRSs are ON and OFF, respectively. Similarly, for L2, the achievable rates of approximately 700 and 370 Mbit/s for cases where the IRSs are ON and OFF, respectively. Lastly, approximately 2.08 and 1.88 Gbit/s data rates are achieved for the UE location L3 when the IRSs are active and inactive, respectively.

The IR band achievable capacity plots for IRS-ON and IRS-OFF cases when the UE is located at points L1, L2, and L3 are depicted in Figure 12.15b. It is important to note that the overall capacity values are higher in the IR band compared to the VL band results due to the higher reflectivity profiles of the coating materials in the IR spectra. For UE location L1, the capacity rates of approximately 1.33 Gbits/s and 237 Mbits/s are achieved when IRSs are ON and OFF, respectively. Moreover, the achievable rates become approximately 2.09 Gbits/s and 884 Mbits/s for IRS active and inactive cases, when the UE is located at L2. Finally, for UE location

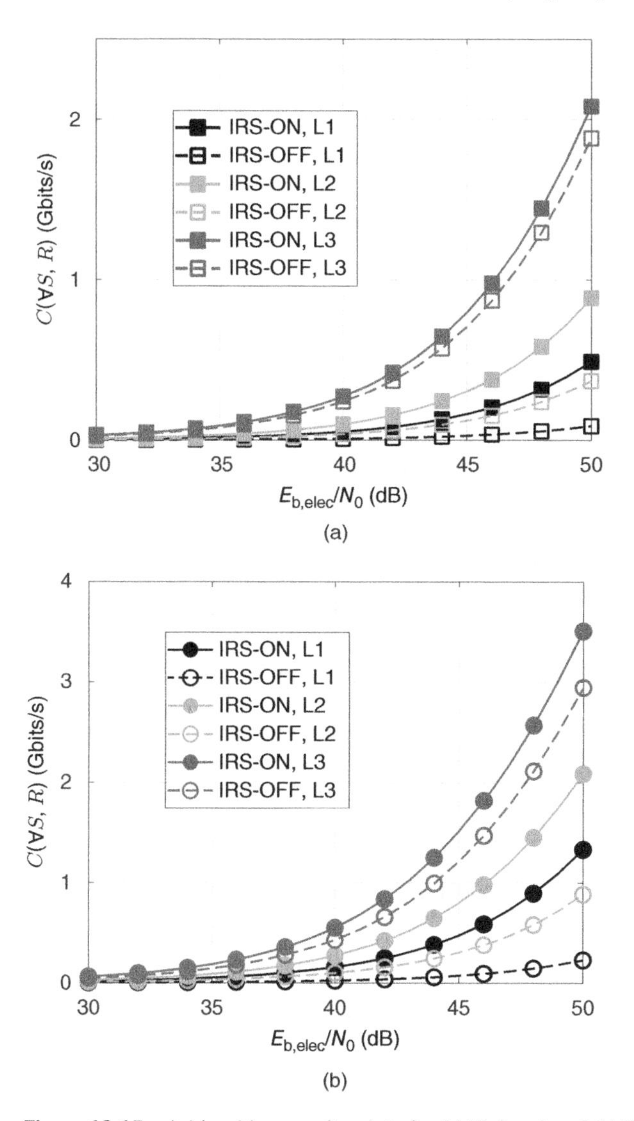

Figure 12.15 Achievable capacity plots for (a) VL band and (b) IR band indoor LiFi for IRS-ON and IRS-OFF cases when the UE is located at L1, L2, and L3.

L3, the achievable rate of 3.51 and 2.94 Gbits/s are obtained for the IRS-ON and IRS-OFF cases, respectively. Please note that based on both VL and IR band plots, the maximum capacity values are achieved for each UE location when the IRSs are activated.

12.4 Challenges and Research Directions

The concept of IRS provides opportunities for achieving unprecedented capabilities when it comes to signal and link manipulation. However, realizing such capabilities is subject to overcoming the challenges associated with the realization of this concept. In this section, we highlight some of the challenges and open research directions related to the integration of IRS in OWC systems and propose a road map for future research directions.

12.4.1 Modelling and Characterization

One of the challenges in the research and implementation of the concept of smart walls in OWC systems is the development of realistic and accurate channel models. More specifically, there is a need to establish realistic models that take into account the type, capabilities, and limitations of different possible IRS structures in order to capture the fundamental behaviour and performance limits of such systems. Moreover, since classical optical channel gain models might not be suitable, it is critical to understand the CIR of different metasurfaces over the range of optical wavelengths. In particular, there is a need to quantify the efficiency and response time for achieving specific functionalities such as amplification factors, absorption capabilities, anomalous reflection. Additionally, it is critical to understand how such systems perform in different mediums such as underwater and outdoor environments with high-ambient noise.

12.4.2 Inter-symbol Interference (ISI)

It has been reported in the previous section that the deployment of the IRSs in typical indoor LiFi networks increases the achievable data rate values significantly both for VL and IR spectra. However, one of the biggest challenges for IRS-aided LiFi arises from the inflated channel time dispersion. The increase in the channel delay spread values could effectively reduce the system performance due to the increased ISI. The ISI could be avoided in multi-carrier transmission methods, especially in orthogonal frequency division multiplexing (OFDM), by the utilization of the cyclic prefix (CP). Furthermore, with the activation of the IRSs, the channel frequency selectivity will also increase which means that the SNR in each subcarrier will significantly differ from each other. Consequently, in order to avoid further channel impairments which could emerge with the utilization of IRSs, suitable CP length selection as well as adaptive bit loading in OFDM are required. A detailed investigation of the optical OFDM systems and their error performances under the frequency selective channel that emerges in IRS-aided LiFi is necessary to obtain better insights on the performance of such systems.

12.4.3 Channel Estimation

The availability of accurate CSI for all the channel paths created by massive numbers of tunable sub-wavelength reflecting elements proves to be a challenging task. Traditional CSI estimation techniques might be unpractical for real-time estimations given the high dimensions of the channel vectors for each network user, particularly if higher-order reflections are considered. There is a need to quantify the trade-off between distributed and centralized CSI acquisition approaches and to come up with intelligent and cost-effective methodologies. Distributed CSI acquisition approaches employ local estimation at each IRS array which requires sensing and processing capabilities. Centralized CSI acquisition approaches, on the other hand, employ a central control unit, typically at the AP. The central unit is responsible for performing the CSI estimation for all network entities, which are then utilized for optimization and reconfiguration decisions that are executed centrally. The obvious advantage of the centralized approach is that the IRS is not required to perform exhaustive sensing and processing, leading to lower-energy consumption and a simpler hardware design. The main challenge of this approach is the signalling overhead, which occurs between the central unit and the IRS.

12.4.4 Real-time Operation

Integrating IRS arrays with a high number of reflecting elements and varying functionalities inevitably results in high-operational complexity which entails increased computational, energy, and overhead cost. The IRS elements need to be reconfigured in real-time to provide precise control over their optical functionalities, which is decided based on the varying system conditions. Such an operation is not trivial and entails different trade-offs. For example, there is a need to develop activation strategies that take into account different metrics, such as throughput and fairness, in order to balance the desired performance enhancement and the associated energy consumption and time delay. Also, how often the reflecting element must be reconfigured must be determined i.e. whether this should be performed with each channel realization or when a change occurs in the user location. Incorporating data-driven optimization tools such as deep learning, reinforcement learning, and federated learning could provide viable solutions to achieve time-efficient optimization.

References

1 Tkach, R.W. (2010). Scaling optical communications for the next decade and beyond. *Bell Labs Tech. J.* 14 (4): 3–9. http://dx.doi.org/10.1002/bltj.20400.

2 Haas, H., Yin, L., Chen, C. et al. (2020). Introduction to indoor networking concepts and challenges in LiFi. *IEEE/OSA J. Opt. Commun. Netw.* 12 (2): A190–A203. http://dx.doi.org/10.1364/JOCN.12.00A190.

3 Haas, H., Yin, L., Wang, Y., and Chen, C. (2016). What is LiFi? *J. Lightw. Technol.* 34 (6): 1533–1544. http://dx.doi.org/10.1109/JLT.2015.2510021.

4 Haas, H., Sarbazi, E., Marshoud, H., and Fakidis, J. (2020). Visible-light communications and light fidelity. In: *Optical Fiber Telecommunications VII*, Chapter 11 (ed. A.E. Willner), pp. 443–493. Academic Press,. ISBN 978-0-12-816502-7. https://doi.org/10.1016/B978-0-12-816502-7.00013-0.

5 Liaskos, C., Nie, S., Tsioliaridou, A. et al. (2018). A new wireless communication paradigm through software-controlled metasurfaces. *IEEE Commun. Mag.* 56 (9): 162–169. http://dx.doi.org/10.1109/MCOM.2018.1700659.

6 Rubin, N.A., Shi, Z., and Capasso, F. (2021). Polarization in diffractive optics and metasurfaces. *Adv. Opt. Photon.* 13 (4): 836–970. http://dx.doi.org/10.1364/AOP.439986.

7 Zaidi, A., Rubin, N.A., Dorrah, A.H. et al. (2021). Generalized polarization transformations with metasurfaces. *Opt. Express* 29 (24): 39065–39078. http://dx.doi.org/10.1364/OE.442844.

8 Suzuki, T. and Kondoh, S. (2018). Negative refractive index metasurface in the 2.0-THz band. *Opt. Mater. Express* 8 (7): 1916–1925. http://dx.doi.org/10.1364/OME.8.001916.

9 Horsley, S.A.R. and Woolley, M. (2020). Zero-refractive-index materials and topological photonics. *Nat. Phys.* 17: 348–355. http://dx.doi.org/10.1038/s41567-020-01082-2.

10 Díaz-Rubio, A., Asadchy, V.S., Elsakka, A., and Tretyakov, S.A. (2017). From the generalized reflection law to the realization of perfect anomalous reflectors. *Sci. Adv.* 3 (8). http://dx.doi.org/10.1126/sciadv.1602714.

11 Ndjiongue, A.R., Ngatched, T., Dobre, O., and Haas, H. (2021). Re-configurable intelligent surface-based VLC receivers using tunable liquid-crystals: the concept. *J. Lightw. Technol.* 39 (10): 3193–3200. http://dx.doi.org/10.1109/JLT.2021.3059599.

12 Li, Y., Li, X., Chen, L. et al. (2017). Orbital angular momentum multiplexing and demultiplexing by a single metasurface. *Adv. Opt. Mater.* 5 (2): 1600502.

13 Aoni, R.A., Rahmani, M., Kamali, K.Z. et al. (2019). High-efficiency visible light manipulation using dielectric metasurfaces. *Sci. Rep.* 9: 6510.

14 Overvig, A.C., Shrestha, S., Malek, S.C. et al. (2019). Dielectric metasurfaces for complete and independent control of the optical amplitude and phase. *Light: Sci. Appl.* 8: 92.

15 Arbabi, A., Horie, Y., Bagheri, M., and Faraon, A. (2015). Dielectric metasurfaces for complete control of phase and polarization with subwavelength spatial resolution and high transmission. *Nat. Nanotechnol.* 10: 937–943.

16 Al-Kinani, A., Wang, C., Zhou, L., and Zhang, W. (2018). Optical wireless communication channel measurements and models. *IEEE Commun. Surv. Tutorials* 20 (3): 1939–1962.

17 Abumarshoud, H., Mohjazi, L., Dobre, O.A. et al. (2021). LiFi through reconfigurable intelligent surfaces: a new frontier for 6G? *IEEE Veh. Technol. Mag.* 2–11. http://dx.doi.org/10.1109/MVT.2021.3121647.

18 Abdelhady, A.M., Salem, A.K.S., Amin, O. et al. (2021). Visible light communications via intelligent reflecting surfaces: metasurfaces vs mirror arrays. *IEEE Open J. Commun. Soc.* 2: 1–20. http://dx.doi.org/10.1109/OJCOMS.2020.3041930.

19 Wang, Q., Wang, Z., and Dai, L. (2015). Multiuser MIMO-OFDM for visible light communications. *IEEE Photon. J.* 7 (6): 1–11. http://dx.doi.org/10.1109/JPHOT.2015.2497224.

20 Yesilkaya, A., Bian, R., Tavakkolnia, I., and Haas, H. (2019). OFDM-based optical spatial modulation. *IEEE J. Sel. Top. Signal Process.* 1. http://dx.doi.org/10.1109/JSTSP.2019.2920577.

21 Yesilkaya, A., Purwita, A.A., Panayirci, E. et al. (2020). Flexible LED index modulation for MIMO optical wireless communications. *Proceedings of IEEE Global Commications Conference (GLOBECOM)*, pp. 1–7, Taipei, Taiwan, December 2020. http://dx.doi.org/10.1109/GLOBECOM42002.2020.9322528.

22 Yesilkaya, A., Basar, E., Miramirkhani, F. et al. (2017). Optical MIMO-OFDM with generalized LED index modulation. *IEEE Trans. Commun.* 65 (8): 3429–3441. http://dx.doi.org/10.1109/TCOMM.2017.2699964.

23 Yesilkaya, A., Cogalan, T., Panayirci, E. et al. (2018). Achieving minimum error in MISO optical spatial modulation. *Proceedings of IEEE Internaitonal Conference on Communications (ICC)*, pp. 1–6, Kansas City, MO, USA, May 2018. http://dx.doi.org/10.1109/ICC.2018.8422436.

24 Fath, T. and Haas, H. (2013). Performance comparison of MIMO techniques for optical wireless communications in indoor environments. *IEEE Trans. Commun.* 61 (2): 733–742. http://dx.doi.org/10.1109/TCOMM.2012.120512.110578.

25 Purwita, A.A., Yesilkaya, A., Tavakkolnia, I. et al. (2019). Effects of irregular photodiode configurations for indoor MIMO VLC with mobile users. *Proceedings of IEEE 30th International Symposium on Personal, Indoor and Mobile Radio Communications (PIMRC)*, pp. 1–7, Istanbul, Turkey, September 2019.

26 Mesleh, R., Elgala, H., and Haas, H. (2011). Optical spatial modulation. *IEEE/OSA J. Opt. Commun. Netw.* 3 (3): 234–244. http://dx.doi.org/10.1364/JOCN.3.000234.

27 Özdogan, Ö., Björnson, E., and Larsson, E.G. (2020). Using intelligent reflecting surfaces for rank improvement in MIMO communications. *arXiv*, 2002.02182.

28 Ibrahim, E., Nilsson, R., and van de Beek, J. (2021). Intelligent reflecting surfaces for MIMO communications in LoS environments. *Proceedings of*

IEEE Wireless Communications and Networking Conference (WCNC), pp. 1–6. http://dx.doi.org/10.1109/WCNC49053.2021.9417270.

29 Mesleh, R.Y., Haas, H., Sinanovic, S. et al. (2008). Spatial modulation. *IEEE Trans. Veh. Technol.* 57 (4): 2228–2241. http://dx.doi.org/10.1109/TVT.2007.912136.

30 Basar, E. (2019). Media-based modulation for future wireless systems: a tutorial. *IEEE Wireless Comm.* 26 (5): 160–166. http://dx.doi.org/10.1109/MWC.2019.1800568.

31 Li, Q., Wen, M., and Di Renzo, M. (2020). Single-RF MIMO: from spatial modulation to metasurface-based modulation. *arXiv preprint arXiv:2009.00789*.

32 Wang, Z., Liu, L., and Cui, S. (2020). Channel estimation for intelligent reflecting surface assisted multiuser communications. *Proceedings IEEE Wireless Communications and Networking Conference (WCNC)* 1–6. http://dx.doi.org/10.1109/WCNC45663.2020.9120452.

33 Abumarshoud, H., Alshaer, H., and Haas, H. (2019). Dynamic multiple access configuration in intelligent LiFi attocellular access points. *IEEE Access* 7: 62126–62141. http://dx.doi.org/10.1109/ACCESS.2019.2916344.

34 Marshoud, H., Sofotasios, P.C., Muhaidat, S. et al. (2017). On the performance of visible light communication systems with non-orthogonal multiple access. *IEEE Trans. Wireless Commun.* 16 (10): 6350–6364. http://dx.doi.org/10.1109/TWC.2017.2722441.

35 Wu, Q., Zhou, X., and Schober, R. (2021). IRS-assisted wireless powered NOMA: do we really need different phase shifts in DL and UL?. *IEEE Wireless Commun. Lett.* 10 (7): 1493–1497. http://dx.doi.org/10.1109/LWC.2021.3072502.

36 Marshoud, H., Muhaidat, S., Sofotasios, P.C. et al. (2018). Optical non-orthogonal multiple access for visible light communication. *IEEE Wireless Commun.* 25 (2): 82–88. http://dx.doi.org/10.1109/MWC.2018.1700122.

37 Yang, G., Xu, X., and Liang, Y.-C. (2020). Intelligent reflecting surface assisted non-orthogonal multiple access. *Proceedings of IEEE Wireless Communications and Networking Conference (WCNC)*, pp. 1–6. http://dx.doi.org/10.1109/WCNC45663.2020.9120476.

38 Abumarshoud, H., Selim, B., Tatipamula, M., and Haas, H. (2021). Intelligent reflecting surfaces for enhanced NOMA-based visible light communications. *arXiv*, 2111.04646.

39 Abumarshoud, H., Chen, C., Islim, M.S., and Haas, H. (2020). Optical wireless communications for cyber-secure ubiquitous wireless networks. *Proc. R. Soc. A* 476 (2242). http://dx.doi.org/10.1098/rspa.2020.0162.

40 Abumarshoud, H., Soltani, M.D., Safari, M., and Haas, H. (2021). Realistic secrecy performance analysis for LiFi systems. *IEEE Access* 9: 120675–120688. http://dx.doi.org/10.1109/ACCESS.2021.3108727.

41 Panayirci, E., Yesilkaya, A., Cogalan, T. et al. (2020). Physical-layer security with optical generalized space shift keying. *IEEE Trans. Commun.* 1. http://dx.doi.org/10.1109/TCOMM.2020.2969867.

42 Yesilkaya, A., Cogalan, T., Erkucuk, S. et al. (2020). Physical-layer security in visible light communications. *Proceedings of the 2nd 6G Wireless Summit (6G SUMMIT)*, pp. 1–5, Levi, Finland, March 2020. http://dx.doi.org/10.1109/6GSUMMIT49458.2020.9083799.

43 Su, N., Panayirci, E., Koca, M. et al. (2021). Physical layer security for multi-user MIMO visible light communication systems with generalized space shift keying. *IEEE Trans. Commun.* 69 (4): 2585–2598. http://dx.doi.org/10.1109/TCOMM.2021.3050100.

44 Qian, L., Chi, X., Zhao, L., and Chaaban, A. (2021). Secure visible light communications via intelligent reflecting surfaces. *arXiv*, 2101.12390.

45 ZEMAX-OpticStudio. https://www.zemax.com/products/opticstudio (accessed 18 November 2021).

46 Channel modelling for LC in TGbb: validation of ray tracing models by measurements. https://mentor.ieee.org/802.11/dcn/20/11-20-1234-00-00bb-channel-modelling-for-lc-in-tgbb-validation-of-ray-tracing-models-by-measurements.pptx, August 2020 (accessed 10 February 2022).

47 Eldeeb, H.B., Uysal, M., Mana, S.M. et al. (2020). Channel modelling for light communications: validation of ray tracing by measurements. *Proceedings of IEEE International Symposium on Communicaiton Systems, Networks and Digital Signal Processing (CSNDSP)*, pp. 1–6. http://dx.doi.org/10.1109/CSNDSP49049.2020.9249565.

48 Eldeeb, H.B., Uysal, M., Mana, S.M. et al. (2020). Channel measurements and ray tracing simulations for MIMO light communication at 200 MHz. *Proceedings of IEEE Photonics Conference (IPC)*, pp. 1–2. http://dx.doi.org/10.1109/IPC47351.2020.9252508.

49 Eldeeb, H.B., Mana, S.M., Jungnickel, V. et al. (2021). Distributed MIMO for Li-Fi: channel measurements, ray tracing and throughput analysis. *IEEE Photon. Technol. Lett.* 33 (16): 916–919. http://dx.doi.org/10.1109/LPT.2021.3072254.

50 Barry, J.R., Kahn, J.M., Krause, W.J. et al. (1993). Simulation of multipath impulse response for indoor wireless optical channels. *IEEE J. Sel. Areas Commun.* 11 (3): 367–379.

51 Barry, J.R. (1994). *Wireless Infrared Communications*. Springer.

52 OSRAM OSCONIQ P 3030, GW QSPPA1.EM High Power LED Datasheet. https://dammedia.osram.info/media/resource/hires/osram-dam-8405096/GW%20QSSPA1.EM.EN.pdf (accessed 18 Novemer 2021).

53 OSRAM TOPLED, SFH4253 High Power Infrared Emitter Datasheet. https://dammedia.osram.info/media/resource/hires/osram-dam-6035009/SFH%204253.EN.pdf (accessed 18 November 2021).

54 Product datasheet - PANEL VAL 600 36 W 3000 K WT. https://docs.rs-online.com/5768/A700000006908990.pdf, May 2020 (accessed 10 February 2022).

55 Kahn, J.M. and Barry, J.R. (1997). Wireless infrared communications. *Proc. IEEE* 85 (2): 265–298.

56 Miramirkhani, F. and Uysal, M. (2015). Channel modeling and characterization for visible light communications. *IEEE Photon. J.* 7 (6): 1–16. http://dx.doi.org/10.1109/JPHOT.2015.2504238.

57 Miramirkhani, F. and Uysal, M. (2020). Channel modelling for indoor visible light communications. *Philos. Trans. R. Soc. A* 378.

58 Gfeller, F.R. and Bapst, U. (1979). Wireless in-house data communication via diffuse infrared radiation. *Proc. IEEE* 67 (11): 1474–1486.

59 Lopez-Hernandez, F.J., Perez-Jimeniz, R., and Santamaria, A. (1998). Monte Carlo calculation of impulse response on diffuse IR wireless indoor channels. *Electron. Lett.* 34 (12): 1260–1262.

60 Lopez-Hernandez, J., Perez-Jimenez, R., and Santamaria, A. (1998). Modified Monte Carlo scheme for high-efficiency simulation of the impulse response on diffuse IR wireless indoor channels. *Electron. Lett.* 34 (19): 1819–1820.

61 Hayasaka, N. and Ito, T. (2007). Channel modeling of nondirected wireless infrared indoor diffuse link. *Elect. Commun. Jpn. (Part I: Commun.)* 90 (6): 9–19. https://doi.org/10.1002/ecja.20352.

62 Kokaly, R.F., Clark, R.N., Swayze, G.A. et al. (2017). USGS spectral library version 7: U.S. Geological Survey Data Series 1035. https://crustal.usgs.gov/speclab/QueryAll07a.php (accessed 18 November 2021).

63 Miramirkhani, F., Uysal, M., and Panayirci, E. (2015). Novel channel models for visible light communications. *Broadband Access Communication Technologies IX*, Volume 9387, pp. 150–162. International Society for Optics and Photonics, SPIE. http://dx.doi.org/10.1117/12.2077565.

64 Mukherjee, S., Das, S.S., Chatterjee, A., and Chatterjee, S. (2017). Analytical calculation of Rician K-factor for indoor wireless channel models. *IEEE Access* 5: 19194–19212. http://dx.doi.org/10.1109/ACCESS.2017.2750722.

65 Office for National Statistics (2013). Health Survey for England 2016 Adult Health Trends. http://digital.nhs.uk/data-and-information/publications/statistical/health-survey-for-england/health-survey-for-england-2016 (accessed 10 February 2022).

66 Fath, T.C.M. (2013). Evaluation of spectrally efficient indoor optical wireless transmission techniques. PhD thesis. The University of Edinburgh.

13

Conclusion

Muhammad Ali Imran[1], Lina Mohjazi[1], Lina Bariah[3], Sami Muhaidat[2,4], Tei Jun Cui[5], and Qammer H. Abbasi[1]

[1]*James Watt School of Engineering, University of Glasgow, Glasgow, UK*
[2]*KU Center for Cyber-Physical Systems, Department of Electrical Engineering and Computer Science, Khalifa University, Abu Dhabi, UAE*
[3]*Technology Innovation Institute, 9639 Masdar City, Abu Dhabi, UAE*
[4]*Department of Systems and Computer Engineering, Carleton University, Ottawa, Canada*
[5]*State Key Laboratory of Millimeter Waves, Southeast University, Nanjing, China*

While the deployment of the fifth-generation (5G) systems is scaling up globally, it is time to look ahead for beyond 5G (B5G) systems. This is mainly driven by the emerging societal trends, calling for fully automated systems and intelligent services supported by extended reality and haptics communications. To accommodate the stringent requirements of their prospective applications, which are data-driven and defined by extremely low-latency, ultra-reliable, fast and seamless wireless connectivity, research initiatives are currently focusing on a progressive roadmap towards the sixth-generation (6G) networks, which are expected to bring transformative changes to this premise. The millimetre-wave (mm-wave) and terahertz (THz) communications are envisioned as key enablers for 6G systems. They are expected to satisfy the stringent requirements of various potential 6G use cases by exploiting higher-frequency bands. However, owing to the severe attenuation and scattering properties, the detrimental effects on communication efficiency remains the grand challenge in wireless communications. Existing solutions mainly rely on device-side approaches, which consider the wireless environment to be uncontrollable and hence, it remains unaware of the on-going communication processes. Metasurfaces have recently emerged as an innovative technology, which is envisioned to revolutionise wireless communications by allowing wireless system designers to fully manipulate the propagation of electromagnetic (EM) waves in a wireless link. This book provides an in-depth

Intelligent Reconfigurable Surfaces (IRS) for Prospective 6G Wireless Networks, First Edition.
Edited by Muhammad Ali Imran, Lina Mohjazi, Lina Bariah, Sami Muhaidat,
Tie Jun Cui, and Qammer H. Abbasi.

treatment of the fundamental physics behind these intelligent planar structures, and details the research roadmap towards their development and hardware and prototyping features. The book also presents the potentials of intelligent surfaces in boosting the performance of future wireless applications in terms of reliability, security, and latency.

This book delves into the very basic fundamentals of intelligent reconfigurable surface (IRS), as an enabler for novel technological trends accompanied with the evolution of 6G networks. In particular, this book sheds lights on the underlying channel models over a wide range of frequency bands for single and multi-IRS scenarios, the hardware architecture, and operational concepts of IRS-enabled wireless systems, paving the way for the smooth comprehension of the preliminary concepts and facilitating the practical realization of efficient IRS systems. Furthermore, the book investigates new paradigms, emerged as a consequence of the recent advancements in these artificial intelligent surfaces, such as the Internet of Meta-Material Things (IoMMT), which concerns the optimal control of the physical propagation between meta-material devices within the internet-of-things (IoT) ecosystem. Motivated by the fact that artificial intelligence (AI) will be the key to orchestrate wireless networks from the core to the edge, a thorough discussion on the utilization of machine learning (ML) algorithms as a tool to enable optimized IRS reconfiguration, and therefore, to realize the full potential of IRS, is presented. Furthermore, aiming to provide a comprehensive study, the book explores promising research directions with emphasis on the key technologies that manifest themselves as enablers for IRS-assisted systems, while at the same time being enabled by IRS. This includes the integration of IRS into integrated terrestrial non-terrestrial networks, mmWave and THz communication, wireless power transfer, imaging and sensing, multi-user communication, as well as optical wireless communication, to name a few.

Despite the fast progression in the research and implementation of IRS-based wireless networks, several critical aspects are identified as key challenges that need to be addressed in order to realize efficient, yet practical IRS systems. Recalling that the 6G vision is to enable large-scale secure connectivity, it is of paramount importance to revisit current security and privacy mechanisms, which are implemented in conventional relaying and direct wireless communication, to validate their ability to cope with IRS systems requirements. In particular, it is essential to study how to leverage the IRS special functionalities in order to serve a massive number of sourced and un-sourced users, while meeting a high level of links security and users' privacy. On the other hand, although channel modelling in IRS is an extensively studied topic, such models are yet to be investigated while considering practical design aspects, e.g. mutual coupling and wideband reflection. It is worthy to highlight that accurate and realistic channel modelling, incorporating the small-scale fading as well as path-loss modelling, is essential to achieve on-demand accurate channel estimation, despite the absence of an radio-frequency (RF) chain at the IRS. In addition to this, being a perfect candidate for facilitating the implementation of reliable integrated terrestrial and non-terrestrial networks (TNTNs), incorporating the ground, air, and space layers, novel challenges pertaining to the enormous amount of data and network traffic produced, exchanged, and managed in both inter- and intra-layer communications, emerge. Within this context, ML is anticipated to be an indispensable tool in IRS-based integrated TNTN, offering smart resource management, access control, and multi-layer communications. Therefore, novel adaptive, centralized, and distributed ML algorithms should be developed in order to cope with increased system complexity resulted from the heterogeneity nature and reconfigurability requirements of IRS-based TNTN. From a hardware perspective, in order to enable self-adaptivity, self-healing, zero-touch IRS-based network, it is essential to equip the IRS with the required sensing elements, to allow these surfaces to sense the environment, and hence, enable self-reconfigurability to adapt to the current environment state. Different than the current proposed implementation frameworks, novel, robust, energy efficient, and cost-effective designs of IRS with sensing features constitutes the future directions of intelligent surface field. Finally, employing IRS to perform multi-user communication comes with additional challenges, related to inter-user interference, resource allocation, and channel estimation. In specific, IRS configuration is generally performed in order to achieve a particular quality-of-service (QoS) requirement for the intended user. However, in a multi-user scenario, users fairness constitutes a major problem and increases the reconfiguration complexity of IRS. This is particularly pronounced when non-orthogonal resource allocation is considered with mobility. Such a challenge can be addressed by splitting the IRS into multiple sub-arrays, in which each sub-array is responsible for serving a particular user. Nevertheless, such an arrangement raise another concerns pertaining to sub-arrays assignment, reconfiguration, and channel estimation. This applies to RF and optical wireless networks.

Index

Intelligent Reconfigurable Surfaces (IRS) for Prospective 6G Wireless Networks, First Edition.
Edited by Muhammad Ali Imran, Lina Mohjazi, Lina Bariah, Sami Muhaidat, Tie Jun Cui, and Qammer H. Abbasi.

o

p

q

r

THE COMSOC GUIDES TO COMMUNICATIONS TECHNOLOGIES

The ComSoc Guide to Next Generation Optical Transport: SDH/SONET/OTN
Huub van Helvoort

The ComSoc Guide to Managing Telecommunications Projects
Celia Desmond

WiMAX Technology and Network Evolution
Kamran Etemad and Ming-Yee Lai

An Introduction to Network Modeling and Simulation for the Practicing Engineer
Jack Burbank, William Kasch, and Jon Ward

The ComSoc Guide to Passive Optical Networks: Enhancing the Last Mile Access
Stephen Weinstein, Yuanqiu Luo, and Ting Wang

Digital Terrestrial Television Broadcasting: Technology and System
Jian Song, Zhixing Yang, and Jun Wang

TV White Space: The First Step Towards Better Utilization of Frequency Spectrum
Ser Wah Oh, Yugang Ma, Edward Peh, and Ming-Hung Tao

Digital Services in the 21st Century: A Strategic and Business Perspective
Antonio Sanchez and Belen Carro

Toward 6G: A New Era of Convergence
Amin Ebrahimzadeh and Martin Maier

VCSEL Industry: Communication and Sensing
Babu Dayal Padullaparthi, Jim A. Tatum and Kenichi Iga

Intelligent Reconfigurable Surfaces (IRS) for Prospective 6G Wireless Networks
Muhammad Ali Imran, Lina Mohjazi, Lina Bariah, Sami Muhaidat, Tie Jun Cui, and Qammer H. Abbasi.

Intelligent Reconfigurable Surfaces (IRS) for Prospective 6G Wireless Networks, First Edition.
Edited by Muhammad Ali Imran, Lina Mohjazi, Lina Bariah, Sami Muhaidat,
Tie Jun Cui, and Qammer H. Abbasi.

Printed and bound by CPI Group (UK) Ltd, Croydon, CR0 4YY
16/04/2025
14658584-0003